ALGORITHMS :
THEIR COMPLEXITY
AND EFFICIENCY

WILEY SERIES IN COMPUTING

Consulting Editor
Professor D. W. Barron, *Department of Mathematics, Southampton University*

ALGORITHMS :
THEIR COMPLEXITY
AND EFFICIENCY

Lydia I. Kronsjö

The Computer Centre
The University of Birmingham

A Wiley–Interscience Publication

JOHN WILEY & SONS

Chichester · New York · Brisbane · Toronto

Library of Congress Cataloging in Publication Data:

Kronsjö, Lydia I.
  Algorithms: their complexity and efficiency.

  (Wiley series in computing)
  'A Wiley–Interscience publication.'
  Includes index.
  1. Electronic digital computers—Programming.
2. Algorithms.   3. Numerical analysis—Data processing.
4. Computational complexity. I. Title.
QA76.6.K76   519.4   78–22090

ISBN 0 471 99752 8

Photosetting by Thomson Press (India) Limited, New Delhi
and printed in Great Britain by The Pitman Press, Bath, Avon.

To
my parents Ivan and Varvara,
my father-in-law Erik,
my husband Tom,
my son Tim.

Contents

Preface

This book is concerned with the study of algorithms and evaluation of their performance. Analysis of algorithms is a new area of research, and emerged as a new scientific subject during the sixties and has been quickly established as one of the most active fields of study, becoming an important part of computer science. The reason for this sudden interest in the study of algorithms is not difficult to trace as the fast and successful development of digital computers and their uses in many different areas of human activity has led to the construction of a great variety of computer algorithms. At present it is often the case that several different algorithms exist for the solution of a single problem or a class of problems and these algorithms need to be carefully analysed in order to provide a basis for selecting the best one for the purpose. In many cases, analysis of algorithms also leads to the revelation of completely new algorithms that are even faster than all algorithms known before. On the other hand, the study of algorithms has brought about many no less startling discoveries of certain natural problems for which all algorithms are inefficient.

The book is intended as a text for an intermediate course in computer science or computational mathematics which focuses on the basic principles and concepts of algorithms. It requires general familiarity with computers, preferably some courses on programming and introductory computer science. Each chapter is devoted to one particular class of problems and their algorithms. Chapter 1 introduces the subject, while Chapters 2, 3, 4, and 5 discuss four different classes of problems that are termed as numerical, i.e. the mathematical problems, solution of which is of a numerical nature. For such problems numerical accuracy of the computed results is of particular importance. Chapter 6 discusses the problem of (asymptotically) fast multiplication of two numbers and is based on the fast Fourier transforms and Chapters 7, 8, and 9 discuss sorting and searching, the most common non-numerical problems encountered in computing.

The exercises at the end of each chapter are used to provide examples as well as to complete or generalize some proofs.

All algorithms in the text are given in a form of a sequence of steps, each xiv

step describing the actions to be undertaken, in natural English. This seems to be a most neutral way of introducing the algorithms. Actual implementation details, e.g. a specific programming language, various programming tricks etc., are left to an interested reader. The reference list at the end of the book contains published sources for the algorithms and theoretical results discussed in the text.

I would like to thank all the people who have critically read various portions of the manuscript and offered many helpful improvements. In particular, I would like to thank Peter Jarratt, Stuart Hollingdale, Michael Atkinson, Nelson Stevens, and Tom Axford. Sincere thanks go to Ilsie Browne for her excellent typing of the manuscript.

L. I. K.

Birmingham, November 1978

Table of Notations

Written	Denotes
$a \in A$	a is contained in the set A
$x \leftarrow y + z$	x is assigned the value $y + z$
$\lceil a \rceil$	the least integer greater than or equal to a
$\lfloor b \rfloor$	the greatest integer less than or equal to b
$\ln a$	natural logarithm
$a \approx b$	a approximately equal to b
$\mathbf{A}^*$	conjugate transpose of the matrix $\mathbf{A}$
$\mathbf{a}^*$	conjugate transpose of the vector $\mathbf{a}$
$\mathbf{A}^\mathsf{T}$	transpose of the real matrix $\mathbf{A}$
$\mathbf{a}^\mathsf{T}$	transpose of the real vector $\mathbf{a}$
$\mu[y]$	mean value of the statistical variable y
$\sigma[y]$	standard deviation of the statistical variable y
$a * b$	a is multiplied by b
$a \equiv b$	a is equivalent to b
$x \gg y$	x is much larger than y

Chapter 1

Introduction

This book will be chiefly concerned with an investigation of algorithms in order to evaluate their performance. This comparatively new field of study is known as algorithmic analysis and forms part of the more general discipline of computer science. In practical terms, a goal of algorithmic analysis is 'to obtain sufficient understanding about the relative merits of complicated algorithms to be able to provide useful advice to someone undertaking an actual computation' (Gentleman, 1973). In broader interpretation, however, algorithmic analysis includes the study of all aspects of performance in computational problem solving, from the preliminary formulation, through the programming stages, to the final task of interpreting the results obtained. We would also like to prove lower bounds on the computation time of various classes of algorithms. In order to show that there is no algorithm to perform a given task in less than a certain amount of time, we need a precise definition of what constitutes an algorithm. First, then, what is an algorithm?

1.1 Definition of an Algorithm

A procedure consisting of a finite set of unambiguous rules which specify a finite sequence of operations that provides the solution to a problem, or to a specific class of problems, is called an algorithm.

Several important features of this definition must now be emphasized.

First, each step of an algorithm must be unambiguous and precisely defined. The actions to be carried out must be rigorously specified for each case.

Secondly, an algorithm must always arrive at a problem solution after a finite number of steps. Indeed, the general restriction of finiteness is not sufficient in practice, as the number of steps needed to solve a specific problem, although finite, may be too large for practicable computation. A useful algorithm must require not only a finite number of steps, but a reasonable number.

Thirdly, every meaningful algorithm possesses zero or more inputs and provides one or more outputs. The inputs may be defined as quantities which are

1

given to the algorithm initially, before it is executed, and the outputs as quantities which have a specified relation to the inputs and which are delivered at the completion of its execution.

Fourthly, it is preferable that the algorithm should be applicable to any member of a class of problems rather than only to a single problem. This property of generality, though not a necessity, is certainly a desirable attribute of a useful algorithm.

Finally, we would like to mention that although the concept of an algorithm is a very broad one, in this book we restrict ourselves to algorithms designed to be executed on a computer. Such an algorithm must be embodied in a computer program (or set of programs), and so in the sequel the two terms will be used interchangeably.

1.2 Measures of Efficiency

It is relatively easy to invent algorithms. In practice, however, one wants not only algorithms, one wants good algorithms. Thus, the objective is to invent good algorithms and prove that they are good. The 'goodness' of an algorithm can be appraised by a variety of criteria. One of the most important is the time taken to execute it. There are several aspects of such a time criterion. One might be concerned with the execution time required by different algorithms for solution of a particular problem on a particular computer. However, such an empirical measure is strongly dependent upon both the program and the machine used to implement the algorithm. Thus, a change in a program may not represent a significant change in the underlying algorithm but may, nevertheless, affect the speed of execution. Furthermore, if two programs are compared first on one machine and then another, the comparisons may lead to different conclusions. Thus, while comparison of actual programs running on real computers is an important source of information, the results are inevitably affected by programming skill and machine characteristics.

A useful alternative to such empirical measurements is a mathematical analysis of the intrinsic difficulty of solving a problem computationally. Judiciously used, such an analysis provides an important means of evaluation the cost of algorithm execution.

The performance time of an algorithm is a function of the size of the computational problem to be solved. However, assuming we have a computer program which eventually terminates, solving a particular problem requires only sufficient time and sufficient storage. Of more general interest are algorithms which can be applied to a collection of problems of a certain type. For these algorithms, the time and storage space required by a program will vary with the particular problem being solved. Consider, for example, the following classes of problems, and note the role of the value of the parameter n.

1. Find the largest in a sequence of n integers.
2. Solve a set of linear algebraic equations $\mathbf{Ax} = \mathbf{b}$, where $\mathbf{A}$ is an $n \times n$ real matrix and $\mathbf{b}$ is a real vector of length n.

3. Let W be a one-dimensional array of n distinct integers. Sort the entries of W into descending order.

4. Evaluate a polynomial $P_n(x) = \sum_{k=0}^{n} a_{n-k}x^k$ at $x = x_0$.

In each of these problems, the parameter n provides a measure of the size of the problem in the sense that the time required to solve the problem, or the storage space required, or both, will increase as n increases.

1.3 Algorithms and Complexity

In order to measure the cost of executing a program, we customarily define a complexity (or cost) function F, where $F(n)$ is either a measure of the time required to execute the algorithm on a problem of size n, or is a measure of the memory space required for such execution. Accordingly, we may speak of either the time complexity function or the space complexity function of the algorithm. We can also refer to either kind of function as simply a complexity (or cost) function of the algorithm. In this book our principal concern will be with time complexity functions, but we shall distinguish, where appropriate, between the time complexity and the computational complexity of an algorithm. The latter term will be used to refer to estimates of the computational power required to solve the problem and will be measured by the number of arithmetic or logical operations required, e.g. the number of multiplications and additions needed to evaluate a polynomial, the number of comparisons between two entries in a sort algorithm, or the number of function evaluations in an iterative root-finding algorithm. In this context, the computational complexity measure will be considered as a branch of the time complexity measure.

In general, the cost of obtaining a solution increases with the problem size, n. If the value of n is sufficiently small, even an inefficient algorithm will not cost much to run; consequently, the choice of an algorithm for a small problem is not critical (unless the problem is to be solved many times). In most cases, however, as n increases one eventually arrives at a situation when the algorithm can no longer be executed in an acceptable period of time. This point is illustrated in Tables 1.1 and 1.2. Table 1.1 shows the growth of some common complexity

Table 1.1 A comparison of the growth of some common complexity functions

Time complexity function	Problem size n			
	10	10^2	10^3	10^4
$\log_2 n$	3.3	6.6	10	13.3
n	10	10^2	10^3	10^4
$n \log_2 n$	0.33×10^2	0.7×10^3	10^4	1.3×10^5
n^2	10^2	10^4	10^6	10^8
n^3	10^3	10^6	10^9	10^{12}
2^n	1024	1.3×10^{30}	$> 10^{100}$	$> 10^{100}$
$n!$	3×10^6	$> 10^{100}$	$> 10^{100}$	$> 10^{100}$

Table 1.2 Limits on sizes of problems which can be solved using algorithms with some common complexity functions (one operation per microsecond, 10^{-6} sec, is assumed)

Time complexity function	Maximum problem size for execution in:		
	1 sec	1 min	1 hour
$\log_2 n$	2^{10^6}	$2^{6 \times 10^7}$	$2^{3.6 \times 10^9}$
n	10^6	6×10^7	3.6×10^9
$n \log_2 n$	62746	2.8×10^6	1.3×10^8
n^2	10^3	7746	60 000
n^3	10^2	3.9×10^2	1.5×10^3
2^n	20	25	32
$n!$	9	11	12

functions as n increases, while Table 1.2 gives the maximum sizes of problems which can be solved using algorithms with these complexity functions, in a specified length of time.

In most cases of practical interest we wish to compare the performance of algorithms for relatively large values of n. In this context, it is convenient to talk about the asymptotic behaviour of the complexity function. The asymptotic complexity function describes the limiting behaviour of the complexity function when the size parameter, n, tends to infinity. The asymptotic (time or space) complexity function ultimately determines the size of problems that can be solved by the algorithm. In terms of the terminology associated with the notion of complexity, we shall say, for example, that the (time) complexity of the algorithm is $O(\ln(n))$, read 'order $\ln(n)$' if the processing by the algorithm of inputs of size n takes a time $C \ln(n)$, for some constant C, as n tends to infinity. Generally throughout the text, in relation to, say, the time complexity of a particular algorithm, any of the following terminologies will be used in an equivalent sense:

(a) the time complexity of the algorithm is of order $\ln(n)$; this can also be written as '$O(\ln(n))$',
(b) the algorithm is executed in $O(\ln(n))$ time,
(c) the amount of work required by the algorithm is proportional to $\ln(n)$ or is '$O(\ln(n))$'.

Where appropriate, the term 'unit of time' will be used in the sense equivalent to the term 'one basic operation' (e.g. Chapter 6). Consequently, the following additional terminology will be used in the sense equivalent to (a), (b) and (c):

(d) the algorithm requires a number of basic operations proportional to $\ln(n)$.

1.4 Numerical Algorithms and Computational Complexity

In this text we distinguish between numerical and non-numerical algorithms. By a numerical algorithm we mean an algorithm that solves a numerical problem such as, for example, the calculation of the roots of an equation or the evaluation of a polynomial for a given argument value. A numerical algorithm produces some numerical output which represents a computed solution of the corresponding numerical problem. Chapters 2 to 5 of the book discuss the numerical algorithms for solving problems of the polynomial evaluation, the calculation of a root of an equation, the solution of linear sets of equations and the calculation of the Fourier transforms, respectively.

Non-numerical algorithms, on the other hand, solve problems, the solution of which, although in some cases expressed by a number (or a set of numbers), is of a non-numerical nature. Two important examples of non-numerical problems are sorting in order a given set of unordered items (note that the items can be expressed in the form of either numerals or sequences of letters), and searching for a specific item in a given set of items. Chapters 7 to 9 of the text are concerned with algorithms for solving such problems. (The problem of fast multiplication of two integers discussed in Chapter 6 is another example of a numerical problem.)

In the complexity analysis of the two distinct types of algorithms, at a higher level of discrimination between various measures of complexity, one selects the measure functions that are most suited for the given type of algorithms. In particular, for the analysis of computational complexity of the numerical algorithms it is convenient to distinguish between (a) algebraic and (b) analytic complexity measures, both of which represent branches of the computational complexity defined earlier.

The objectives of algebraic complexity studies are manifold: e.g. (i) to find out how many arithmetic operations are used by a given algorithm, (ii) to try to discover how many arithmetic operations are needed to solve a given problem, and (iii) to answer the question of what is the best way to solve a given problem in terms of arithmetic operations. The evaluation of a polynomial at a given point(s) and the solution of linear sets of equations by direct methods are examples of problems studied in terms of algebraic complexity. A common feature of the computational processes used for solving such problems is that only a finite number of steps is needed to achieve an answer. Hence, the number of arithmetic operations required is also finite. Analytic complexity, on the other hand, addresses the question of how much computation has to be performed to obtain a result with a given degree of accuracy, and focuses on those computational processes which in a certain sense never end. Iterative processes provide an obvious example. Here, the process is interrupted at some point and, if the current value of the result lies within some error bound, that value is accepted; otherwise the computation is continued until a satisfactory result is obtained. In this case one estimates the number of arithmetic operations per iteration step, and a 'best' algorithm in terms of the computational complexity is identified by the fewest total number of arithmetic operations required to achieve an approximate result computed to some prescribed level of accuracy.

1.5 Numerical Algorithms and Numerical Accuracy

In analysing algorithms for solving numerical problems, the accuracy of the computed results is another very important criterion for distinguishing between 'good' and 'bad' algorithms. The need to examine the accuracy of mathematical computations arises from the fact that a computer is a finite machine; it is capable of representing numbers only to a finite number of digit positions. As a result, most numbers, and even integers, if they are too long for the computer to represent exactly, are rounded. This means that; usually, only a finite approximation to a number can be represented in a computer. Thus, if care is not exercised, some numerical algorithms implemented on a computer may produce approximations to the true results that are wildly inaccurate. The study of mathematical methods for solving problems numerically and ascertaining the bounds of errors in the results is a very well developed field, known as numerical analysis. The foundations of modern mathematical error analysis were laid down in the celebrated paper of von Neuman and Goldstine (1947), although it was not until the mid to late fifties that this area of study became really active.

One of the most important ideas in error analysis is that of backward error analysis. It was proposed by Givens (1954) and brilliantly utilized by Wilkinson (1963) in the analysis of the rounding errors in algebraic processes. The question asked in backward error analysis is: What problem have we solved exactly and how far is this problem from the one we set out to solve? This may be contrasted with usual forward error analysis where the question asked is: By how much does the calculated answer differ from the true answer? The advantage of backward error analysis is that it is often easier to perform and the answers are often more useful. Applying backward analysis, Wilkinson (1965) was able to give a complete *a priori* analysis of the solution of a system of linear algebraic equations by Gaussian elimination with pivoting.

1.6 Types of Algorithm Analysis

It is convenient to distinguish between the analysis of a particular algorithm and the analysis of a class of algorithms. In analysing a particular algorithm we would probably investigate its most important characteristics, such as time complexity (i.e. how many times each part of the algorithm is likely to be executed) and space complexity (i.e. how much computer storage space it is likely to need). In analysing a class of algorithms, on the other hand, we study the entire family of algorithms for the solution of a particular problem and seek to identify the 'best possible' algorithm of the family. If such an algorithm has not yet been found, then attempts are made to establish bounds on the computational complexity of the algorithms in the class.

Analyses of the first type have been made since the earliest days of computer programming. Analyses of the second type were, with a few exceptions, not embarked upon until much later. It is easy to see that analyses of the second type are far more powerful and useful since they study many algorithms at the same time. It has always been a dream of mankind to find the best (in mathematical terms, optimal) ways of solving the many problems with which he is faced. Computer algorithms are just another area of human activity where the optimal ways for solving problems are eagerly sought. Instead of analysing each algorithm that is devised, it is obviously better to prove once and for all that a particular algorithm is the 'best possible'. However, it so happens that many of the chosen algorithms turn out to be the 'best' only in a very narrow sense; very small changes in the definition of the optimal criteria can significantly affect the choice of what algorithm is indeed the 'best possible'.

Consider, for example, one particular case: how to compute x^n with the fewest possible multiplications. This problem was first raised by Arnold Scholtz in 1937. Let us consider x^{31}. It cannot be calculated in fewer than 8 multiplications, i.e.

$$x; x^2; x^4; x^8; x^{16}; x^{24}; x^{28}; x^{30}; x^{31}.$$

It can however be done in only 6 arithmetic operations if division is allowed:

$$x; x^2; x^4; x^8; x^{16}; x^{32}; x^{31}.$$

Other examples of the dependence of complexity estimates on underlying assumptions are these of the different estimates of Pan (1966), Motzkin (1955), and Winograd (1967) on the number of multiplications necessary to evaluate polynomials, and the different estimates of Klyuyev and Kokovkin–Shcherbak (1965), Strassen (1969) and Miller (1973) on the number of arithmetic operations necessary to solve n linear equations in n unknowns. Generally, analyses of the second type are difficult and relatively few results have been obtained to date. The problem of computing x^n with fewest multiplications, for instance, is far from being solved. An engaging discussion of this problem is given in Knuth (1969).

Another example is that of sorting. The need to sort a given group of elements into some prescribed order arises quite frequently in programming practice, and sorting algorithms have been studied intensively for a number of years. However, very few optimal parameters have yet been established, and even then only for problems of quite small size. For instance, for sorting algorithms which are based on comparisons of the elements, the exact value of the optimal number of comparisons, $C(n)$, is known only for $n \leqslant 12$ and $n = 20, 21$. The difficulty in performing analyses of the second type is the reason why such analyses have not superseded analyses of the first type. In fact, in a certain sense, the latter analyses are even more important in practice since they can be designed to measure all the relevant factors about the performance of a particular algorithm.

1.7 Bounds on Complexity and Models of Computation

The complexity analysis among other things, is concerned with obtaining upper and lower bounds on the performance of algorithms (or classes of algorithms) that solve various problems.

The existence of complexity bounds for the algorithms available can serve as a basis for classifying problems. For example, Jakeman (1976) has suggested that four distinct classes of problems can be identified in this way:

First, there are those problems for which complexity bounds can be determined, using appropriate theorems, without the actual construction of an algorithm for the problems. Any proofs of the theorems in such a case are thus non-constructive and do not give any indication as to whether or how an algorithmic solution can be obtained. A discussion of some complexity theorems of this kind, e.g. those concerned with some finite combinatorial problems, can be found in Rabin (1974).

Secondly, there are those problems for which lower bounds on complexity have been derived, but which are unattainable by available algorithms. A notable example of this type is matrix multiplication, where a minimum bound of $O(n^2)$ has been established but the fastest available methods are $O(n^{2.8})$. This discrepancy suggests that the bound $O(n^2)$ is not sharp enough.

Thirdly, there are problems for which

   (i)   lower bounds on complexity are known and
   (ii) algorithms which attain these bounds can be constructed but
   (iii) the algorithms do not possess numerical stability.

Fast methods of matrix multiplication belong to this category. In fact, using the modern approach of interpreting the operation of matrix multiplication as the computation of bilinear forms, Miller (1973) has shown that for any algorithm that multiplies two $n \times n$ matrices using only addition, subtraction and multiplication, to be strongly stable in the sense of the ideas drawn from backward error analysis, the algorithm has to be essentially equivalent to the usual algorithm, and hence must involve $O(n^3)$ operations.

Fourthly, and finally, there are problems which satisfy criteria (i) and (ii) above, and for which, in addition, numerically stable algorithms do exist. In this category, for example, belong the generally available numerical methods which are stable and attain well-known bounds of complexity. One particularly remarkable method of this class is the Fast Fourier Transform (FFT) algorithm for computing the discrete finite Fourier transforms of functions.

Algorithms and their complexity are normally studied under assumption of a specific (precisely defined) model of a computing device that is capable of executing the algorithm in hand. A basic step in a computation has also to be precisely defined. Perhaps the most important motivation for formal models of computation is the desire to discover the inherent computational difficulty of various problems.

Unfortunately, there is no one computational model which is suitable for all situations, and so for each problem one selects an appropriate model which for the given situation reflects as accurately as possible the actual computation process on a real computer.

This approach is adopted in the present text. In each chapter we outline the main features of the computational model suitable for the problem in hand. The models vary from chapter to chapter because of the wide range of the problems considered. However one basic property that justifies and guides our choice of models is the fact that every model is defined in such a way that any computational process which can be carried out using such a model can also be carried out by a real computer.

In other words, we assume that none of our computational models possess such features that are not available on a modern real computer.

In terms of optimality, the goal of complexity analysis is to show that in order to solve a certain problem or class of problems computationally, one requires a certain number of operations of a certain type. This is a very difficult objective; for the majority of practical problems we still have to rely on experience to judge the goodness of an algorithm. However, in some cases useful results have been established. For example, we can now strictly bound from below the number of arithmetic operations required to perform certain calculations and in the course of the book we shall present some basic results of this nature. For instance, we shall show in the chapter on polynomial evaluation that the evaluation of an nth degree polynomial requires at least n multiplications and the same number of additions, unless certain conceptually different conditions are imposed on forming the polynomial coefficients. In other cases, indeed, it is possible to show that for some problems we shall never find optimal algorithms because they simply do not exist. For example, Gentleman (1969) has surveyed the field of numerical algorithms and selected areas for which 'good' algorithms do or do not exist. He refers to areas that rely heavily on matrix operations, such as solving linear systems of equations, linear programming, linear least squares fitting and the calculation of eigenvalues and eigenvectors, as areas for which good, reliable algorithms are available. At the other end he lists such problem areas as numerical differentiation, statistical analyses, pattern recognition, classification and clustering on a set of observations. In the case of numerical differentiation the problem is inherently unstable and so no good computational methods can be expected to exist at all. In solving problems of a stochastic nature, in Gentleman's opinion a human touch will always be needed to check the meaningfulness of any computational model that may be constructed for the purpose of solving such a problem.

Chapter 2

Evaluation of Polynomials

The evaluation of a polynomial is one of the basic and most widely encountered operations in practical computation. Indeed, some of those who use the extensive libraries of computer programs that are now available may not realize how ubiquitous this operation is, not only in contexts directly associated with the evaluation of polynomials, but also in problems of computing transcendental or more complicated algebraic expressions, for example, trigonometric or logarithmic expressions, or the ratio of polynomials. More often than not a polynomial of some kind is used to approximate such expressions and the sine, cosine, exponential and logarithmic functions represent familiar examples.

With the foregoing in mind, it is not surprising that the problem of the accurate and efficient numerical evaluation of a polynomial has received considerable attention. In this chapter we shall explore the results obtained from these studies, and use them to illustrate some of the general ideas that are involved in the analysis of numerical algorithms. The operation we have to consider, then, is that of computing the value of a polynomial expression

$$P_n(x) = a_0 x^n + a_1 x^{n-1} + \ldots + a_n \tag{1}$$

for a particular value, say x_0, of the argument x. We shall refer to this, for brevity, as evaluation of the polynomial at a fixed point x and it will be assumed throughout this chapter, with the exception of the final section, that both the argument and the coefficients are real numbers.

2.1 Polynomial Forms and Their Evaluation Algorithms

Any particular polynomial may be expressed in a variety of ways. Each of them represents the polynomial exactly but where numerical evaluation is concerned the various forms may have different properties. We now examine alternative methods of evaluating a polynomial, depending on the form in which it is presented.

10

1. *The Power Form*

The most common form of a polynomial is the Power form given by 2–(1). To evaluate (1) term by term requires $(2n - 1)$ multiplications and n additions. On a modern computer, the operations of addition and subtraction are of equal complexity, and so the term 'additions' is used to embrace any combination of addition and subtraction operations.

Fortunately, however, we can rewrite (1) in the so called *Horner*, or *nested*, form given by

$$P_n(x) = (\ldots(a_0 x + a_1)x + \ldots + a_{n-1})x + a_n. \tag{2}$$

This immediately suggests the recursive evaluation algorithm:

$$
\begin{aligned}
V_n &= a_0, \\
V_k &= xV_{k+1} + a_{n-k}, \qquad\qquad k = n-1, n-2, \ldots, 0 \\
P_n(x) &= V_0
\end{aligned}
\tag{3}
$$

This algorithm requires n multiplications and n additions and is the one most frequently used. We shall refer to (3) as the Horner computational algorithm, which has to be distinguished from the Horner polynomial form given by (2).

2. *The Root Product Form*

The Root Product form of a polynomial is given by

$$P_n(x) = a_0 \prod_{i=1}^{n} (x - \gamma_i), \qquad \text{where } P_n(\gamma_i) = 0. \tag{4}$$

The roots γ_i's can be real or complex, though for our purpose here we may assume the form with all the roots real.

Important cases of polynomials in this form arise, for example, in statistics. The computational algorithm based on (4) may be expressed as follows:

$$
\begin{aligned}
V_n &= a_0, \\
V_k &= (x - \gamma_{k+1})V_{k+1}, \qquad\qquad k = n-1, n-2, \ldots, 0 \\
P_n(x) &= V_0.
\end{aligned}
\tag{5}
$$

It can be seen that the evaluation of this form requires n multiplications and n additions, the same as is required by the Horner algorithm.

3. *The Newton Form*

The Newton polynomial form is given by

$$
\begin{aligned}
P_n(x) = c_0(x - \beta_n)(x - \beta_{n-1})\ldots(x - \beta_1) &+ c_1(x - \beta_{n-1})\ldots(x - \beta_1) \\
&+ c_{n-1}(x - \beta_1) + c_n.
\end{aligned}
\tag{6}
$$

The computational algorithm based on (6) may be stated as follows:

$$V_n = c_0,$$
$$V_k = (x - \beta_{k+1})V_{k+1} + c_{n-k}, \qquad k = n-1, n-2, \ldots, 0 \qquad (7)$$
$$P_n(x) = V_0.$$

The algorithm requires n multiplications and $2n$ additions.

This polynomial form arises in the problem of interpolating a function given in tabular form. It should be noted that this form does not usually occur directly but is arrived at after some manipulation on the direct interpolation polynomial. Suppose, for example, that the Lagrange interpolating polynomial is used to interpolate the given function, $f(x)$. If we take the case of three points $(f(x_k), x_k, k = 0, 1, 2)$, corresponding to $n = 2$, for which the Lagrange polynomial interpolates the function, we have

$$P_2^{(L)}(x) = f_0 \frac{(x - x_1)(x - x_2)}{(x_0 - x_1)(x_0 - x_2)} + f_1 \frac{(x - x_0)(x - x_2)}{(x_1 - x_0)(x_1 - x_2)}$$
$$+ f_2 \frac{(x - x_0)(x - x_1)}{(x_2 - x_0)(x_2 - x_1)}, \qquad (8)$$

where $P_2^{(L)}(x)$ denotes the Lagrange interpolating polynomial of degree 2.

This polynomial is easily obtained but it is clear that the form is computationally uneconomical, if used to evaluate the polynomial. Indeed, for general n, as in $P_2^{(L)}(x)$, $(2n^2 + n - 1)$ multiplications, $(n + 1)$ divisions and $(2n^2 + n + 2)$ additions are required and there is an obvious need to seek a simpler form. This may be done using the following approach.

In the above example, in addition to $P_2^{(L)}(x)$ we consider also $P_1^{(L)}(x)$, where

$$P_1^{(L)}(x) = f_0 \frac{x - x_1}{x_0 - x_1} + f_1 \frac{x - x_0}{x_1 - x_0}.$$

Now examine the difference

$$P_2^{(L)}(x) - P_1^{(L)}(x) = f_0 \frac{(x - x_1)}{(x_0 - x_1)} \left(\frac{x - x_2}{x_0 - x_2} - 1 \right) + f_1 \frac{(x - x_0)}{(x_1 - x_0)} \left(\frac{x - x_2}{x_1 - x_2} - 1 \right)$$
$$+ f_2 \frac{(x - x_0)(x - x_1)}{(x_2 - x_0)(x_2 - x_1)}.$$

We observe that $[P_2^{(L)}(x) - P_1^{(L)}(x)]$, which is a second degree polynomial in x, vanishes at $x = x_0$ and x_1, and must therefore be a multiple of $(x - x_0)(x - x_1)$.

We therefore set

$$P_2^{(L)}(x) - P_1^{(L)}(x) = \alpha_2(x - x_0)(x - x_1),$$

which gives

$$P_2^{(L)}(x) = \alpha_2(x - x_0)(x - x_1) + P_1^{(L)}(x), \qquad (9)$$

and similarly, since $P_0^{(L)}(x) = f_0$, we have

$$P_1^{(L)}(x) - P_0^{(L)}(x) = f_0 \frac{x - x_1}{x_0 - x_1} + f_1 \frac{x - x_0}{x_1 - x_0} - f_0$$

$$= f_0 \left(\frac{x - x_1}{x_0 - x_1} - 1 \right) + f_1 \frac{x - x_0}{x_1 - x_0}.$$

This polynomial vanishes when $x = x_0$, and so we can set

$$P_1^{(L)}(x) - P_0^{(L)}(x) = \alpha_1(x - x_0),$$

giving

$$P_1^{(L)}(x) = \alpha_1(x - x_0) + P_0^{(L)}(x). \tag{10}$$

Substituting from (10) into (9), we find

$$P_2^{(L)}(x) = \alpha_2(x - x_0)(x - x_1) + \alpha_1(x - x_0) + \alpha_0, \tag{11}$$

which is the Newton form of the polynomial.

4. *The Orthogonal Form*

The orthogonal polynomial form is given by

$$P_n(x) = \sum_{k=0}^{n} b_k Q_k(x), \tag{12}$$

where the orthogonal polynomials, $\{Q_i(x), i = 0, 1, \ldots, n\}$, satisfy a recurrence relation

$$Q_{i+1}(x) = (A_i x + B_i)Q_i(x) - C_i Q_{i-1}(x),$$

with $A_i \neq 0$, $Q_0(x) = 1$, $Q_{-1}(x) = 0$,
and where A_i, B_i and C_i may depend on i but not on x.

The computational algorithm for (12) is

$$
\begin{aligned}
V_n &= b_n, \\
V_{n-1} &= (A_{n-1}x + B_{n-1})V_n + b_{n-1}, \\
V_k &= (A_k x + B_k)V_{k+1} - C_k V_{k+2} + b_k, \quad k = n-2, \ldots, 0 \quad (13) \\
P_n(x) &= V_0.
\end{aligned}
$$

In the general situation this scheme of evaluation requires $(3n - 1)$ multiplications and $(3n - 1)$ additions.

Among the classical orthogonal polynomials, the Chebyshev polynomials, $T_n(x)$, turn out to be the best choice on the grounds of efficiency of evaluation. In this case (12) becomes

$$P_n(x) = \sum_{k=0}^{n} b_k T_k(x), \tag{14}$$

and we have $B_i = 0$, $C_i = 1$ for $i \geq 0$, $A_i = 2$ for $i \geq 1$, and $A_0 = 1$.

One computational algorithm based on (14) is due to Clenshaw (1955) and is given by

$$
\begin{aligned}
Y &= 2x \\
V_n &= b_n \\
V_{n-1} &= YV_n + b_{n-1} \\
V_k &= YV_{k+1} - V_{k+2} + b_k, \qquad k = n-2, \ldots, 1 \qquad (15) \\
P_n(x) = V_0 &= xV_1 - V_2 + b_0.
\end{aligned}
$$

We shall refer to this algorithm as the Chebyshev–Clenshaw algorithm.

The evaluation of a general polynomial as a weighted sum of Chebyshev polynomials using algorithm (15) requires $(n+1)$ multiplications and $2n$ additions.

A further computational algorithm based on (14) is due to Bakhvalov (1971) and is given by

$$
\begin{aligned}
D_0 &= b_0 - b_2/2 \\
D_k &= (b_k - b_{k+2})/2, \qquad k = 1, \ldots, n-2 \\
D_{n-1} &= b_{n-1}/2 \\
V_n &= b_n/2 \\
Y &= 2x \\
V_{n-1} &= YV_n + D_{n-1} \\
V_k &= (YV_{k+1} - V_{k+2}) + D_k, \qquad k = n-2, \ldots, 0 \qquad (16) \\
P_n(x) &= V_0.
\end{aligned}
$$

This algorithm will be referred to as the Chebyshev–Bakhvalov algorithm. Algorithm (16) requires $(n+1)$ multiplications, $(n+1)$ divisions by a factor of 2, and $3n - 1$ additions.

Compared to (15), algorithm (16) is less efficient in terms of the number of arithmetic operations involved, but for certain polynomials, its use may be preferable to that of (15) on the grounds of achievable numerical accuracy.

In using the general recursive formulae for computing V_k in algorithms (15) and (16), it will be noted that the parameter k assumes values from $n-2$ to 0. This 'reverse' ordering of the computational sequence is dictated by the following observation. When adding a large sequence of numbers on a computer under conditions of normalized floating-point arithmetic, it is desirable to add the smaller numbers first, since at each step of the computation the local round-off error is proportional to the partial accumulated sum.

For 'smooth' polynomials the coefficients b_k in (14) have a tendency to decrease with k increasing and it is therefore desirable to have the order of accumulation of the summation terms as given in (15) and (16).

In the case of the algorithms discussed earlier which were based on other polynomial forms, similar orderings for the computational sequences were used in the interests of uniform presentations.

2.2 Preprocessing the Coefficients

Of the computational algorithms based on the various polynomial forms and considered so far, the Horner algorithm requires for a polynomial of degree n the smallest number of operations of multiplication and addition, namely n of each. The question may be raised however whether or not it is possible to devise an improved form and corresponding computational algorithm which will require fewer additions and/or fewer multiplications than the Horner algorithm. The idea of establishing lower bounds on the number of arithmetic operations required to evaluate a polynomial was originally introduced by Ostrowski (1954). He showed that at least n multiplications and n additions are necessary to evaluate nth degree polynomial for $n \leqslant 4$, a crucial point in his proof being the underlying assumption that the polynomial coefficients were not in any way artificially transformed for the purpose of minimizing the number of multiplications and additions.

Since then, under the same assumption, this result has been proved true for all non-negative values of n. However, in 1955, Motzkin introduced the idea of 'preprocessing' the polynomial coefficients for the purpose of designing evaluation algorithms which may require fewer multiplications and/or additions than, say, the Horner algorithm. Such preprocessing of the coefficients will clearly involve additional arithmetic operations, but if the same polynomial is to be evaluated at many points, the overall savings on the number of arithmetic operations may be significant.

To illustrate the idea of coefficient preprocessing we consider a simple example of a fourth degree polynomial which follows from the form proposed by Motzkin (1955) and is described by Todd (1955).

We make use of the identity

$$
\begin{aligned}
a_0 x^4 &+ a_1 x^3 + a_2 x^2 + a_3 x + a_4 \\
&\equiv a_0 [(x(x + \alpha_0) + \alpha_1)(x(x + \alpha_0) + x + \alpha_2) + \alpha_3] \\
&= a_0 x^4 + a_0 (2\alpha_0 + 1) x^3 + a_0 (\alpha_1 + \alpha_2 + \alpha_0 (\alpha_0 + 1)) x^2 \\
&\quad + a_0 ((\alpha_1 + \alpha_2)\alpha_0 + \alpha_1) x + a_0 (\alpha_1 \alpha_2 + \alpha_3).
\end{aligned}
\tag{1}
$$

Equating coefficients of corresponding powers of x and solving the resulting equations, we obtain:

$$
\alpha_0 = \frac{a_1 - a_0}{2a_0}, \qquad \alpha_1 = \frac{a_3}{a_0} - \alpha_0 \frac{a_2}{a_0} + \alpha_0^2 (\alpha_0 + 1),
$$

$$
\alpha_2 = \frac{a_2}{a_0} - \alpha_0 (\alpha_0 + 1) - \alpha_1, \qquad \alpha_3 = \frac{a_4}{a_0} - \alpha_1 \alpha_2.
\tag{2}
$$

Thus, if we first compute α_i, $i = 1, \ldots, 4$ the evaluation algorithm becomes:

$$
\begin{aligned}
V_1 &= (x + \alpha_0) x \\
V_2 &= (V_1 + \alpha_1)(V_1 + x + \alpha_2) \\
V_3 &= V_2 + \alpha_3 \\
F_4(x) &= V_4 = a_0 V_3,
\end{aligned}
\tag{3}
$$

where $F_4(x)$ has been written instead of the usual $P_4(x)$ to emphasize that the polynomial is expressed in the form (1).

This algorithm entails 3 multiplications and 5 additions, whereas the standard Horner algorithm would require 4 multiplications and 4 additions.

On a digital computer, multiplication is usually a more complex operation than addition, so even this modest saving is worthwhile if the polynomial is to be evaluated many times.

For polynomials of higher degree, the savings due to preprocessing their coefficients become steadily more significant as n increases. For instance, a polynomial of degree six

$$P_6(x) = a_0 x^6 + \ldots + a_6$$

may be evaluated by expressing it in the form

$$F_6(x) = [(((x + \alpha_0)x + \alpha_1)(x + \alpha_2) + \alpha_3 + (x + \alpha_0)x + \alpha_1 + \alpha_4)$$
$$\times (((x + \alpha_0)x + \alpha_1)(x + \alpha_2) + \alpha_3) + \alpha_5]a_0,$$

and using the following algorithm (Knuth, 1969):

$$V_1 = (x + \alpha_0)x + \alpha_1$$
$$V_2 = V_1(x + \alpha_2) + \alpha_3$$
$$V_3 = (V_2 + V_1 + \alpha_4)V_2 + \alpha_5$$
$$F_6(x) = V_4 = a_0 V_3. \tag{4}$$

This requires 4 multiplications and 7 additions as opposed to the standard Horner algorithm with 6 operations of each, multiplication and addition.

The operation of computing the parameters $\alpha_0, \ldots, \alpha_r$ in terms of the coefficients $a_0, \ldots, a_n$ has become generally known as *preprocessing* the polynomial coefficients, and the polynomial $F_n(x)$ expressed in terms of the α_i corresponding to $P_n(x)$ is known as *the polynomial with preprocessed coefficients*. In distinction to the polynomial forms considered earlier, polynomials with preprocessing coefficients do not arise 'naturally' but are obtained in an artificial way, with the sole purpose of facilitating their economical evaluation.

2.3 On Optimality of the Algorithms for Polynomial Evaluation. Optimality of the Horner Algorithm

The question of the economical evaluation of a polynomial can now be raised in its most general form. Given a polynomial expression, $P_n(x) = \sum_{k=0}^{n} a_{n-k}x^k$, what is the minimum number of multiplications and additions that is required to evaluate such an expression at a fixed point using (a) the 'natural' coefficients, and (b) the preprocessed coefficients?

The answer to the first question is given by the following theorem:

Theorem 2.1 Any algorithm for evaluating the polynomial $P_n(x)$ without preprocessing the coefficients involves at least n multiplication operations and at least n addition operations.

The proof of the theorem has been given by several authors, see for example, Belaga (1961), Pan (1966), Reingold and Stocks (1972), and Knuth (1973).

It also has been shown that the Horner algorithm is an optimal algorithm and, for that matter, the only algorithm which optimises the number of additions and multiplications required (Borodin, 1971).

We shall now show that the Horner algorithm is optimal with respect to the number of multiplications required, provided no preprocessing of the coefficients is made. For this it is necessary to specify a definite computational model to evaluate a polynomial.

The problem of polynomial evaluation belongs to a class of problems where certain dominant parameters may be identified in relation to which the computational complexity can be measured. For polynomials such a dominant parameter is its degree. For problems of this type, the computation can be performed using the model known as the straight-line program, once the dominant parameter has been fixed.

A straight-line program may be defined as a sequence of statements, $s_1, s_2, \ldots, s_r$, where each statement is of the form

$$\lambda_i = \lambda_j \circ \lambda_k, \qquad 0 \leqslant j, k < i \tag{1}$$

where '$\circ$' denotes one of the operations: addition, subtraction, multiplication or division and $\lambda_i, \lambda_j, \lambda_k$ denote various parameters and variables associated with the particular problem.

In the case of polynomial evaluation, we shall assume that in model (1), '$\circ$' denotes an operation of addition or multiplication; each of λ_j and λ_k denotes one of the following:

    (i) the polynomial coefficient, $a_k, k = 0, \ldots, n,$

    (ii) the argument, x,

    (iii) a previous variable in the straight-line program, where $j, k < i,$

    (iv) an arbitrary rational constant,

and λ_i denotes a new variable in the straight-line program.

Further, we shall call the step λ_i from (1) an 'active step in a_m' if at least one of λ_j and λ_k is either a_m (it may also be a linear function of a_m, i.e. $\beta a_m + \gamma$, where β and γ are real constants, and $\beta \neq 0$) or an active step in a_m, and the other is *not* a constant.

As an example of a straight-line program to evaluate a polynomial, consider:

$$P_5(x) = ((((a_0 x + a_1)x + a_2)x + a_3)x + a_4)x + a_5.$$

Set $\lambda_{-6} = a_5, \qquad \lambda_{-5} = a_4, \qquad \lambda_{-4} = a_3, \qquad \lambda_{-3} = a_2$

        $\lambda_{-2} = a_1, \qquad \lambda_{-1} = a_0, \qquad \lambda_0 = x,$

and define a sequence, $\lambda_1, \ldots, \lambda_{10}$, by

$$\lambda_1 = \lambda_0 \times \lambda_{-1} \qquad \lambda_2 = \lambda_1 + \lambda_{-2} \qquad \lambda_3 = \lambda_2 \times \lambda_0$$

$$\lambda_4 = \lambda_3 + \lambda_{-3} \qquad \lambda_5 = \lambda_4 \times \lambda_0 \qquad \lambda_6 = \lambda_5 + \lambda_{-4}$$

$$\lambda_7 = \lambda_6 \times \lambda_0 \qquad \lambda_8 = \lambda_7 + \lambda_{-5} \qquad \lambda_9 = \lambda_8 \times \lambda_0$$

$$\lambda_{10} = \lambda_9 + \lambda_{-6}.$$

Then the sequence $\{\lambda_k, k = 1, \ldots, 10\}$ defines the Horner straight-line program for evaluating a general polynomial of degree 5.

The steps λ_1, λ_3, λ_5, λ_7 and λ_9 are the multiplication steps active in a_0; the steps λ_2, λ_4, λ_6, λ_8 and λ_{10} are the addition steps active in a_1, a_2, a_3, a_4, a_5, respectively.

An example of a straight-line program which computes a polynomial with preprocessed coefficients is given in Section 2.5.1.

Now, we wish to establish the fact that in order to evaluate the general polynomial of degree n,

$$P(x; a_0, \ldots, a_n) = a_0 x^n + a_1 x^{n-1} + \ldots + a_{n-1} x + a_n,$$

using the computational model (1), at least n multiplication operations are required which actively involve the $\{a_k, k = 0, \ldots, n\}$.

To do that we prove a somewhat more general theorem.

Theorem 2.2 Let $\mathbf{U} = \{u_{ij}\}$, $0 \leqslant i \leqslant m$, $0 \leqslant j < n$, be an $(m + 1) \times (n + 1)$ matrix of real numbers, of rank $n + 1$; and let $\mathbf{V} = (v_0, v_1, \ldots, v_m)$ be a column vector of real numbers. Then, any polynomial straight-line program which computes

$$P(x; a_0, \ldots, a_n) = \sum_{i=0}^{m} (u_{i0} a_0 + u_{i1} a_1 + \ldots + u_{in} a_n + v_i) x^i$$

$$= (\mathbf{U}\mathbf{a} + \mathbf{V})^T \mathbf{x}, \tag{2}$$

where $\mathbf{a} = (a_0, \ldots, a_n)$ and $\mathbf{x} = (x^0, x, x^2, \ldots, x^m)$ are the column vectors and $m \geqslant n$,

involves at least n multiplication steps active in the $\{a_k, k = 0, \ldots, n\}$.

Here, the matrix $\mathbf{U}$ and the vector $\mathbf{V}$ are chosen arbitrarily, while the straight-line program and its variables, $\lambda_i, \lambda_j, \lambda_k$, will generally depend on the chosen $\mathbf{U}$ and $\mathbf{V}$. The proof establishes the fact that no matter how $\mathbf{U}$ and $\mathbf{V}$ are chosen, it is impossible to compute $P(x; a_0, \ldots, a_n)$ without doing n active in a_k multiplication steps in the corresponding straight-line program. Note that the fact that $\mathbf{U}$ is of rank $(n + 1)$ implies that $m \geqslant n$.

Proof. The theorem holds for $n = 0$, since in this case we have

$$P(x; a_0) = \sum_{i=0}^{m} (u_{i0} a_0 + v_i) x^i \tag{3}$$

and for $m = 0$ (recall that by definition $m \geqslant n$), the corresponding straight-line program, whatever form it may be given, will, in respect to the a_0, only involve a multiplication step which multiplies a_0 by a constant, u_{00}.

Next, for $n = 1$ we have

$$P(x; a_0, a_1) = \sum_{i=0}^{m} (u_{i0} a_0 + u_{i1} a_1 + v_i) x^i$$

and, in particular, for $m = n = 1$:

$$P(x; a_0, a_1) = (u_{00} a_0 + u_{01} a_1 + v_1) + (u_{10} a_0 + u_{11} a_1 + v_1)x. \qquad (4)$$

Now, the rank of matrix $\mathbf{U} = \begin{bmatrix} u_{00} & u_{01} \\ u_{10} & u_{11} \end{bmatrix}$ is assumed to be equal to 2, so at least three of the entries u_{ij} must be non-zero. It follows that evaluation of $P(x; a_0, a_1)$ in (4) involves at least one multiplication step which is active either in a_0 or in a_1.

We shall now use the method of induction.

Let $n > 1$ and let a polynomial straight-line program which computes $P(x; a_0, a_1, \ldots, a_n)$ be given. Further, let $\lambda_{i_q} = \lambda_{j_q} \times \lambda_{k_q}$ be the first multiplication step active in one of $a_0, \ldots, a_n$. Without loss of generality, we may assume that λ_{j_q} involves a_0. Thus, λ_{j_q} has the form $\beta a_0 + \alpha$, where β and α are real constants, $\beta \neq 0$. Now change step i_q to $\lambda_{i_q} = \kappa \times \lambda_{k_q}$ where κ is an arbitrary real number. Add further steps to calculate

$$\lambda = (\kappa - \alpha)/\beta.$$

These new steps involve only additions and constant multiplication by suitable new constants.

Finally, replace a_n everywhere in the straight-line program by this new variable λ. We thus get the straight-line program which calculates

$$R(x; a_0, \ldots, a_{n-1}) = P\left(x; a_0, \ldots, a_{n-1}, \frac{\kappa - \alpha}{\beta}\right),$$

and this straight-line program has one less multiplication step active in the $\{a_k, k = 0, \ldots, n\}$, since a_n has been eliminated.

Now, by virtue of (2), the polynomial R may be expressed as

$$R(x; a_0, \ldots, a_{n-1}) = \sum_{i=0}^{m} \left(u_{i0} a_0 + u_{i1} a_1 + \ldots + u_{in-1} a_{n-1} + u_{in} \frac{\kappa - \alpha}{\beta} + v_i \right) x^i$$

$$= (\mathbf{U}_1 \mathbf{a}_1 + \mathbf{V}_1)^{\mathrm{T}} \mathbf{x}, \qquad (5)$$

where we may denote

$$\mathbf{U}_1 = \begin{bmatrix} u_{00} & u_{01} & \ldots & u_{0n-1} \\ u_{10} & u_{11} & \ldots & u_{1n-1} \\ \vdots & & & \\ u_{m0} & u_{m1} & \ldots & u_{mn-1} \end{bmatrix}, \quad \mathbf{a}_1 = \begin{bmatrix} a_0 \\ a_1 \\ \vdots \\ a_{n-1} \end{bmatrix},$$

$$\mathbf{V}_1 = \begin{bmatrix} u_{0n} \dfrac{\kappa - \alpha}{\beta} + v_0 \\[2mm] u_{1n} \dfrac{\kappa - \alpha}{\beta} + v_1 \\[2mm] \vdots \\[2mm] u_{mn} \dfrac{\kappa - \alpha}{\beta} + v_m \end{bmatrix}, \quad \mathbf{x} = \begin{bmatrix} x^0 \\ x^1 \\ x^2 \\ \vdots \\ x^m \end{bmatrix}.$$

It follows from (5) that R satisfies the general form (2) assumed by the Theorem. Hence, by induction on n, its straight-line program involves at least $n - 1$ multiplication steps active in $\{a_k, k = 0, \ldots, n\}$ and so the original straight-line program which computes P has at least n multiplication steps active in $\{a_k; k = 0, \ldots, n\}$.

This completes the proof.

From Theorem 2.2, the standard Horner algorithm and the proof of its optimality in terms of the number of multiplications required follows as a special case, where the matrix $\mathbf{U}$ is set to be the identity matrix, $\mathbf{I}$, and the vector $\mathbf{V}$ is set to be the zero-vector.

Returning to the Theorem 2.1, we may say that since the Horner algorithm is an optimal algorithm for evaluating a polynomial with 'natural' coefficients and since this algorithm requires n multiplication operations to evaluate a polynomial of degree n, it follows that n multiplications is the minimum number of these operations which may be required by any algorithm for evaluating a polynomial with 'natural' coefficients. The proof of the optimal number of addition operations required may be developed in a similar way.

2.4 The Belaga Theorems

We now turn to the second question, namely that of determining the minimum number of arithmetic operations required to evaluate a polynomial with pre-processed coefficients. For this we require the following theorems:

Theorem 2.3 For any polynomial $P_n(x)$ of degree n, there exists a computational evaluation scheme that requires $\lfloor (n + 1)/2 \rfloor + 1$ multiplications and $n + 1$ additions.

Theorem 2.4 No evaluation scheme exists with less than $\lfloor (n + 1)/2 \rfloor$ multiplications or with less than n additions.

It has not yet been shown whether an evaluation scheme exists with exactly $\lfloor (n + 1)/2 \rfloor$ multiplications and n additions. Theorems 2.3 and 2.4 were first formulated by Belaga (1958). The first of the theorems is proved using a constructive approach, namely a concrete evaluation scheme is suggested which requires the number of multiplications and additions as stated in the theorem. We shall explore this scheme in greater detail. Theorem 2.4 establishes lower bounds on the number of multiplications and additions, under the assumption that some preprocessing of the coefficients is allowed without cost. Proofs of the theorem have been elaborated mainly by Winograd (1970) who in the course of proof uses some deep results from topology and by Knuth (1969) who argues the facts using the so called 'degrees of freedom' from number theory. Some readers might wish to consult a somewhat simpler proof due to Reingold and Stocks (1972) that is based on the notion of algebraic independence of the preprocessed parameters. It is felt, however, that a careful treatment of the proof of Theorem 2.4 is beyond the scope of the present text and hence interested readers are referred to the sources of information just mentioned. We shall now turn to the scheme for evaluation of a polynomial that has been suggested by Belaga. It is a general scheme which evaluates any polynomial of degree $n \geqslant 4$ with prior preprocessing of the coefficients and is given as:

$$V_1 = (x + \alpha_1)x,$$
$$V_2 = (V_1 + x + \alpha_2)(V_1 + \alpha_3) + \alpha_4,$$
$$V_k = V_{k-1}(V_1 + \alpha_{2k-1}) + \alpha_{2k} \qquad k = 3, \ldots, \lfloor n/2 \rfloor,$$
$$P_n(x) = F_n(x) = \begin{cases} \alpha_{n+1} V_{\lfloor n/2 \rfloor}, & n \text{ even} \\ \alpha_{n+1} x V_{\lfloor n/2 \rfloor}, & n \text{ odd} \end{cases} \quad n \geqslant 4. \tag{1}$$

We can see that the Belaga computational scheme requires $\lfloor (n+1)/2 \rfloor + 1$ multiplications and $2\lfloor n/2 \rfloor + 1$ additions.

A method for computing the parameters α_i in scheme (1) in terms of $(a_0, \ldots, a_n)$ where $P_n(x) = \sum_{k=0}^{n} a_k x^{n-k}$, was given by Cheney (1962).

From (1) we can write

$$P_n(x) = a_0 x^n + a_1 x^{n-1} + \ldots + a_n = \begin{cases} \alpha_{n+1} V_m, \\ \alpha_{n+1} x V_m \end{cases}, \qquad m = \left\lfloor \frac{n}{2} \right\rfloor. \tag{2}$$

Without restricting the generality of the method, we may assume that $a_0 = \alpha_{n+1} = 1$, and consider

$$V_m = x^{2m} + c_1 x^{2m-1} + c_2 x^{2m-2} + \ldots + c_{2m}$$
$$= (x^{2m-2} + b_1 x^{2m-3} + b_2 x^{2m-4} + \ldots + b_{2m-2})(x^2 + \alpha_1 x + \alpha_{2m-1})$$
$$+ \alpha_{2m}, \tag{3}$$

where $c_j = a_j, \quad j = 1, \ldots, 2m, \quad m = \left\lfloor \frac{n}{2} \right\rfloor.$

Equating like powers of x in (2) gives:

$$c_1 = \alpha_1 + b_1.$$

Setting $\alpha_1 = \dfrac{1}{m}(c_1 - 1)$, obtain $b_1 = \dfrac{(m-1)c_1 + 1}{m}$. This choice satisfies the requirement of the polynomial being monic.

$$c_2 = \alpha_{2m-1} + \alpha_1 b_1 + b_2.$$

Solve the equation for b_2.

$$c_k = \alpha_{2m-1} b_{k-2} + \alpha_1 b_{k-1} + b_k, \qquad k = 3, \ldots, 2m - 2.$$

Solve each equation for b_k.

$$c_{2m-1} = \alpha_{2m-1} b_{2m-3} + \alpha_1 b_{2m-2}.$$

Solve the equation for α_{2m-1}.

At this point, substitute the value of α_{2m-1} into the expressions for $b_{2m-2}, \ldots,$ b_2 and obtain their values.

Finally, solve the equation

$$c_{2m} = \alpha_{2m-1} b_{2m-2} + \alpha_{2m} \qquad \text{for } \alpha_{2m}. \tag{4}$$

We have now obtained values for $\alpha_n = \alpha_{2m}$ and $\alpha_{n-1} = \alpha_{2m-1}$, and also

$$\alpha_1 = \frac{1}{m}(c_1 - 1) = \frac{1}{m}(a_1 - 1).$$

By replacing m by $m - 1$ in (3) and solving the system obtained in the manner similar to the above, we obtain α_{n-2} and α_{n-3}. The process is repeated until all α_j are computed. Since the Cheney scheme for determining α_j in terms of the $\{a_0, \ldots, a_n\}$ is valid for any polynomial of degree n, it may be considered as a constructive proof of Theorem 2.3.

As an example of computing parameters of the Belaga form for a polynomial originally given in the power form, consider

$$P_7(x) = x^7 - x^6 + 8x^5 - 4x^4 + 6x^3 + 2x^2 - 5x + 1. \tag{5}$$

From (2) and (3) we get

$$V_3 = x^6 - x^5 + 8x^4 - 4x^3 + 6x^2 + 2x - 5$$
$$= (x^4 + b_1 x^3 + b_2 x + b_4)(x^2 + \alpha_1 x + \alpha_5) + \alpha_6$$

with $c_1 = -1$, $c_2 = 8$, $c_3 = -4$, $c_4 = 6$, $c_5 = 2$, $c_6 = -5$.

We now solve the system (4) for obtaining the values for α_6, α_5 (and α_1). We have

$(c_1) \quad -1 = \alpha_1 + b_1.$ We set $\alpha_1 = -\frac{2}{3}$, and get $b_1 = -\frac{1}{3}$.

$(c_2) \quad 8 = \alpha_5 + \alpha_1 b_1 + b_2,$ giving $b_2 = 7.778 - \alpha_5$.

$(c_3) \quad -4 = \alpha_5 b_1 + \alpha_1 b_2 + b_3,$ giving $b_3 = 1.185 - 0.333\alpha_5$.

$(c_4) \quad 6 = \alpha_5 b_2 + \alpha_1 b_3 + b_4,$ giving $b_4 = 6.790 - 8\alpha_5 + \alpha_5^2$.

$(c_5) \quad 2 = \alpha_5 b_3 + \alpha_1 b_4,$ giving $\alpha_5^2 - 6.519\alpha_5 + 6.527 = 0$,

                            and $(\alpha_5)_1 = 5.284$, $(\alpha_5)_2 = 1.236$.

$(c_6) \quad -5 = \alpha_5 b_4 + \alpha_6,$ giving two values for α_6,

                            $(\alpha_6)_1 = 34.952$ and $(\alpha_6)_2 = -3.059$,

                            corresponding to $(\alpha_5)_1$ and $(\alpha_5)_2$,
                            respectively.

At this stage two sets of the parameters values have been obtained:

(i) $\alpha_6 = 34.952$, $\alpha_5 = 5.284$, $\alpha_1 = -0.667$,
      $b_1 = -0.333$, $b_2 = 2.494$, $b_3 = -0.576$, $b_4 = -7.561$;

and

(ii) $\alpha_6 = -3.059$, $\alpha_5 = 1.236$, $\alpha_1 = -0.667$,
      $b_1 = -0.333$, $b_2 = 6.542$, $b_3 = 0.773$, $b_4 = -1.570$.

The calculations have now to be proceeded for each of the two sets independently. So, using set (i) we obtain

$$V_2 = x^4 - 0.333x^3 + 2.494x^2 - 0.576x - 7.561$$
$$= (x^2 + d_1 x + d_2)(x^2 + \alpha_1 x + \alpha_3) + \alpha_4.$$

The system (4) for obtaining the values for α_3 and α_4 is given by:

$(b_1) - 0.333 = \alpha_1 + d_1,$	giving $d_1 = \frac{1}{3} = 0.333.$
$(b_2)\quad 2.494 = \alpha_3 + \alpha_1 d_1 + d_2,$	giving $d_2 = 2.716 - \alpha_3.$
$(b_3) - 0.576 = \alpha_1 d_2 + \alpha_3 d_1,$	giving $\alpha_3 = 1.235.$
	We then get $d_2 = 1.481,$ and
	$\alpha_2 = d_2 = 1.481.$
$(b_4) - 7.561 = \alpha_3 d_2 + \alpha_4,$	giving $\alpha_4 = -9.390.$

This completes the calculation for set (i).

Now, using set (ii) we obtain

$$V_2 = x^4 - 0.333x^3 + 6.542x^2 + 0.773x - 1.570$$
$$= (x^2 + g_1 x + g_2)(x^2 + \alpha_1 x + \alpha_3) + \alpha_4.$$

The system (4) for obtaining the values for α_3 and α_4 is given by:

$(b_1)\quad -0.333 = \alpha_1 + g_1,$	giving $g_1 = \frac{1}{3} = 0.333.$
$(b_2)\quad 6.542 = \alpha_3 + \alpha_1 g_1 + g_2,$	giving $g_2 = 6.764 - \alpha_3.$
$(b_3)\quad 0.773 = \alpha_3 g_1 + \alpha_1 g_2,$	giving $\alpha_3 = 5.282.$
	We thus find $g_2 = 1.481,$ and
	$\alpha_2 = g_2 = 1.481.$
$(b_4)\quad -1.570 = \alpha_3 g_2 + \alpha_4,$	giving $\alpha_4 = -9.398.$

This completes the calculations for set (ii).

We have obtained two sets of the Belaga parameters:

α_1	α_2	α_3	α_4	α_5	α_6	α_7
$-\dfrac{2}{3}$	1.481	1.235	-9.390	1.236	-3.059	1
$-\dfrac{2}{3}$	1.481	5.282	-9.398	5.284	34.952	1

giving two variants of the Belaga algorithm for the $P_7(x)$:

(I) $\quad V_1 = (x - \frac{2}{3})x,$

$\qquad V_2 = (V_1 + x + 1.481)(V_1 + 1.235) - 9.390,$

$\qquad V_3 = V_2(V_1 + 1.236) - 3.059,$ $\hfill$ (6)

$\qquad P_7(x) = xV_3,$

and

(II) $V_1 = (x - \frac{2}{3})x,$

$V_2 = (V_1 + x + 1.481)(V_1 + 5.282) - 9.398,$

$V_3 = V_2(V_1 + 5.284) + 34.952,$ (7)

$P_7(x) = xV_3.$

In the case of the example at hand, we observe that the parameters of the Belaga form for a given polynomial are not unique. The other more serious practical flaw in the Belaga algorithm arises from the fact that some of the α_j may be complex for a polynomial with real 'natural' coefficients.

As an example of such a case consider the polynomial

$$P_6(x) = x^6 + x^5 + x^4 + x^3 + x^2 + x + 1.$$ (8)

The Belaga coefficients of this polynomial are:

$\alpha_1 = 0 \quad \alpha_2 = 0 \quad \alpha_3 = 1 - q$

$\alpha_4 = 0 \quad \alpha_5 = q \quad \alpha_6 = 1 \quad \alpha_7 = 1,$

where

$$q = \tfrac{1}{2}(1 \pm i\sqrt{3}),$$

and the Belaga algorithm is given by

$V_1 = x^2,$

$V_2 = (V_1 + x)(V_1 + (1 - q)),$ (9)

$P_6(x) = V_3 = V_2(V_1 + q) + 1.$

Pan (1959) has proposed another form for an economical evaluation of a polynomial. His form is valid for any polynomial of degree $n \geqslant 5$, and the computational algorithm is given as:

$W_0 = (x + \lambda_0)x,$

$W_1 = W_0 + x,$

$V_0 = x,$

$V_k = V_{k-1}((W_0 + \lambda_{4k-3})(W_1 + \lambda_{4k-2}) + \lambda_{4k-1}) + \lambda_{4k},$

$$P_n(x) = F_n(x) = \begin{cases} \lambda_n V_k & n = 4k + 1, \\ \lambda_n x V_k + \lambda_{n-1} & n = 4k + 2, \\ \lambda_n(V_k(W_0 + \lambda_{n-2}) + \lambda_{n-1}) & n = 4k + 3, \\ \lambda_n x(V_k(W_0 + \lambda_{n-3}) + \lambda_{n-2}) + \lambda_{n-1} & n = 4k + 4, \end{cases} \quad k \geqslant 1.$$ (10)

The Pan algorithm requires $\lfloor n/2 \rfloor + 2$ multiplications and $n + 1$ additions. For n odd, this gives the same number of arithmetic operations as the Belaga algorithm, and for n even, it gives one extra multiplication.

It is worth noting here that Motzkin (1955) was the first to announce, though without proof, a result which implied that the values of a polynomial $P(x)$ can be computed from an algorithm employing $\lfloor n/2 \rfloor + 2$ multiplications. A limit on the number of additions was not given.

The Pan parameters are, again, not uniquely determined. For example, Rice (1965) gives the following three variants of the Pan form for the polynomial

$$x^7 - 2x^5 + x^3 - 0.1x$$

λ_0	λ_1	λ_2	λ_3	λ_4	λ_5	λ_6	λ_7
$-1/3$	-1.15	-0.481	0.179	0.223	-0.035	0.0078	1
$-1/3$	-0.268	-0.481	-0.264	0.018	-0.917	0.016	1
$-1/3$	-0.804	-0.481	-0.087	-0.043	-0.381	-0.016	1

However, the great advantage of the Pan form over the Belaga form is that for a polynomial with real 'natural' coefficients, all the parameters in (10) are always real.

More recent methods by Rabin and Winograd (1971) show how to achieve $\lfloor n/2 \rfloor + O(\log_2 n)$ multiplication/division operations and $n + O(n)$ addition operations using only rational preprocessing of coefficients. Eve (1964) has also suggested schemes for rational preprocessing of coefficients.

A non-uniqueness of the preprocessed coefficients may at times be considered as a useful attribute since it gives extra freedom of choice among available computational schemes in specific cases. Winograd (1974) gives an example of such a polynomial:

$$\frac{1}{81}x^4 + \frac{5}{81}x^3 + \frac{4}{27}x^2 + \frac{11}{27}x + \frac{5}{9} \tag{11}$$

for evaluation of which along with the traditional Horner scheme,

$$\left(\left(\left(\frac{1}{81}x + \frac{5}{81}\right)x + \frac{4}{27}\right)x + \frac{11}{27}\right)x + \frac{5}{9}, \tag{11A}$$

we can consider using one of the following preprocessed schemes:

(Todd–Motzkin) $\dfrac{1}{81}((x(x+2)+21)(x(x+2)+x-15)+360),$ (11B)

(Rabin–Winograd) $\left(\dfrac{1}{3}x\left(\dfrac{1}{3}x-\dfrac{2}{3}\right)+\dfrac{85}{81}\right)\left(\dfrac{1}{3}x\left(\dfrac{1}{3}x-\dfrac{2}{3}\right)+x-\dfrac{59}{81}\right)+\dfrac{8660}{6561},$

(11C)

(Rabin–Winograd) $\left(\dfrac{1}{3}x\left(\dfrac{1}{3}x+\dfrac{1}{3}\right)+\dfrac{25}{27}\right)\left(\dfrac{1}{3}x\left(\dfrac{1}{3}x+\dfrac{1}{3}\right)+\dfrac{1}{3}x-\dfrac{1}{27}\right)+\dfrac{430}{729}.$

(11D)

Here, inspecting expression (11B) we see that if it happens to be used for the computation of the polynomial values for small x, an uncomfortable problem in rounding errors may occur. Noting further that the algorithms of this type might well be used for building up, say, the elementary function procedures, such as $\sin x$, where it is quite common to call the routine for small x the scheme (11B) will prove not very suitable in practice.

On the other hand, the use of scheme (11C) or, even better, scheme (11D) greatly reduces the rounding error problems incurred by algorithm (11B) since the former have evenly-balanced coefficients.

2.5 Polynomials of Degree Less Than or Equal to 6

It is probably the case that the majority of polynomials that arise in practice are of degree 6 or less and we therefore examine in detail some schemes for the economical evaluation of these polynomials. It is not necessary, however, to consider polynomials of degree 2 or 3, since no advantage over the Horner scheme can be obtained by preprocessing the coefficients in these cases. Accordingly, we restrict consideration in turn to polynomials of degree 4, 5, and 6.

2.5.1 Polynomials of Degree Four

The Belaga theorem on the evaluation of a polynomial with preprocessed coefficients shows that the lowest degree polynomial which yields some savings in the number of multiplications as compared with the Horner scheme, is the polynomial of degree four. The least number of multiplications required for this polynomial is 3. Some computational schemes which achieve this lower bound are given below.

Consider

$$P_4(x) = a_0 x^4 + a_1 x^3 + a_2 x^2 + a_3 x + a_4. \tag{1}$$

The Motzkin–Todd form has already been discussed (see formulae $2.2 - (1) - (2) - (3)$) and we now present it in a slightly different form from $2.2 - (1)$:

$$P_4(x) = M_4(x) = \alpha_0 \big[[x(x + \alpha_0) + \alpha_1] [x(x + \alpha_0) + \alpha_1 + x + \alpha_2] + \alpha_3 \big] \tag{2}$$

where

$$\alpha_0 = \tfrac{1}{2}(a_1/a_0 - 1), \qquad \beta = a_2/a_0 - \alpha_0(\alpha_0 + 1),$$
$$\alpha_1 = a_3/a_0 - \alpha_0\beta, \qquad \alpha_2 = \beta - 2\alpha_1,$$
$$\alpha_3 = a_4/a_0 - \alpha_1(\alpha_1 + \alpha_2).$$

We give also a straight-line arithmetic program for the form (2) as an illustration of this modelling technique applied to the problem of evaluating a polynomial with preprocessed coefficients.

The program is as follows:

$$V_1 = x + \alpha_0$$
$$V_2 = V_1 x$$
$$V_3 = V_2 + \alpha_1$$
$$V_4 = V_3 + x$$
$$V_5 = V_4 + \alpha_2 \qquad\qquad (3)$$
$$V_6 = V_5 V_3$$
$$V_7 = V_6 + \alpha_3$$
$$V_8 = V_7 \alpha_4$$

It can immediately be seen that 3 multiplications and 5 additions are required to evaluate the Motzkin–Todd form.

We also observe that in order to be able to execute the statement $V_6 = V_5 V_3$, the algorithm requires one storage operation for the quantity V_3. Clearly, an accurate estimate of the time efficiency of the algorithm would require the storage operation to be taken into account together with the number of multiplications and additions.

In future, where appropriate, we shall indicate the number of storage operations, along with the account of basic arithmetic operations.

It is also worth noting that polynomials arising in practice often have a rather small leading coefficient, so that the division by a_0 above leads to numerical instability. In such a case it is usually preferable to replace x by $x^4\sqrt{|a_0|}$ as the first step and a similar transformation may be applied to polynomials of higher degree. This idea is due to Fike (1967).

Knuth has suggested the following small variation of the Motzkin–Todd form:

$$P_4(x) = K_4(x) = a_0 \left[\left[x(x + \alpha_0) + \alpha_1 \right] \left[x(x + \alpha_0) + \alpha_1 - x + \alpha_2 \right] + \alpha_3 \right] \qquad (4)$$

where

$$\alpha_0 = \tfrac{1}{2}(a_1/a_0 + 1), \qquad\qquad \beta = a_2/a_0 - \alpha_0(\alpha_0 - 1),$$
$$\alpha_1 = \alpha_0\beta - a_3/a_0, \qquad\qquad \alpha_2 = \beta - 2\alpha_1,$$
$$\alpha_3 = a_4/a_0 - \alpha_1(\alpha_1 + \alpha_2).$$

Although the Knuth algorithm requires the same number of operations as the Motzkin–Todd, they may differ significantly in the accuracy of the computed results. The Knuth algorithm may perform better in some cases while being worse in other. Further details on numerical accuracy are given in Section 2.6. We observe that compared with the Horner algorithm, both the Motzkin–Todd and the Knuth algorithms achieve a saving of 1 multiplication at a price of 1 extra addition and 1 storing operation.

Theorem 2.4 leaves open the question of whether or not it is possible to evaluate a fourth-degree polynomial with 3 multiplications and 4 additions.

28

A general algorithm for doing this is not known; on the other hand, there is as yet no proof of its impossibility.

2.5.2 Polynomials of Degree Five

An economical form of evaluating a general polynomial of degree five may be obtained using the rule

$$P_5(x) = P_4(x)x + a_5, \tag{1}$$

where

$$P_4(x) = a_0 x^4 + a_1 x^3 + a_2 x^2 + a_3 x + a_4$$

and evaluating $P_4(x)$ by one of the schemes given in the previous section. This algorithm requires 4 multiplications and 6 additions together with 1 storage operation.

The usual Pan form is given as:

$$P_5(x) = A_5(x) = [[[x(x + \lambda_0) + \lambda_1][x(x + \lambda_0) + x + \lambda_2] + \lambda_3]x + \lambda_4]\lambda_5 \tag{2}$$

where the λ_j are determined from:

$$2\lambda_0 + 1 = \frac{a_1}{a_0}$$

$$\lambda_1 + \lambda_0(\lambda_0 + 1) + \lambda_2 = \frac{a_2}{a_0}$$

$$\lambda_1(\lambda_0 + 1) + \lambda_0\lambda_2 = \frac{a_3}{a_0}$$

$$\lambda_1\lambda_2 + \lambda_3 = \frac{a_4}{a_0}$$

$$\lambda_4 = \frac{a_5}{a_0}$$

$$\lambda_5 = a_0.$$

The algorithm based on (2) requires 1 extra storage operation compared with algorithm (1), and thus may not be considered as a particularly attractive algorithm for the fifth-degree polynomial.

Knuth has suggested the following alternative form:

$$P_5(x) = V_5(x) = [[[(x + \alpha_0)^2 + \alpha_1](x + \alpha_0)^2 + \alpha_2](x + \alpha_3) + \alpha_4]\alpha_5 \tag{3}$$

where α_0 is obtained as a root of the cubic equation

$$40z^3 - 24a_1 z^2 + (4a_1^2 + 2a_2)z + (a_3 - a_2 a_1) = 0$$

(this equation always has at least one real root, since the polynomial on the left approaches $+\infty$ for large positive z, and it approaches $-\infty$ for large negative z. It may also happen that all three roots are real; in this case, the form (3) is not uniquely determined),

$$\alpha_3 = a_1 - 4\alpha_0,$$
$$\alpha_1 = a_2 - 4\alpha_0\alpha_3 - 6\alpha_0^2,$$
$$\alpha_2 = a_4 - \alpha_0(\alpha_0\alpha_1 + 4\alpha_0^2\alpha_3 + 2\alpha_1\alpha_3 + \alpha_0^3),$$
$$\alpha_4 = a_5 - \alpha_3(\alpha_0^4 + \alpha_1\alpha_0^2 + \alpha_2)$$
$$\alpha_5 = a_0.$$

The algorithm based on (3) requires 1 fewer addition than algorithms (1) and (2), although it uses 3 storage operations as compared with 1 in algorithm (1) and 2 in algorithm (2).

2.5.3 Polynomials of Degree Six

The Pan form for this case is as follows:

$$P_6(x) = A_6(x) = [[[x(x + \alpha_0) + \alpha_1 - x + \alpha_3][x(x + \alpha_0) + \alpha_1 + x + \alpha_2]$$
$$+ \alpha_4][x(x + \alpha_0) + \alpha_1] + \alpha_5]\alpha_6 \tag{1}$$

where it can be verified that

$$\alpha_0 = a_1/3 \text{ and}$$
$$\alpha_1 = (a_5 - \alpha_0 a_4 + \alpha_0^2 a_3 - \alpha_0^3 a_2 + 2\alpha_0^5)/(a_3 - 2\alpha_0 a_2 + 5\alpha_0^3).$$

As can be seen, the Pan method requires that the denominator does not vanish, i.e. that

$$a_3 - 2a_2\frac{a_1}{3a_0} + 5\left(\frac{a_1}{3a_0}\right)^3 \neq 0. \tag{2}$$

Indeed, in practice this quantity should not be so small that α_1 becomes un-computably large.

The remaining α's may be determined from the equations

$$\beta_1 = 2\alpha_0, \quad \beta_2 = a_2 - \alpha_0\beta_1 - \alpha_1,$$
$$\beta_3 = a_3 - \alpha_0\beta_2 - \alpha_1\beta_1, \quad \beta_4 = a_4 - \alpha_0\beta_3 - \alpha_1\beta_2,$$
$$\alpha_3 = \tfrac{1}{2}(\beta_3 - (\alpha_0 - 1)\beta_2 + (\alpha_0 - 1)(\alpha_0^2 - 1)) - \alpha_1,$$
$$\alpha_2 = \beta_2 - (\alpha_0^2 - 1) - \alpha_3 - 2\alpha_1,$$
$$\alpha_4 = \beta_4 - (\alpha_2 + \alpha_1)(\alpha_3 + \alpha_1), \quad \alpha_5 = a_6 - \alpha_1\beta_4.$$
$$\alpha_6 = a_0.$$

The algorithm based on (1) requires 4 multiplications and 8 additions plus 2 storage operations.

We also note that since the use of the Pan algorithm is restricted to condition (2), it cannot be considered to be completely general for a polynomial of degree five.

The Knuth form, which is slightly more economical than the Pan form, is given by

$$P_6(x) = K_6(x)$$
$$= [[[x(x + \alpha_0) + \alpha_1](x + \alpha_2) + \alpha_3 + x(x + \alpha_0) + \alpha_1 + \alpha_4]$$
$$\times [[x(x + \alpha_0) + \alpha_1](x + \alpha_2) + \alpha_3] + \alpha_5]\alpha_6 \qquad (3)$$

where

$$\beta_1 = \tfrac{1}{2}(a_1 - 1), \quad \beta_2 = a_2 - \beta_1(\beta_1 + 1), \quad \beta_3 = a_3 - \beta_1\beta_2,$$
$$\beta_4 = \beta_1 - \beta_2, \quad \beta_5 = a_4 - \beta_1\beta_3,$$

and β_6 is a real root of the cubic equation

$$2y^3 + (2\beta_4 - \beta_2 + 1)y^2 + (2\beta_5 - \beta_2\beta_4 - \beta_3)y + (a_5 - \beta_2\beta_5) = 0. \qquad (4)$$

Defining

$$\beta_7 = \beta_6{}^2 + \beta_4\beta_6 + \beta_5, \quad \beta_8 = \beta_3 - \beta_6 - \beta_7,$$

we have finally

$$\alpha_0 = \beta_2 - 2\beta_6, \quad \alpha_2 = \beta_1 - \alpha_0, \quad \alpha_1 = \beta_6 - \alpha_0\alpha_2,$$
$$\alpha_3 = \beta_7 - \alpha_1\alpha_2, \quad \alpha_4 = \beta_8 - \beta_7 - \alpha_1, \quad \alpha_5 = a_6 - \beta_7\beta_8.$$
$$\alpha_6 = a_0$$

Clearly the Knuth form is also not uniquely determined in the case when equation (4) possesses three real roots.

No algorithm to evaluate a sixth-degree polynomial in 4 multiplication and 6 addition operations is yet known.

It is obvious that the savings in computer time in evaluation polynomial forms with preprocessed coefficients, as compared with evaluation the forms with 'natural' coefficients, become important only on a computer for which the operation of multiplication is significantly more expensive than that of addition. Some interesting observations on the efficiency of the Horner and the Pan algorithms when using different computers have been made by Rice (1972):

Degree of polynomial	Evaluation algorithm	CDC 3600	Computer 360/50	360/75
		(time is given in microseconds)		
4	Horner	44.60	113.5	12.44
	Belaga[a]	51.97	98.9	11.23
5	Horner	55.25	141	15
	Pan	58.37	122	13.8
6	Horner	66	169	18.5
	Pan	73	136	15.6

[a] No Pan algorithm exists for a polynomial of degree $n < 5$.

2.6 Accuracy of the Numerical Solution and Conditioning of the Problem

The problem of assessing the accuracy of computed results depends critically on the exact way in which numbers are represented within the computer.

For our needs, we shall distinguish between mathematical (or round-off) errors and physical (or inherent) errors in the data. The round-off errors are usually small and what is very important can be kept under control, in the sense that it is usually possible to specify upper bounds for the absolute values of such errors. The physical errors, on the other hand, may be quite large and are outside our control. We must first define the various types of error with which we shall be concerned.

If x is the true value of the number, and X an approximation to the true value, then we define

$$e = X - x \qquad \text{as the } error \qquad (1)$$

$$|e| = |X - x| \qquad \text{as the } absolute\ error \text{ and} \qquad (2)$$

$$\varepsilon = \frac{e}{x} = \frac{X - x}{x} \qquad \text{as the } relative\ error; \qquad (3)$$

we can also write $X = x + e$ and

$$X = x(1 + \varepsilon). \qquad (4)$$

2.6.1 Floating-point Number Representation and Summary of Some Basic Results on Floating-point Arithmetic

In the majority of modern digital computers, numbers are stored in the so-called floating-point (fl) form, i.e.

$$x = 2^b a \qquad (5)$$

where b (the exponent) is a positive or negative integer (or zero) and a (the mantissa) is a fraction in the range $-\frac{1}{2} \geqslant a > -1$ or $\frac{1}{2} \leqslant a < 1$, and of fixed length t digits. If we assume that each arithmetic operation is carried out internally to double precision and the result subsequently rounded to t significant digits, then denoting the floating-point digital representation of the number by the symbol fl or by the bar, we can write

$$\bar{x} = \text{fl}(x) = x(1 + \varepsilon), \qquad |\varepsilon| \leqslant 2^{-t}, \qquad (6)$$

where $|\varepsilon|$ denotes the absolute value of the relative error. The results given below are due to Wilkinson (1963).

Assuming that the input data is read in exactly, we have

$$\text{fl}(x_1 \pm x_2) = (x_1 \pm x_2)(1 + \varepsilon), \qquad |\varepsilon| \leqslant 2^{-t} \qquad (7)$$

$$\text{fl}(x_1 x_2) = (x_1 x_2)(1 + \varepsilon), \qquad |\varepsilon| \leqslant 2^{-t} \qquad (8)$$

$$\text{fl}(x_1/x_2) = (x_1/x_2)(1 + \varepsilon), \qquad |\varepsilon| \leqslant 2^{-t} \qquad (9)$$

Next, we shall state bounds for the errors involved in the calculation of extended products, sums and inner-products, since these will be used in various analyses throughout the text.

For the extended product $x_1 x_2 \ldots x_n$, we have

$$\mathrm{fl}(x_1 x_2 \ldots x_n) = x_1 x_2 \ldots x_n (1 + E), \tag{10}$$

where

$$|E| < (n - 1)2^{-t_1} \tag{11}$$

and t_1, for compactness of presentation, is defined by the relation

$$2^{-t_1} = (1.06)2^{-t}$$

so that

$$t_1 = t - 0.08406.$$

Much the same result may be obtained for extended sequences of multiplication and divisions. We have

$$\mathrm{fl}(x_1 x_2 \ldots x_m / y_1 y_2 \ldots y_n) = (x_1 x_2 \ldots x_m / y_1 y_2 \ldots y_n)(1 + E) \tag{12}$$

where

$$|E| < (m + n - 1)2^{-t_1} \tag{13}$$

It may be observed from (11) and (13) that for n (and $n + m$) appreciably smaller than 2^t, the computed results, given by (10) and (12), have a low relative error, E.

This observation is not, however, valid for extended sequences of additions, and here we have

$$\mathrm{fl}(x_1 + x_2 + \ldots + x_n) = x_1(1 + \eta_1) + x_2(1 + \eta_2) + \ldots x_n(1 + \eta_n), \tag{14}$$

where

$$|\eta_r| < (n + 1 - r)2^{-t_1}, \quad r = 1, \ldots, n. \tag{15}$$

Finally, for the inner product with accumulation we have

$$\sum_{i=1}^{n} x_i y_i = \left[\sum_{i=1}^{n} x_i y_i (1 + e_i) \right](1 + e) \tag{16}$$

where

$$|e_r| < \tfrac{3}{2}(n + 2 - r)2^{-2t_2}, \quad 2^{-t_2} = (1.06)2^{-2t} \tag{17}$$
$$|e| < 2^{-t}$$

2.6.2 Numerical Accuracy and Conditioning

As we have just seen a computer representation of any number consists of a finite number of digits only.

This limited accuracy in the representation of numbers in a computer gives rise to a phenomenon known as the numerical (or induced) instability. Numerical instability is associated with a particular computational scheme. The instability manifests itself in a growth of round-off errors during the computation.

A computational scheme which may generate an uncontrolled growth of round-off errors is said to be numerically unstable, and a computational scheme which controls the growth of round-off errors is described as numerically stable.

Another concept which relates to the fact that for some forms there occurs a much higher loss of accuracy during evaluation than one would normally expect is that of ill-conditioning. Ill-conditioning is associated with a particular formulation of the mathematical problem. It exists independently of round-off effects and its nature is unaffected by various schemes for the control of round-off errors. Suppose that the Horner form of a polynomial which has been obtained as a mathematical model describing a certain physical process, has one (or more) coefficient, a, which is uncertain. By uncertainty in this context we mean that the coefficient is known only to lie within some interval $(a_h - \varepsilon_1, a_h + \varepsilon_2)$. The value of the polynomial, p, taken at some fixed point will lie in the corresponding interval $(p - \eta_1, p + \eta_2)$. The quantities η_1 and η_2 are functions of ε_1 and ε_2 and hence one of two possible cases will occur:

  (i) for 'small' ε_1 and ε_2,
     the quantities η_1 and η_2 will be 'small' as well;
  (ii) for 'small' ε_1 and ε_2,
     the quantities η_1 and η_2 will be very 'large'.

(We use the terms 'small' and 'large' in the relative sense, hence the quotation marks.)

In case (i) we say that the particular polynomial form (i.e. the Horner form in our example) 'defines the polynomial with the accuracy of the coefficients'. Computationally, it means that provided a stable computational algorithm is used, a specified value of the polynomial may be computed with some degree of accuracy. In case (ii) we say that the particular polynomial form does not define the polynomial with the accuracy of its coefficients. We also say that this polynomial form is ill-conditioned. For such a form, however carefully a computational evaluation algorithm may be chosen, it cannot alleviate the effects of ill-conditioning in the form. There is as yet no general theory about the circumstances which give rise to ill-conditioning in various problems; each must be analysed individually.

The problem of ill-conditioning in the evaluation of polynomials has been studied by Rice (1965). He first observed that the ill-conditioning is basically a relative concept and that in relation to this phenomenon one is concerned with the question of how different the computed result is from what one would expect it to be, allowing normal round-off effects. He then introduced a quantitative measure of degree of ill-conditioning of a polynomial form at a point and in an interval. This measure relates the number of arithmetic operations required by a particular algorithm for evaluating a polynomial at a fixed point, a number of

34

Table 2.1

Conditioning of the form	Polynomial form	Stability of the algorithm
May be ill-conditioned	Horner	May be unstable
Always well-conditioned	Chebyshev	Always stable
Comparatively well-conditioned	Root Product	Always stable
Well-conditioned variants of Newton's form are obtainable, given a standard Newton's form, which displays symptoms of ill-conditioning, however at a price of involving more parameters and more operations for evaluation as compared with the standard form	Newton	Stability depends on the choice of the interpolation points. More or less equal distribution of these points over the interval of evaluation will usually give satisfactory stability. The interpolation points can also be chosen so as to guarantee maximum stability
May be ill-conditioned, by empirical evidence. Conjecture: If the polynomial is such that its Belaga and/ or Pan coefficients are small then it is well-conditioned	$\left\{\begin{array}{l}\text{Belaga}\\[1em]\text{Pan}\end{array}\right.$	May be very unstable May be extremely unstable

digits used in a floating-point arithmetic to the number of accurate significant digits in the value or in the specific norm of the polynomial. Finally, he tested a number of polynomials experimentally for their conditioning.

The numerical stability of various evaluation algorithms has been studied by several authors. The Horner algorithm, in particular, has been examined very thoroughly (see, for example, Fox and Mayers (1968)). Other algorithms for polynomial evaluation have been analysed by Mesztenyi and Witzgall (1967), Hart (1968), Bakhvalov (1971), Newbery (1974).

In Table 2.1 we present a summary of the results of these studies.

No computational algorithm based on an ill-conditioned form will accurately evaluate the polynomial since ill-conditioning is a function of the way in which a problem is formulated.

One may search for a different better-conditioned polynomial form instead but this is a numerically very delicate problem. This follows from the observation that a particular polynomial form may be ill-conditioned not only in terms of its evaluation at a specified argument but also in terms of the problems associated with the transformation of one polynomial form into another. For example, the form may be ill-conditioned in terms of its root-finding. On the other hand, if the evaluation algorithm based on a particular polynomial form is numerically unstable, while the form itself is well-conditioned, then it may be preferable to transform the form in question to some other form for which a stable evaluation algorithm is available. For example, in some cases it is advantageous to transform the conventional power form into an orthogonal form (e.g. the Chebyshev form) and use this to evaluate the polynomial. Another way of putting the matter is to say that one should first ensure that the chosen polynomial form is well-conditioned and then select an evaluation algorithm that is numerically stable.

The cases when the algorithm based on the Chebyshev polynomial form will not perform better than the Horner algorithm has been studied by Newbery (1974). He has shown that when the Horner coefficients of the polynomial are of uniform sign or of strictly alternating sign, then the Horner algorithm, in terms of the upper bound on the computational error, may indeed be better— or, at least, not worse—than the algorithm based on the Chebyshev form. In other words, the transformation of the Horner form to the Chebyshev form in these two cases would not guarantee more accurate computations. Bakhvalov (1971) has given an instance when it is advantageous to transform a given Horner polynomial form into a Chebyshev form.

2.6.3 Error Analysis of Evaluation Algorithms

Assuming that the polynomial forms are well-conditioned, we now present some results of the studies of round-off error effects in the evaluation of polynomials at a fixed point, using various computational algorithms.

We consider

$$P_n(x) = \sum_{k=0}^{n} a_{n-k} x^k = \sum_{k=0}^{n} b_k T_k(x) = Q_n(x) \tag{1}$$

It is important to bear in mind that the form $Q_n(x)$ assumes that x is given in the interval $[-1, 1]$ since the Chebyshev polynomials, $T_k(x)$, are defined in this interval only. The transformation when considering the form $Q_n(x)$ does not result in any loss of generality and may be readily carried out, cf. Newbery (1974).

We assume, further, that the a_{n-k}, b_k and x have round-off errors only, that is

$$
\begin{aligned}
\bar{a}_{n-k} &= \mathrm{fl}(a_{n-k}) = a_{n-k}(1 + \varepsilon_{n-k}), \\
\bar{b}_k &= \mathrm{fl}(b_k) = b_k(1 + \varepsilon_k), \\
\bar{x} &= \mathrm{fl}(x) = x(1 + \delta),
\end{aligned}
\tag{2}
$$

where $|\varepsilon_{n-k}|$, $|\varepsilon_k|$, $|\delta|$ do not exceed 2^{-t}.

Let the exact value $P(x)$ of the polynomial with coefficients a_r at the point x be

$$P(x) = \sum_{k=0}^{n} a_{n-k} x^k. \tag{3}$$

and the exact value $\bar{P}(x)$ of the polynomial with coefficients $\bar{a}_r$ at the point x be given by

$$\bar{P}(x) = \sum_{k=0}^{n} \bar{a}_{n-k} x^k. \tag{4}$$

If $\bar{a}_{n-k} = a_{n-k}(1 + \varepsilon_{n-k})$ and $|\varepsilon_{n-k}| \leqslant \varepsilon < 2^{-t}$ then the difference

$$\bar{P}(x) - P(x) = \sum_{k=0}^{n} a_{n-k} \varepsilon_{n-k} x^k,$$

giving

$$|\bar{P}(x) - P(x)| \leqslant \varepsilon \Sigma |a_{n-k}| \, |x^k|. \tag{5}$$

For all polynomials this bound can be attained by taking $\varepsilon_{n-k} = \varepsilon \operatorname{sign}(a_{n-k} x^k)$.

Now, let the exact value $P(\bar{x})$ of the polynomial with coefficients a_r at the point $\bar{x} = x(1 + \delta)$ be given by

$$P(\bar{x}) = \sum_k a_{n-k} x^k (1 + \delta)^k. \tag{6}$$

Then the error

$$P(\bar{x}) - P(x) = \sum_k k \delta a_{n-k} x^k, \tag{7}$$

which is obtained by retaining in the expression for $P(\bar{x})$ the first order terms only. We have

$$|P(\bar{x}) - P(x)| \leqslant \sum_k k \varepsilon |a_{n-k}| \, |x^k| \qquad \text{when } |\delta| \leqslant \varepsilon < 2^{-t},$$

and the bound can be attained only if all $a_{n-k} x^k$ are of the same sign. Note though that in this case,

$$\frac{|P(\bar{x}) - P(x)|}{|P(x)|} \leqslant n\varepsilon, \tag{8}$$

and hence the result then has a low relative error.

The computed value $\overline{P(\bar{x})} = \mathrm{fl}(\bar{P}(\bar{x}))$ obtained using floating-point arithmetic for the polynomial with coefficients $\bar{a}_r$ at the point $\bar{x}$ is given by

$$\overline{P(\bar{x})} = \mathrm{fl}(\bar{P}(\bar{x})) = \sum_k \bar{a}_{n-k} \bar{x}^k (1 + \varepsilon_k)$$

with $|\varepsilon_k| \leqslant (2k + 1)2^{-t}$ to first order.

Retaining the first order terms only, we get

$$|\overline{P(\bar{x})} - P(x)| \leqslant \Sigma |a_{n-k}| \, |x^k| (3k + 2)2^{-t}$$
$$\leqslant (3n + 2)2^{-t} \Sigma |a_{n-k}| \, |x^k| \tag{9}$$

with the same assumptions about $\bar{a}_r$ and $\bar{x}$ as above.

We note from (9) that for all polynomials and all x this bound is merely $(3n + 2)$ times as large as the bound resulting from the errors in the coefficients alone and the latter was attainable. This factor is rarely if ever, attainable.

We now follow a more detailed analysis of round-off error accumulation in various evaluation algorithms.

First consider the Horner algorithm as given by $2.1 - (3)$. By virtue of assumptions (2), we have

$$\bar{V}_n = \text{fl}(a_0) = a_0(1 + \varepsilon_0) \tag{10}$$

and
$$\begin{aligned}
\bar{V}_k &= \text{fl}(\bar{x}\bar{V}_{k+1} + \bar{a}_{n-k}) \\
&= [x(V_{k+1} + T_{k+1})(1 + \delta)(1 + \alpha_k) + a_{n-k}(1 + \varepsilon_{n-k})](1 + \beta_k) \\
&= xV_{k+1}(1 + \delta + \alpha_k + \beta_k) + xT_{k+1} + \alpha_{n-k}(1 + \varepsilon_{n-k} + \beta_k)
\end{aligned} \tag{11}$$

where T_{k+1} denotes the absolute error in the computed value of V_{k+1}, and $|\delta|, |\alpha_k|, |\beta_k|, |\varepsilon_{n-k}| \leqslant 2^{-t}$, $k = n - 1, \dots, 0$.

Here, as before, assuming that the errors are small, we have ignored terms containing powers of errors higher than unity. It follows that

$$T_n = |\bar{V}_n - V_n| \leqslant |a_0|\varepsilon \tag{12}$$
$$\begin{aligned}
T_k &= |\bar{V}_k - V_k| = |\text{fl}(\bar{x}\bar{V}_{k+1} + \bar{a}_{n-k}) - (xV_{k+1} + a_{n-k})| \\
&\leqslant |x|T_{k+1} + 3|x||V_{k+1}|\varepsilon + 2|a_{n-k}|\varepsilon
\end{aligned} \tag{13}$$

where $\varepsilon = \max\{|\alpha_k|, |\beta_k|, |\delta|, |\varepsilon_{n-k}|\}$ and $\varepsilon \leqslant 2^{-t}$

Hence we have

$$T_k \leqslant |x|T_{k+1} + (3|x||V_{k+1}| + 2|a_{n-k}|)\varepsilon$$

which shows that the accuracy of the Horner algorithm is quite sensitive to the magnitude of x. This observation was made by Newbery (1974).

Next, for the Root Product algorithm we have:

$$\begin{aligned}
\bar{V}_n &= \text{fl}(a_0) = a_0(1 + \varepsilon_0) \\
\bar{V}_k &= \text{fl}([\bar{x} - \bar{\gamma}_{k+1}]\bar{V}_{k+1}) \\
&= \text{fl}([x(1 + \delta) - \gamma_{k+1}(1 + w_{k+1})](V_{k+1} + T_{k+1})) \\
&= [x(1 + \delta) - \gamma_{k+1}(1 + w_{k+1})](1 + \beta_k)(V_{k+1} + T_{k+1})(1 + \alpha_k) \\
&= x(V_{k+1} + T_{k+1})(1 + \delta)(1 + \beta_k)(1 + \alpha_k) - \gamma_{k+1}(V_{k+1} + T_{k+1}) \\
&\quad \times (1 + w_{k+1})(1 + \beta_k)(1 + \alpha_k)
\end{aligned} \tag{14}$$

Proceeding as before, by first retaining only linear error terms in (14) and then considering the differences between the exact and computed values of V_j's we get for the errors

$$\begin{aligned}
T_n &= |\bar{V}_n - V_n| \leqslant |a_0|\varepsilon \\
T_k &= |\bar{V}_k - V_k| \leqslant |x - \gamma_{k+1}|T_{k+1} + 3|(x - \gamma_{k+1})V_{k+1}|\varepsilon
\end{aligned} \tag{15}$$

where $\varepsilon \leqslant 2^{-t}$.

Further, using

$$|V_k| = |x - \gamma_{k+1}||V_{k+1}| \tag{16}$$

we can write

$$T_k \leqslant 3|V_k|\varepsilon + \frac{|V_k|}{|V_{k+1}|} T_{k+1},\qquad (17)$$

and for the relative error we get

$$\frac{T_k}{|V_k|} \leqslant \frac{T_{k+1}}{|V_{k+1}|} + 3\varepsilon \qquad (18)$$

We see that the evaluation algorithm 2.1–(5), based on the Root Product form is completely stable.

It should however be kept in mind that although theoretically transformation into the Root Product form is always possible, finding the roots may be an unstable process itself, notwithstanding availability of a whole arsenal of techniques for polynomial root searching, and, thus, the transformation may be of little practical help.

For the Newton algorithm given by 2.1–(7), similar analysis gives the following relations for the errors

$$T_n = |\bar{V}_n - V_n| \leqslant |a_0|\varepsilon$$
$$T_k = |\bar{V}_k - V_k| \leqslant |x - \beta_{k+1}|T_{k+1} + (4|(x - \beta_{k+1})V_{k+1}| + 2|a_{n-k}|)\varepsilon \qquad (19)$$

where $\varepsilon \leqslant 2^{-t}$.

The relations obtained show that the numerical stability of this evaluation algorithm depends on the choice of the interpolation points, β_k. More or less equal distribution of the β_k over the interval of evaluation will usually give satisfactory stability. The points β_k can also be chosen so as to guarantee maximum stability. If there are no zeros in the interval of evaluation, the stability will be complete, cf. Mesztenyi and Witzgall (1967).

The upper bound on the round-off error accumulated in the Chebyshev–Clenshaw algorithm given by 2.1–(15) may be derived using the following results due to Newbery (1974):

$$Q_n(x) = xV_1 - V_2 + b_0 \qquad (20)$$

and

$$|\bar{V}_1 - V_1| \leqslant \frac{\sigma\left(\sum_k |\bar{b}_k|\right) n[1 + (1 + 2|x|)n(n+1)/2]}{1 - \sigma(1 + 2|x|)n(n+1)/2} \qquad (21)$$

$$|\bar{V}_2 - V_2| \leqslant \frac{\sigma\left(\sum_k |\bar{b}_k|\right)(n-1)[1 + (1 + 2|x|)n(n-1)/2]}{1 - \sigma(1 + 2|x|)n(n-1)/2} \qquad (22)$$

where $\sigma = \varepsilon(2 + \varepsilon)$, $|\varepsilon| \leqslant 2^{-t}$.

We proceed using relations (20)–(22) as follows:

$$T_0 = |\bar{V}_0 - V_0| = |\text{fl}(\bar{x}\bar{V}_1 - \bar{V}_2 + \bar{b}_0) - (xV_1 - V_2 + b_0)| \qquad (23)$$

where

$$\bar{V}_0 = \mathrm{fl}(\bar{x}\bar{V}_1 - \bar{V}_2 + \bar{b}_0)$$
$$= \mathrm{fl}(x(1 + \delta)(V_1 + T_1) - (V_2 + T_2) + b_0(1 + \varepsilon_0))$$
$$= [[x(1 + \delta)(V_1 + T_1)(1 + \alpha_0) - (V_2 + T_2)](1 + \beta_0) + b_0(1 + \varepsilon_0)]$$
$$\times (1 + \xi_0)$$
$$= x(V_1 + T_1)(1 + \delta)(1 + \alpha_0)(1 + \beta_0)(1 + \xi_0) - (V_2 + T_2)(1 + \beta_0)$$
$$\times (1 + \xi_0) + b_0(1 + \varepsilon_0)(1 + \xi_0) \tag{24}$$

Using arguments similar to the Horner algorithm analysis, we derive

$$T_0 \leqslant |x - 1|T + (4|xV_1| + 2|V_2| + 2|b_0|)\varepsilon. \tag{25}$$

Recalling that here $-1 \leqslant x \leqslant 1$, we conclude from (25) that the Chebyshev–Clenshaw algorithm is always numerically stable. An analysis similar to that for the Chebyshev–Clenshaw algorithm can be readily carried out for the Chebyshev–Bakhvalov algorithm.

Bakhvalov (1971) has given an upper bound on the round-off error for the Chebyshev–Bakhvalov algorithm:

$$|\overline{Q_n(\bar{x})} - Q_n(x)| \leqslant Cq(x,n)\left(\sum_k |\bar{b}_k|\right)\log n\varepsilon \tag{26}$$

where $q(x,n) = \min\left\{\dfrac{1}{\sqrt{(1 - x^2)}}, n\right\}, \quad -1 \leqslant x \leqslant 1,$.

and C is an absolute constant.

Finally, following familiar techniques, and after somewhat tedious arithmetic manipulations with various expressions, we obtain the error estimates for the Belaga and Pan algorithms. So, for the Belaga algorithm given by 2.4–(1) the error satisfy

$$T_1 = |\bar{V}_1 - V_1| \leqslant 4(|x|^2 + |\alpha_1 x|)\varepsilon, \tag{27}$$

$$T_2 = |\bar{V}_2 - V_2| \leqslant (2|V_1| + |x| + |\alpha_2| + |\alpha_3|)T_1 + (5|V_1|^2 + 6|xV_1|$$
$$+ 5|\alpha_2 V_1| + 6|\alpha_3 V_1| + 7|\alpha_3 x| + 6|\alpha_2 \alpha_3| + 2|\alpha_4|)\varepsilon, \tag{28}$$

$$T_{k+1} = |\bar{V}_{k+1} - V_{k+1}| \leqslant (|V_1| + |\alpha_{2k+1}|)T_k + (3|V_k V_1| + 4|V_k||x|^2$$
$$+ 4|V_k||\alpha_1 x| + 4|V_k \alpha_{2k+1}| + 2|\alpha_{2k+2}|)\varepsilon \tag{29}$$

where $\varepsilon \leqslant 2^{-t}$.

From (29) we observe that if we assume validity of the Rice conjecture, which states that a well-conditioned Belaga polynomial form has 'small' coefficients, α_j, then the numerical stability of the Belaga algorithm is very highly sensitive to the magnitude of x, since the error, T_k, grows as

$$T_k \leqslant P_2(|x|)T_{k-1} + A\varepsilon, \quad \varepsilon \leqslant 2^{-t}, \tag{30}$$

where $P_2(x)$ is a polynomial of second degree, A depends on x and the Belaga coefficients as defined by (29)

For the Pan algorithm given by 2.4–(10) we first get

$$E_0 = |\bar{W}_0 - W_0| \leqslant 4(|x|^2 + |\lambda_0 x|)\varepsilon, \tag{31}$$

$$E_1 = |\bar{W}_1 - W_1| \leqslant 1.25 E_0 + 2|x|\varepsilon. \tag{32}$$

Now, noting that

$$\bar{V}_0 = \mathrm{fl}(x) = x(1 + \delta), \quad |\delta| \leqslant 2^{-t},$$

we obtain an estimate for the error of the general term, given by

$$T_k = |\bar{V}_k - V_k| \leqslant P_4(|x|)T_{k-1} + B\varepsilon, \quad \varepsilon \leqslant 2^{-t} \tag{33}$$

where $P_4(x)$ is a fourth degree polynomial and B is a function of x and the Pan coefficients, which can readily be derived.

From (33) we again conclude that the Pan algorithm is extremely sensitive to the magnitude of x and may be very unstable indeed.

2.7 Evaluation of the Derivatives of a Polynomial

The need to evaluate the derivatives of a polynomial at a specific point arises in many contexts, a typical example being that of finding the roots of polynomial equations. The usual procedure in such cases is to use Horner's algorithm repetitively and we shall now show that to evaluate all the derivatives in this way requires $\frac{1}{2}n(n + 1)$ multiplications and the same number of additions.

Given a polynomial of degree n,

$$P_n(x) = a_0 x^n + a_1 x^{n-1} + \ldots + a_n$$

and recalling the Horner or 'nested' form represented by

$$P_n(x) = \underbrace{(\ldots((a_0 x + a_1)x + a_2)x + \ldots + a_{n-1})x + a_n}_{n-1 \text{ brackets}},$$

we write for the first derivative

$$P_n'(x) = na_0 x^{n-1} + (n-1)a_1 x^{n-2} + \ldots + a_{n-1},$$

and then present the latter expression in the form:

$$\begin{aligned}
&[a_{n-1} + x(a_{n-2} + x(a_{n-3} + \ldots + x(a_1 + a_0 x)\ldots))] \\
&+ x[a_{n-2} + x(a_{n-3} + \ldots + x(a_1 + a_0 x)\ldots)] \\
&+ x^2[a_{n-3} + x(a_{n-4} + \ldots + x(a_1 + a_0 x)\ldots)] \\
&+ \ldots + x^{n-2}[a_1 + a_0 x] + x^{n-1}[a_0].
\end{aligned} \tag{1}$$

In (1), each of the expressions in the square brackets is computed as an intermediate result in the process of evaluation the polynomial $P_n(x)$ by the Horner algorithm.

Denoting the values in the square brackets of (1) by b_i, we can write

$$P_n'(x) = b_0 x^{n-1} + b_1 x^{n-2} + \ldots + b_{n-3} x^2 + b_{n-2} x + b_{n-1}$$

which gives, using the Horner scheme,

$$P'_n(x) = (\ldots((b_0x + b_1)x + b_2)x + \ldots + b_{n-2})x + b_{n-1}.$$

We see that all the b's have been computed while evaluating the polynomial itself and a further application of the same procedure suffices to evaluate the derivative. The process can, obviously, be extended to compute in turn the higher derivatives, $P_n^{(i)}(x)$ for $i = 2, \ldots, n$.

Hence the algorithm which evaluates both a polynomial and its first derivative may be expressed as follows:

$$V_n = a_0$$
$$V_k = xV_{k+1} + a_{n-k}, \qquad k = n - 1, \ldots, 0$$
$$W_{n-1} = xV_n + V_{n-1}$$
$$W_k = xW_{k+1} + V_{k-1}, \qquad k = n - 2, \ldots, 1 \qquad (2)$$
$$P'_n(x) = W_1,$$
$$P_n(x) = V_0.$$

We see that algorithm (2) uses $2n - 1$ multiplications and the same number of additions, of which $n - 1$ are required for the evaluation of the derivative.

Algorithm (2) can readily be extended to evaluate all the derivatives of $P_n(x)$ and this would require

$$n + (n - 1) + (n - 2) + \ldots + 1 = \tfrac{1}{2}n(n + 1)$$

multiplications and the same number of additions.

Kirkpatrick (1972) has shown that provided the operation of division is excluded, a minimum of at least $2n - 1$ additions is required to evaluate both a polynomial of degree n and its first derivative. The minimum number of multiplications, however, does not need to be as large as $2n - 1$ and in fact can be reduced to $n + 1$, cf. Borodin (1973).

An alternative and a more economical way to compute the derivatives has been proposed by Shaw and Traub (1974). They consider the so called normalized derivatives which are defined as $P_n^{(s)}(x)/s!$, where $P_n(x)$ is, as usual, a polynomial of degree n. The algorithm computes all the normalized derivatives of a polynomial in a total of $3n - 2$ operations of multiplication and division and requires the same number of additions as in the 'repeated Horner' algorithm.

Let

$$P_n(x) = \sum_{i=0}^{n} a_{n-1}x^i, \qquad (3)$$

then the algorithm is given as follows:

$$V_i^{-1} = a_{i+1}x^{n-i-1}, \qquad i = 0, 1, \ldots, n - 1$$
$$V_j^j = a_0x^n, \qquad j = 0, 1, \ldots, n$$
$$V_i^j = V_{i-1}^{j-1} + V_{i-1}^j, \qquad j = 0, 1, \ldots, n - 1 \qquad (4)$$
$$\qquad\qquad\qquad\qquad i = j + 1, \ldots, n$$

where the V_i^j in terms of (3) may be expressed as

$$V_i^j = x^{n-i+j} \sum_{k=j}^{i} \binom{k}{j} a_{i-k} x^{k-j},$$

the $\binom{k}{j}$ being binomial coefficients.

We thus get

$$V_n^j = x^j \sum_{k=j}^{n} \binom{k}{j} a_{n-k} x^{k-j} = \frac{P_n^{(j)}(x)}{j!} x^j, \quad j = 0, 1, \ldots, n-1 \tag{5}$$

where $P_n^{(0)}(x) = P_n(x)$.

To estimate the number of arithmetic operations involved, we observe that $x^2, \ldots, x^n$ may be obtained by $n-1$ multiplications and $a_0 x^n, \ldots, a_{n-1} x$ may be obtained by n further multiplications. Hence, the V_n^j may be obtained in $2n-1$ multiplications. The normalized derivatives

$$\frac{P_n^{(j)}(x)}{j!} = \frac{V_n^j}{x^j} \quad j = 1, \ldots, n-1 \tag{6}$$

can then be calculated in $n-1$ divisions.

(Note that $\dfrac{P_n^{(n)}(x)}{n!} = a_0$ and so need not be computed.)

Thus, the algorithm given by (4) and (6) yields 'point' values of all the normalized derivatives and the polynomial itself in $3n-2$ operations of multiplication and division, and $\frac{1}{2}n(n+1)$ operations of addition.

It may also be shown that the calculation of the

$$V_n^j = \frac{x^j P_n^{(j)}(x)}{j!}, \quad j = 0, \ldots, n$$

in $2n-1$ multiplications as accomplished in the above algorithm optimizes the number of multiplications required by any algorithm for solving the problem, cf. Borodin and Munro (1975).

Woznaikowski (1974a) has shown that the algorithm given by (4) and (6) is numerically stable. In particular, he proved that under conditions of normalized floating-point arithmetic, the computed values of the V_i^j's are given by

$$\bar{V}_i^j = \mathrm{fl}(V_i^j) = x^{n-i+j} \sum_{k=j}^{i} \binom{k}{j} a_{i-k} (1 + \eta_{i,k}^j) x^{k-j}$$

where

$$|\eta_{i,-1}^{-1}| \leqslant (n-i-1)\varepsilon \qquad\qquad i = 0, \ldots, n-1$$
$$|\eta_{j,j}^{j}| \leqslant n\varepsilon \qquad\qquad j = 0, \ldots, n$$

$$|\eta_{i,k}^j| \leqslant \{2k + 1 - \delta_{i,k} - j + (n - i + j)\}\varepsilon,$$

$$j = 0,\ldots,n$$
$$i = j + 1,\ldots,n$$
$$k = j,\ldots,i$$

and $\delta_{i,k}$ denotes the Kronecker symbol.

2.8 Evaluation of Polynomials with a Complex Argument and Complex Coefficients

We now turn, in conclusion, to the problem of evaluating of polynomial

$$P(z) = \sum_{k=0}^{n} a_{n-k} z^k$$

for a complex value of the argument z_0. The coefficients a_{n-k} are assumed to be real. The operations of addition and multiplication of complex numbers can be reduced to a sequence of arithmetic operations on real numbers in accordance with the following scheme:

numbers

real + complex uses 1 addition

complex + complex uses 2 additions

real × complex uses 2 multiplications

complex × complex uses 4 multiplications, 2 additions

or

3 multiplications, 5 additions.

In the scheme, the operation of computing the product of two complex numbers in 3 real multiplications and 5 additions is less obvious than the rest of the scheme. Such an operation, however, may be easily observed from the identity (here $i = \sqrt{-1}$)

$$(a + ib)(c + id) = a(c + d) - (a + b)d + i(a(c + d) + (b - a)c)$$
$$\text{also} \equiv ac - bd + i((a + b)(c + d) - ac - bd).$$

It appears that the identity was first suggested by Winograd (1970).

Now, one computational algorithm for evaluation of $P(z)$ at a fixed point $z = z_0$ is based on the following representation of the polynomial

$$P(z) = \sum_{k=0}^{n} a_{n-k} z^k = g(z)(z - z_0) + V_n \tag{1}$$

where

$$g(z) = V_0 z^{n-1} + V_1 z^{n-2} + V_2 z^{n-3} + \ldots + V_{n-1}$$

and V_n is a constant.

(The process of obtaining (1) is generally referred to as the synthetic division of $P(z)$ by $(z - z_0)$ and $g(z)$ is called the reduced degree polynomial.)

From (1) it follows that provided the coefficients of $g(z)$ and the constant V_n are known, the polynomial value $P(z_0) = V_n$.

Hence, the evaluation algorithm is given by

$$
\begin{aligned}
V_0 &= a_0 \\
V_k &= V_{k-1}z_0 + a_k, \qquad k = 1, \ldots, n \\
P(z_0) &= V_n.
\end{aligned}
\tag{2}
$$

It requires either $4n - 2$ multiplications and $3n - 2$ additions or $3n - 1$ multiplications and $6n - 5$ additions and is essentially the same as the Horner algorithm.

An alternative procedure for evaluating $P(z)$ is based on division of the polynomial by a quadratic factor, i.e. the synthetic division of the form

$$
P(z) = \sum_{k=0}^{n} a_{n-k}z^k = g(z)d(z) + V_{n-1}z + V_n
\tag{3}
$$

where the divisor is

$$
d(z) = (z - z_0)(z - \bar{z}_0)
$$

and the quotient is

$$
g(z) = V_0 z^{n-2} + V_1 z^{n-3} + \ldots + V_{n-2}
$$

Let, for convenience, the fixed point z_0 at which the value of $P(z)$ is searched, be expressed as $z_0 = x_0 + iy_0$. Then equating like powers of z in expression (3) we obtain

$$
\begin{aligned}
a_0 &= V_0, \\
a_1 &= V_1 - 2x_0 V_0, \\
a_k &= V_k - 2x_0 V_{k-1} + (x_0^2 + y_0^2)V_{k-2}, \qquad k = 2, \ldots, n-2, \\
a_{n-1} &= \quad\ -2x_0 V_{n-2} + (x_0^2 + y_0^2)V_{k-3} + V_{n-1}, \\
a_n &= \qquad\qquad\quad (x_0^2 + y_0^2)V_{n-2} + V_n.
\end{aligned}
\tag{4}
$$

From the set (4) the following computational algorithm may be constructed

$$
\begin{aligned}
V_0 &= a_0, \qquad p = x_0 + x_0, \qquad q = x_0^2 + y_0^2 \\
V_1 &= V_0 p + a_1, \\
V_k &= V_{k-1}p - V_{k-2}q + a_k, \qquad k = 2, \ldots, n-1, \\
V_n &= -V_{n-2}q + a_n, \\
P(z_0) &= V_n z_0 + V_{n-1}.
\end{aligned}
\tag{5}
$$

It is readily deduced that this algorithm requires $2n + 2$ multiplications and $2n + 1$ additions. The algorithm is due to Knuth (1965) and it is an improvement over the Horner-like algorithm for $n \geq 3$.

Finally, in the case of a polynomial with complex coefficients evaluated for a complex argument, the first computational algorithm given by (2), requires obviously either $4n$ multiplications and $4n$ additions or 3n multiplications and $7n - 3$ additions, while the Knuth algorithm, given by (5), becomes less interesting with either $4n + 10$ multiplications and the same number of additions or with $4n + 9$ multiplications and $4n + 13$ additions.

2.9 Final Comments

A concept of preprocessing the polynomial coefficients has been introduced in this chapter. It is an important concept which the investigations concerning the efficient evaluation schemes for polynomials have brought to light. This concept is finding more and wider applications than the theory originally have indicated. As Winograd remarks, the method of preprocessing is generally useful wherever one has the problem of continually re-evaluating a set of expressions where some of the parameters remain fixed. Hoffman and Winograd (1970) have studied the problem of finding the minimum path between any two nodes in a graph, and, by means of preprocessing, succeeded in reducing the number of additions from $O(n^3)$ to $O(n^{5/2})$.

We have concentrated on efficient evaluation of a polynomial at a given point. Other problems involving polynomials include efficient evaluation of a polynomial at more than one point, e.g. in a process of finding polynomial roots, polynomial interpolation, polynomial division, polynomial multiplication. The latter is closely related to the problem of efficient multiplication of matrices; for example, given an $m \times n$ matrix $X = (x_{ij})$ and an $n \times r$ matrix $Y = (y_{ij})$, their product $Z = XY$ means that

$$Z = (z_{ik}), \quad \text{where } z_{ik} = \sum_{1 \le j \le n} x_{ij} y_{jk}, \quad 1 \le i \le m, \quad 1 \le k \le r.$$

This equation may be regarded as the computation of mr simultaneous polynomials in $mn + nr$ variables. So, what economical ways exist to perform such a computation? A comprehensive reference to recent results in this area may be found in Borodin and Munro (1975).

Exercises

2.1 Discuss the problem of efficient evaluation of a polynomial. Give examples of different evaluation algorithms. Comment on the aspects of numerical accuracy and speed of evaluation.

2.2 State in your own words Belaga's theorem on the evaluation of polynomials with preprocessed coefficients.

2.3 The polynomial $x^2 + y^2 + z^2 + t^2 - 2(xz - yt)$ can obviously be evaluated with 7 multiplications. By suitable factoring, show how the evaluation can be accomplished with only 2 multiplications.

2.4 Consider the polynomial $P_n(x) = \sum_{k=0}^{n} a_{n-k} x^k$ with real coefficients and a real argument x. Let the time complexity function $F(n)$ of an algorithm to evaluate $P_n(x)$ be the function such that $F(n)$ is the maximum number of multiplications required to evaluate $P_n(x)$ for any set of $\{a_{n-k}\}$. Suppose it is known that $a_{n-k} = 0$ for all odd/even k. Construct an algorithm to take advantage of this restriction and analyse its complexity.

2.5 Prove that any algorithm (using only $+, -, \times$) which computes $\sum_{k=0}^{n} a_{n-k}x^k$ requires at least n addition operations.

2.6 Write a program which evaluates the polynomial $P_n(x) = \sum_{k=0}^{n} a_{n-k}x^k$ and all its derivatives by a repetitive use of the Horner scheme.

2.7 Prove that the algorithm of Exercise 2.7 requires $\frac{1}{2}n(n+1)$ operations of multiplication and the same number of additions.

2.8 Estimate the total number of arithmetic operations required by each of the algorithms 2.8–(2) and 2.8–(5) in the case of a polynomial with complex coefficients, evaluated for a real argument.

2.9 Find a straight-line program for the algorithm of 2.7–(4), given $n = 6$.

Chapter 3

Iterative Processes

3.1 Definition of an Iterative Process

The idea of *iteration* (from the Latin *iteratio* = repetition) is one of the most important concepts used in the numerical methods for solving different mathematical problems. It may also be called *successive approximation*. Let us illustrate the idea on the example of solving an equation of the form:

$$x = F(x). \tag{1}$$

We assume that the function $F(x)$ is differentiable.

In an iterative method we start with some initial approximation x_0, which for the majority of equations may be quite crude, and thereafter form a sequence of numbers from the following rule:

$$x_1 = F(x_0), \qquad x_2 = F(x_1), \dots. \tag{2}$$

Each calculation of the type

$$x_{n+1} = F(x_n) \tag{3}$$

is called an *iteration*.

If the sequence of numbers obtained converges to a limiting value α, then

$$\lim_{n \to \infty} F(x_n) = F(\alpha). \tag{4}$$

Hence, $x = \alpha$ satisfies the equation $x = F(x)$.

We may say that with n increasing, the number x_n tends to the root α. An algorithm based on an iterative process is assumed to be completed when some prescribed accuracy for the approximation to the root is achieved.

In this chapter we shall be concerned with iterative methods for computing approximate solutions of the non-linear equation

$$f(x) = 0 \tag{5}$$

where attention is restricted to functions of a real variable that are real, single-valued, and possess a certain number of continuous derivatives in the neighbour-

47

hood of a real root, α, of (5). One important subclass of non-linear equations is the set of polynomial equations, and iterative methods constitute a powerful group of methods for determining the roots of polynomials. It may be noted, however, that one frequently needs to determine both the real and complex roots of such equations or to determine all the roots simultaneously (e.g. when stability problems are considered). Also, polynomial equations have a special structure as compared with general non-linear equations. These considerations call for methods which are specially suitable for solving polynomial equations and a number of special techniques have in fact been developed which are not suitable for the general equations. These specialized techniques will not be considered here. The evaluation of complex and multiple roots of a general non-linear equation and the solution of systems of non-linear equations will not be discussed here either.

3.2 The Bisection Method

Probably the most primitive procedure for approximating a real root of an equation is the following bisection method. Let a and b be two points such that $f(a)f(b) < 0$. Then the equation $f(x) = 0$ has at least one real root in (a, b). For simplicity, let us assume that $f(a) < 0$, $f(b) > 0$. A new approximation is now obtained by computing $(a + b)/2$ and finding the sign of $f((a + b)/2)$. If $f((a + b/2) = 0$, a root has been found, otherwise the following changes are made:

$$\text{if } f\left(\frac{a + b}{2}\right) > 0 \quad \text{then} \quad \frac{a + b}{2} \text{ replaces } b,$$

$$\text{if } f\left(\frac{a + b}{2}\right) < 0 \quad \text{then} \quad \frac{a + b}{2} \text{ replaces } a.$$

The process is then repeated. We can see that the root will either be found by this procedure or be known to lie on an interval of length $|b - a|/2^m$ after the bisection operation has been applied m times. We may then take the approximate value of the root to be the midpoint of this last interval; this estimate can have a maximum error of $\frac{1}{2}|b - a|/2^m$, and to achieve this accuracy we need just m evaluations of $f(x)$. It is interesting to observe that in this procedure the accuracy to which the root may be found is limited only by the accuracy with which $f(x)$ may be evaluated.

This method is guaranteed to *converge*. On the other hand, since no use is made of the structure of $f(x)$, indeed it is necessary only to be able to decide on the sign of $f(x)$, we can expect that a large number of function evaluations would be required to get a reasonably accurate result. At each iteration we gain in accuracy one binary digit. Since $10^{-1} \approx 2^{-3.3}$, we, thus, gain on average one decimal digit in 3.3 iterations. It is natural to expect that the convergence could be speeded up if the iterative method used were to take account of the structure of $f(x)$, and possibly its derivative or derivatives.

3.3 The Order of Convergence of an Iterative Process

We shall now study iterative methods of the general form

$$x_{n+1} = \phi(x_n, x_{n-1}, \ldots, x_{n-m+1}), \tag{1}$$

where ϕ is known as the *iteration function*.

The convergence of an iterative method, which is of paramount importance in the theory and application of such methods is studied by means of error analysis. Provided the method converges, the next question we may ask is: How *fast* does the method converge? One way to interpret this general question is to ask *how many* iterations are required to achieve a result with a prescribed accuracy. The answer to the latter question gives some indication of the comparative 'goodness' of the method.

In order to expound the convergence properties of different iterative methods we now define *the order of convergence of an iterative method*.

Definition Let $x_0, x_1, x_2, \ldots$ be a sequence of values obtained using (1) which converges to α. Let further e_n be the error in the iterate x_n, that is $e_n = |x_n - \alpha|$.
If there exists a number p and *a* constant $C \neq 0$ such that

$$\lim_{n \to \infty} \frac{e_{n+1}}{e_n^p} = C, \tag{2}$$

then p is called the order of convergence of the sequence and C its asymptotic error constant.
For $p = 1$, 2, or 3 the convergence is said to be linear, quadratic, or cubic, respectively. The corresponding iteration function $\phi(x_n, x_{n-1}, \ldots, x_{n-m+1})$ is then said to be of order p.

The order p may be considered as a suitable measure of the improvement in accuracy of the approximation to the root which is gained at one iteration. Assume, for example, that the magnitude of C in (2) is unity. Let x_n agree with the exact solution, α, to γ significant figures. Then, for n sufficiently large, x_{n+1} approximates α to $p\gamma$ significant figures. It is also clear that the improvement in accuracy will be most rapid when p is large and C small.

We now go on to a discussion of various methods which yield the general form (1).

3.4 The Newton–Raphson Method. Convergence

In order to study how to derive an iterative method to obtain a root of $f(x) = 0$, the function $y = f(x)$ is represented graphically in Fig. 3.1. Now, the root α is represented by the point of intersection of the curve and the x-axis. Hence, if an iterative method is being designed to approximate this root, one basic idea might be to replace the curve by a suitable straight line whose intersection

50

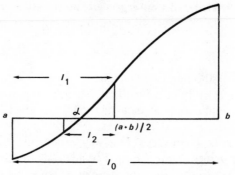

Fig. 3.1. Illustration to the bisection method.
$l_0, l_1, l_2, \ldots$ is a sequence of diminishing intervals, $l_k = |b - a|/2^k$, containing the root of
$$f(x)$$

with the x-axis can be easily computed. Starting with an arbitrary initial approximation to α, namely x_0, we then calculate an iterative sequence

$$x_0, x_1, x_2, \ldots .$$

Clearly, we will get different iterative methods depending on how the direction of the straight line is chosen. One natural choice for the direction of the straight line is that of the tangent to the curve at the point in question, as indicated in Fig. 3.2. The iterative method based on this idea is the celebrated *Newton–Raphson method*. It is defined as follows:

$$x_{n+1} = \phi(x_n) = x_n - \frac{f(x_n)}{f'(x_n)}, \qquad x_0 \text{ arbitrary.} \tag{1}$$

To study the convergence properties of this method, we assume that $f(x)$ is

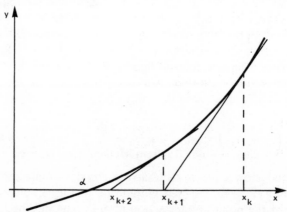

Fig 3.2. The Newton–Raphson method for a convex
function

twice differentiable, and we keep in mind our earlier general assumption that the root α is simple, that is $f'(\alpha) \neq 0$, from which it follows that $f'(x) \neq 0$ for all x in some vicinity of the root α.

We now establish the relation between the errors e_n and e_{n+1} under the above assumptions.

The Taylor series expansion about x_n gives

$$0 = f(\alpha) = f(x_n) + (\alpha - x_n)f'(x_n) + \tfrac{1}{2}(\alpha - x_n)^2 f''(\xi), \quad \xi \in (x_n, \alpha).$$

It follows that

$$\frac{f(x_n)}{f'(x_n)} + \alpha - x_n = \alpha - x_{n+1} = -\frac{1}{2}(\alpha - x_n)^2 \frac{f''(\xi)}{f'(x_n)} \tag{2}$$

Thus

$$e_{n+1} = \frac{1}{2} e_n^2 \frac{f''(\xi)}{f'(x_n)} \tag{3}$$

and if $x_n \to \alpha,$ we have

$$\frac{e_{n+1}}{e_n^2} \to \frac{1}{2} \frac{f''(\alpha)}{f'(\alpha)}. \tag{4}$$

Since e_{n+1} is approximately proportional to e_n^2, we say that the Newton–Raphson method has quadratic convergence or that the method is of second order.

We have shown that provided the Newton–Raphson method converges, then its convergence is of order 2.

Let us now study the conditions under which the method converges. Assume that I is a vicinity of the root α, such that

$$\frac{1}{2}\left|\frac{f''(y)}{f'(x)}\right| \leqslant M \qquad \text{for all } x, y \in I \tag{5}$$

If $x_n \in I$, then it follows from (3) that $|e_{n+1}| \leqslant M e_n^2$ which can be written as $|Me_{n+1}| \leqslant (Me_n)^2$. Assume that $|Me_0| < 1$ and that the interval $[\alpha - |e_0|, \alpha + |e_0|]$ is contained in I. It can then be shown by induction that $x_n \in I$ for all n and that

$$|e_n| \leqslant \frac{1}{M}(Me_0)^{2^n} \qquad \text{for } n > 0. \tag{6}$$

It follows that provided x_0 is chosen sufficiently close to α, that is provided

$$|Me_0| = M|x_0 - \alpha| < 1,$$

the Newton–Raphson method always converges.

In relation to a poor choice of starting values, we would like to point out a hazard which may occur when using the Newton–Raphson method. It is illustrated in Fig. 3.3. In the situation shown, the subsequent iterations will

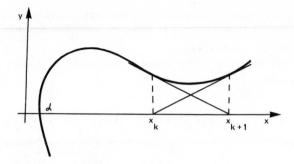

Fig. 3.3. A case of non-convergence of the Newton–
Raphson method. The iterations after $(k + 1)$st will
produce two points, x_k and x_{k+1}, alternatively, without
any further progress in approaching a root α

produce two points, x_k and x_{k+1}, alternatively, without any further progress in
approaching a root α.

3.5 The Secant Method. Convergence

If in the Newton–Raphson method we replace the derivative $f'(x_n)$ by the
difference ratio

$$\frac{f(x_n) - f(x_{n-1})}{x_n - x_{n-1}},$$

we derive another iterative method for finding a root of $f(x) = 0$,
namely

$$x_{n+1} = x_n - f(x_n) \frac{x_n - x_{n-1}}{f(x_n) - f(x_{n-1})}, \quad f(x_n) \neq f(x_{n-1}). \tag{1}$$

The method is known as the Secant iterative method. It is one of the oldest
methods known for the solution of $f(x) = 0$. After some period of neglect, its
use has recently been revived since the method is particularly convenient for
use with digital computers. The following geometric interpretation may be
given to the method: x_{n+1} is determined as the abissa of the intersection
between the secant through $(x_{n-1}, f(x_{n-1}))$ and $(x_n, f(x))$ and the axis. The
method is illustrated in Fig. 3.4.

To study the relation between the errors at two consecutive steps in the
secant method, consider the Newton interpolation formula for the points
x_n and x_{n-1}:

$$f(x) = f(x_n) + (x - x_n) \frac{f(x_n) - f(x_{n-1})}{x_n - x_{n-1}} + \frac{1}{2}(x - x_{n-1})(x - x_n) f''(\xi), \tag{2}$$

where $\xi \in$ the smallest interval containing x, x_{n-1}, x_n.
The Secant method is obtained by ignoring the remainder term in this formula,

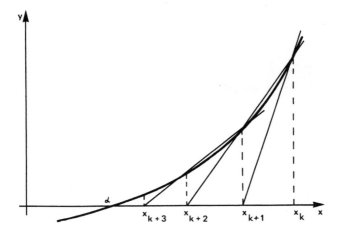

Fig. 3.4. The Secant method for a convex function

and x_{n+1} satisfies the equation

$$0 = f(x_n) + (x_{n+1} - x_n)\frac{f(x_n) - f(x_{n-1})}{x_n - x_{n-1}}. \tag{3}$$

Now setting $x = \alpha$ in (2) and using equation (3), we get, since $f(\alpha) = 0$:

$$(\alpha - x_{n+1})\frac{f(x_n) - f(x_{n-1})}{x_n - x_{n-1}} + \frac{1}{2}(\alpha - x_n)(\alpha - x_{n-1})f''(\xi) = 0. \tag{4}$$

By virtue of the mean value theorem we have

$$\frac{f(x_n) - f(x_{n-1})}{x_n - x_{n-1}} = f'(\xi'), \tag{5}$$

where $\xi' \in$ the smallest interval containing x_{n-1} and x_n, from which it follows that

$$e_{n+1} = \left|\frac{f''(\xi)}{2f'(\xi')}\right| e_n e_{n-1}. \tag{6}$$

We conclude that the secant method converges for sufficiently good initial values, x_0 and x_1, provided that $f'(\alpha) \neq 0$ and $f(x)$ is twice continuously differentiable.

Assuming now that the secant method converges, then for very large n, we may take $\xi \approx \alpha$, $\xi' \approx \alpha$, and hence

$$|e_{n+1}| \approx C|e_n||e_{n-1}|, \quad \text{where } C = \left|\frac{1}{2}\frac{f''(\alpha)}{f'(\alpha)}\right|. \tag{7}$$

We note that the constant C is the same as in the Newton–Raphson method. To establish the order of convergence, p, of the secant method we make the fol-

54

lowing conjecture:

$$|e_{n+1}| \approx K|e_n|^p,$$
$$|e_n| \approx K|e_{n-1}|^p.$$

Substituting the proposed relation between the errors into (7) we get

$$K|e_n|^p \approx C|e_n|K^{-1/p}|e_n|^{1/p}.$$

This relation is valid only if $p = 1 + (1/p)$, that is $p = \frac{1}{2}(1 \pm \sqrt{5})$, and if $C = K^{1+(1/p)} = K^p$.
We thus have

$$|e_{n+1}| \approx C^{1/p}|e_n|^p \quad \text{for} \quad p = \frac{1}{2}(1 \pm \sqrt{5}). \tag{8}$$

Now, the value $p = \frac{1}{2}(1 - \sqrt{5}) = -0.618\ldots$ implies that the error of the current iteration, $|e_{n+1}|$, depends inversely on the error of the preceding iteration, $|e_n|$. Since the error $|e_n|$, in general, is very small, for example it is much smaller than unity, its reciprocal, $|e_n|^{-1}$ may be large, which in turn implies that the error $|e_{n+1}|$ will grow uncontrollably with n increasing. This contradicts the assumption on which the proof is based, that the method is convergent. Hence, the proof is valid only for $p = \frac{1}{2}(1 + \sqrt{5})$.

We conclude that

$$|e_{n+1}| \approx C^{1/p}|e_n|^p, \quad p = \frac{1}{2}(1 + \sqrt{5}) = 1.618\ldots, n \gg 1.$$

We shall confine ourselves to this heuristic discussion of the order of convergence; a rigorous proof of the above result has been given by Ostrowski (1966). We wish once again to emphasize that a good initial approximation to the root is essential for the secant method to converge. It is not difficult to find examples of functions for which the method would give a divergent sequence of approximations.

One such example is shown in Fig. 3.5.

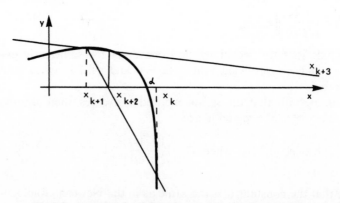

Fig 3.5. An example of divergence of the Secant method

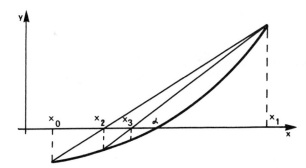

Fig. 3.6. The method of False Position for a convex
function

3.6 Method of False Position (*Regula Falsi*)

A simple modification of the Secant method will produce a method which is always convergent. Suppose that the two initial approximations can be chosen so that the two function values have opposite signs. It is possible then to generate a sequence of values which always possesses this property. The value of x_2 is found as the intersection between the chord joining $f(x_0)$ and $f(x_1)$, and the x-axis. Then choosing x_2 and x_i, where i is either 0 or 1, so as to make $f(x_2)f(x_i) < 0$, the above procedure is repeated to obtain x_3 and so on. The method of false position is given as

$$x_{n+1} = x_n - f(x_n) \frac{x_n - x_{n-1}}{f(x_n) - f(x_{n-1})}, \text{ where } f(x_n)f(x_{n-1}) < 0.$$

An illustration of the method is shown in Figure 3.6.

The penalty involved in the modification of the method is to sacrifice the high rate of convergence of the Secant method. As soon as an interval is reached on which the function is convex or concave, thereafter one of the end-points of this interval is always retained and this feature slows down convergence to first order. This is, of course, the price one would expect to have to pay for a guaranteed convergence.

Various modifications of the *Regula Falsi* basic scheme which lead to a considerable enhancement in speed of convergence without the guarantee of convergence being lost, are discussed in an excellent paper by Jarratt (1973).

3.7 Formulation of the Optimality Problem

Using the order p as a measure of the relative efficiency of an iterative method (e.g. p shows how fast the method converges) one intuitively feels that the Newton–Raphson method will be 'faster' than the Secant method in terms of the number of iterations required to arrive at a given accuracy in a root. Simple examples show this to be true at least for some problems (e.g. the well-separated simple roots of well-conditioned polynomials). It is not difficult, however, to find examples which contradict the above conjecture. Consider the problem of finding a root of $g(x) = 0$ where

$$g(x) = x - \cos\left(\frac{0.785 - x\sqrt{(1 + x^2)}}{1 + 2x^2}\right).$$

To compute the root $x = 0.97989961$ requires 8 iterations by the Newton–Raphson method and 5 iterations by the Secant method. (As a matter of interest, it takes 5 iterations to compute the same root by the Traub method, to be described in Section 3.9.4.) In our case the initial approximations were taken as $x_0 = 0, x_1 = 0.1$ (the Secant method needs two starting values).

Such 'unexpected behaviour' of iterative methods is explained by the fact that the order of a method is a property local to the neighbourhood of a root. It therefore measures how good a method is when it is near convergence, while the total number of iterations depends on the initial approximations to the root.

It is also important to realise that the time taken to execute one iteration on a digital computer varies from method to method. Among other things, this time depends on the number of function evaluations per iteration and on the number of arithmetic operations necessary to combine the function values to form the iterative formula. It, thus, makes more sense to compare the computational efficiency of various methods by measuring how much computation must be done to compute an approximation to the root to a prescribed numerical accuracy. Thus, we may say that the efficiency of the root-finding iterative methods can be discussed in terms of

(a) the number of iteration steps required versus the numerical accuracy achieved; and
(b) the enhancement of accuracy per single iteration step versus the iteration 'cost'.

The 'optimal' algorithm may then be thought of as that for either

(a) the iterative method which would guarantee a prescribed accuracy after the fewest number of iteration steps; or
(b) the iterative method which would need the minimum amount of work, e.g. the number of arithmetic operations per iteration step, required to produce an approximation of a prescribed accuracy.

These two forms of optimal model are not necessarily contradictory or independent; indeed if one could find an iterative method which combined the least possible amount of work per iteration and the fewest number of iteration steps to give a numerical approximation of prescribed accuracy, then such a method would clearly be ideal. The optimal requirements formulated above, however, prompt the observation that any reasonable efficiency measure of an iterative method should be a function both of the parameters set by condition (a) and of those set by condition (b). This suggests that an efficiency measure should be maximized for these parameters.

The complete optimization problem for the universe of root searching iterative techniques has yet to be solved. We shall now study some recent developments in this area.

First, it is convenient to distinguish two groups of iterative methods. One group consists of those methods that are designed to solve root-finding problems where enough is known about the root location, to enable one to choose starting values such that the convergence of the method adopted would not be likely to present any difficulties. Then with every new iteration step, one 'marches' closer and closer to the root until a sufficiently accurate approximation is obtained.

The Newton–Raphson and Secant methods are examples of this group.

In cases where the *a priori* information on the location of the root is poor, some guarantee of the convergence of the method becomes of paramount importance. This has led to the development of a number of algorithms which are relatively insensitive to the choice of starting values. Most of these so called 'robust' methods are based on the technique of searching until two values of x are found for which $f(x)$ possesses opposite signs. The root is then bracketed between these values. The iterations are then repeated in such a way that such a bracket is always preserved. The bisection algorithm described in Section 3.2 is an example of such a method.

For methods in this group, provided the tolerance ε is known, an estimate of the upper bound can be given of the number of iterations required to achieve an adequate solution. For instance, in the bisection method, starting with some initial root interval $[a, b]$, one repeatedly halves it. After n iterations, it is easy to see that the root will lie in an interval of length $|b - a|/2^n$. We deduce that the method guarantees an approximation to the root, with a tolerance $\pm \varepsilon$, after $n = \log_2 |(b - a)/\varepsilon|$ iterations. Clearly this does not take into account round-off errors which may make the computed result somewhat worse. We shall refer to such an estimate of the number of iterations required as the guaranteed convergence number of the iterative method. The efficiency of the different methods of the group may then be compared in terms of the guaranteed convergence number.

We shall now investigate further methods for finding a zero of a function defined on an interval.

3.8 Iterative Methods for Finding Zeros of Functions That Change Sign in an Interval. Dekker's Algorithm

Bisection and the method of False Position are two examples of iterative methods suitable for finding a zero of a function $f(x)$ defined in an interval $[a, b]$ and such that $f(a)f(b) < 0$. Using these basic techniques, several hybrid algorithms have been developed, — a number of them quite recently, — which for the majority of problems encountered in practice converge faster than either

58

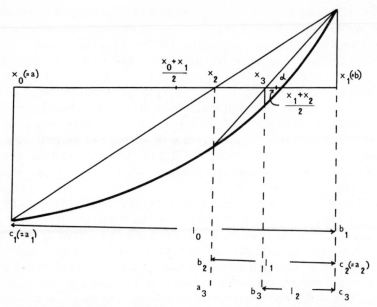

Fig. 3.7. The Dekker method for a convex function. At each iteration b_i is the best approximation to the root so far, a_i is the previous value of b, i.e. $a_i = b_{i-1}$, and the root must lie between b_i and c_i. Initially $a = c$

of the two basic methods. We shall describe one such method due to Dekker (1969). (Variants of the method have been proposed by van Wijngaarden, Zonneveld, and Dijkstra (1963), Wilkinson (1967), Peters and Wilkinson (1969).)

The Dekker method is illustrated in Fig. 3.7. A typical step involves three points a, b, and c such that $f(b)f(c) \leqslant 0$, $|f(b)| \leqslant |f(c)|$, and a may coincide with c. The points a, b, and c change during the algorithm, but there should be no confusion if subscripts are omitted; b is the best approximation so far to the root, a is the previous value of b, and the root must lie between b and c. Initially $a = c$.

The sequence of calculations corresponding to the situation shown in Fig. 3.7 is:

Compute $f(a)$ and $f(b)$;

set $x_0 = c_1 = a_1 = a$ and $x_1 = b_1 = b$
$$\text{since } |f(b)| \leqslant |f(a)|.$$

Compute $m_1 = \dfrac{b_1 + c_1}{2}$ and $s_1 = b_1 - \dfrac{f(b_1)(b_1 - a_1)}{f(b_1) - f(a_1)}$;

set $x_2 = s_1$ since s_1 lies between m_1 and b_1;

set $b_2 = x_2$ and $c_2 = x_1$ since $f(x_2)f(x_1) \leqslant 0$ and $|f(x_2)| \leqslant |f(x_1)|$;

set $a_2 = b_1$.

Compute $m_2 = \dfrac{b_2 + c_2}{2}$ and $s_2 = b_2 - \dfrac{f(b_2)(b_2 - a_2)}{f(b_2) - f(a_2)}$;

set $x_3 = s_2$ since s_2 lies between b_2 and m_2;

set $b_3 = x_3$ and $c_3 = x_1$ since $f(x_3)f(x_1) \leqslant 0$ and $|f(x_3)| \leqslant |f(x_1)|$;

set $a_3 = b_2$,

     etc.

We shall now give the complete form of Dekker's algorithm. Let $[a, b]$ be the initial interval.

Initial step

$$
\left.
\begin{aligned}
\text{Set } x_0 &= a, \\
x_1 &= b, \\
a_1 &= a, \\
c_1 &= a, \\
b_1 &= b,
\end{aligned}
\right\}
\qquad \text{if} \quad |f(b)| \leqslant |f(a)|,
$$

$$
\left.
\begin{aligned}
x_0 &= b, \\
x_1 &= a, \\
a_1 &= a, \\
c_1 &= a, \\
b_1 &= b,
\end{aligned}
\right\}
\qquad \text{otherwise.}
$$

General Iteration Step

Define the linear interpolation formula, for $a \neq b$, by

$$
\begin{aligned}
l = l(b, a) = b - \frac{f(b)(b - a)}{f(b) - f(a)}, && \text{if } f(b) \neq f(a), \\
= \infty, && \text{if } f(b) = f(a) \neq 0, \\
= b, && \text{if } f(b) = f(a) = 0.
\end{aligned}
$$

Let further

$$
h = h(b, c) = b + \text{sign}(c - b)\delta(b)
$$

where $\delta(x)$ is a positive tolerance function δ satisfying $0 < \tau \leqslant \delta(x)$ and τ is a given positive constant (for instance, $\delta(x) \equiv \tau$ may define an absolute tolerance τ and $\delta(x) = \varepsilon|x| + \tau$ may define a relative tolerance ε when $|x|$ is large),

$$
m = m(b, c) = \tfrac{1}{2}(b + c)
$$

and $v = v(l,b,c) = l$ if l is between $h(b,c)$ and $m(b,c)$,

$\quad\quad\quad\quad\quad\;\; = h(b,c)$ if $|l - b| \leqslant \delta(b)$,

$\quad\quad\quad\quad\quad\;\; = m(b,c)$ otherwise.

The new iterate x_i is then calculated by

$$x_i = v(\lambda_i, b_{i-1}, c_{i-1})$$

where $\lambda_i = l(b_{i-1}, a_{i-1})$.

Furthermore, let k be the largest (non-negative) integer satisfying $k < i$ and $f(x_k)f(x_i) \leqslant 0$.

Then, b_i, c_i and a_i are defined by

$$\left.\begin{array}{l} b_i = x_i, \\ c_i = x_k, \\ a_i = b_{i-1}, \end{array}\right\} \quad \text{if} \quad |f(x_i)| \leqslant |f(x_k)|,$$

$$\left.\begin{array}{l} b_i = x_k, \\ c_i = x_i, \\ a_i = x_i, \end{array}\right\} \quad\quad\quad \text{otherwise.}$$

Termination Step

Let n be the smallest positive integer satisfying

$$|b_n - c_n| \leqslant 2\delta(b_n).$$

Then the algorithm terminates for $i = n$ and delivers two (distinct) real numbers $x = b_n$ and $y = c_n$,
satisfying

$$f(x)f(y) \leqslant 0,$$
$$|f(x)| \;\leqslant |f(y)|,$$
$$|x - y| \;\leqslant 2\delta(x).$$

Assuming that $f(x)$ is continuous, the first condition ensures that there exists a zero, α, of $f(x)$ in the closed interval with endpoints x and y, the second condition that x is the 'best' approximation to α, the third condition that the required tolerance has been reached.

The order of convergence of Dekker's algorithm is equal to $1.618\ldots$, and according to Brent (1971a), is much faster than bisection on well-behaved functions, e.g. polynomials of moderate degree with well separated zeros.

In fact, the algorithm has proved satisfactory in most practical cases. However, an unfortunate drawback is that its guaranteed convergence number is $n = |(b - a)/\varepsilon|$ which compares unfavourably with the bisection method. Indeed Brent gives an example when this full number of iterations is required. Such slow convergence may occur in problems where, at each iteration, the point computed using linear interpolation (the Secant method) has to be taken as the next approximation to the root. It may then happen that at each step, the length of the root interval is reduced by no more than an amount equal to ε, the prescribed absolute tolerance. The number of iterations required to achieve an approximation to the root within this prescribed tolerance is then given by the relation

$$|b - a - n\varepsilon| \leqslant \varepsilon$$

giving in turn

$$n = \left| \frac{b-a}{\varepsilon} \right|. \tag{1}$$

Brent has suggested a modification to the Dekker algorithm which ensures that a bisection is done at least once in every $2 \log_2 |(b - a)/\varepsilon|$ consecutive iterations with the result that the modified algorithm is never much slower than bisection.

Another modification of the Dekker algorithm due to Brent (1971 b) combines linear interpolation and inverse quadratic interpolation with bisection. This modified algorithm retains the same order of convergence as Dekker's algorithm and guarantees convergence to the root within a tolerance ε in $n = [\log_2 |(b - a)/\varepsilon| + 1]^2 - 2$ iterations.

In 1974 Bus and Dekker developed a new algorithm for finding the zero of a function within a given interval which uses both bisection and rational approximation technique. Iterative methods in which a rational function is fitted through previously computed values were first studied by Jarratt and Nudds (1965) and Jarratt (1966a, 1966b). For example, a useful practical technique may be obtained using the three-parameter bilinear form

$$g(x) = \frac{x - r}{px + q} \tag{2}$$

such that $g(x) = f(x_v)$, $v = i, i - 1, i - 2$, where x_v are previously computed approximations to the root.

Using bisection and a rational function of form (2), Bus and Dekker have derived an algorithm within an order of convergence equal to 1.84 and a guaranteed convergence number equal to $n = 4 \log_2 |(b - a)/\varepsilon|$.

We would also like to mention an algorithm due to Krautstengel (1968). It is based on the use of linear interpolation (*Regula Falsi*) and linear extrapolation (Secant method) and is illustrated in Fig. 3.8. The class of functions to which the algorithm may usefully be applied is restricted by rather severe conditions which require that depending on whether the function is convex or concave on $[a, b]$, one of the following conditions holds

(i) $\dfrac{f(b)}{b - a} \leqslant f'(b),$

(ii) $-\dfrac{f(a)}{b - a} \leqslant f'(a).$ \hfill (3)

62

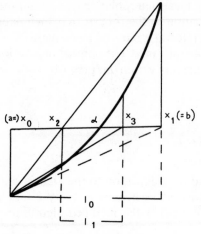

Fig. 3.8. The Krautstengel method for a
convex function

Table 3.1 A summary of main characteristics of some methods for finding a root of a
function which changes sign in an interval

Method	Order of convergence	Guaranteed convergence number
Bisection	1	$\log_2 \left\| \dfrac{b-a}{\varepsilon} \right\|$
Secant and *Regular Falsi*	1.618	$\left\| \dfrac{b-a}{\varepsilon} \right\|$
Dekker's (bisection + Secant)	1.618	$\left\| \dfrac{b-a}{\varepsilon} \right\|$
Dekker–Brent (bisection + Secant; with bisection ensured in every $2 \log_2 \left\| \dfrac{b-a}{\varepsilon} \right\|$ consecutive iterations)	1.618	$\log_2 \left\| \dfrac{b-a}{\varepsilon} \right\|$
Brent–Dekker (bisection + Secant + inverse quadratic interpolation)	1.618	$\left[\log_2 \left\| \dfrac{b-a}{\varepsilon} \right\| + 1 \right]^2 - 2$
Bus–Dekker (bisection + rational approximation)	1.84	$4 \log_2 \left\| \dfrac{b-a}{\varepsilon} \right\|$

The illustration shown in Fig. 3.8 is given for a convex function that satisfies

$$f(a) < 0, \quad f(b) > 0,$$
$$f'(x) > 0, \quad f''(x) > 0, \quad x \in [a,b]$$

and $-\dfrac{f(a)}{b-a} \leqslant f'(a)$.

It is shown by Krautstengel that when used on functions that satisfy conditions (3), the algorithm converges with the order $1 + \sqrt{2}$. The guaranteed convergence number of the algorithm has not yet been determined.

Table 3.1 gives a brief summary of the characteristics of the robust iteration methods discussed in this and the preceding Sections.

3.9 Optimality of Root-Finding Algorithms for Functions That Change Sign in an Interval

We have seen that in root-finding methods that are suitable for functions which change sign in an interval, the basic iteration step consists of calculating a value of $f(x)$ at some point in the interval and comparing it with the previously obtained value or values. With such methods, therefore, the number of iteration steps is usually the same as the number of function evaluations required. Furthermore, the guaranteed convergence of the method gives the maximum number of function evaluations required to achieve an *a priori* prescribed numerical accuracy of the approximation to the root, assuming a given initial root interval. This number may therefore be considered as one possible criterion with respect to which an 'optimal' method can be formulated. That is to say, the 'optimal' method can be identified by the minimum value of the maximum number of function evaluations required to obtain an approximation to the root within a fixed tolerance ε and starting with a fixed initial root interval. This problem belongs to the general class of optimization problems known as minimax problems and we shall refer to it as the guaranteed convergence minimax problem. It still awaits complete solution.

Another optimization problem can be formulated as follows:

Assuming that the initial root interval and the number of function evaluations are given, find the iterative method which computes the best numerical approximation to the root.

This latter problem can also be described as that of minimizing the maximum length of the interval which can be guaranteed to include the desired zero, after a fixed number of function evaluations, performed sequentially. It thus is another minimax problem and we shall refer to it as the numerical accuracy minimax problem.

It has been shown that bisection is an optimal algorithm in the sense of the numerical accuracy minimax problem, for finding zeros of functions which change sign in an interval, cf. Brent (1971a). For functions which satisfy the

Lipschitz condition with a known constant K, i.e.

$$|f(x') - f(x'')| \leqslant K|x' - x''|, \qquad x', x'' \in [a,b] \tag{1}$$

the optimality of bisection has been proved by Sukharev (1976) and Chernous'ko (1968).

Assuming an initial root interval $[a,b]$ and $f(a)f(b) < 0$, the optimal algorithm may be given in the form:

$$a_0 = a,$$

$$b_0 = b,$$

$$x_1 = \frac{a_0 + b_0}{2},$$

$$a_k = \begin{cases} a_{k-1}, & \text{if } f(x_k) > 0, \\ x_k - \dfrac{f(x_k)}{K}, & \text{if } f(x_k) < 0, \end{cases} \tag{2}$$

$$b_k = \begin{cases} x_k - \dfrac{f(x_k)}{K}, & \text{if } f(x_k) > 0, \\ b_{k-1}, & \text{if } f(x_k) < 0, \end{cases}$$

$$x_{k+1} = \frac{a_k + b_k}{2}, \qquad k = 1, 2, \ldots.$$

We note that the current root intervals are computed using condition (1).

Sukharev (1976) has also shown that for the class of functions which satisfy condition (1), the bisection method given by (2) is optimal in the sense of the so-called W-criterion. The W-criterion is defined as

$$W(X^r, Y^r) = \min_{x \in [a,b]} \quad \max_{f \in L, f(X^r) = Y^r} |f(x)| \tag{3}$$

where X^r and Y^r denote the sets $\{x_i, i = 0, 1, \ldots, r\}$

and $\{y_i = f(x_i), i = 0, 1, \ldots, r\}$ respectively,

and '$f \in L, f(X^r) = Y^r$' reads as 'all the functions which satisfy the Lipschitz condition and such that $f(x_i) = y_i, i = 0, 1, \ldots, r$'.

A 'W-optimal' root-finding algorithm guarantees that after n iteration steps there will be found a point, x_{n+1}, at which the absolute value of the function will not exceed some tolerance w, and also that there exists no other algorithm which, with the same number of iteration steps, would guarantee the point, x_{n+1}, at which the absolute value of the function would not exceed any $w' < w$.

However, bisection does not yield optimality conditions for various suitably restricted classes of allowable functions. For example, for convex functions, Bellman and Dreyfus (1962) (see also Gross and Johnson (1959)) have developed an algorithm which is 'better' than bisection in the sense of the numerical accuracy minimax problem.

3.10 Complexity Parameters and Efficiency Measures

We now proceed to study the computational complexity of the general class of iterative methods given by equation 3.3–(1) and identify possible optimal characteristics of an iterative process from the point of view of the 'lowest cost' of an iteration step. The analysis in the first instance concerns iterative methods which are suitable when good initial approximations to the root are known.

In order to study the computational complexity of these methods, some measure of the efficiency of an iterative method must be introduced which can then be used to compare different methods. Here, we shall follow the work of Traub and his co-workers. Consider the flow diagrams of the basic step of the following two iterative methods.

For the Newton–Raphson method we have:

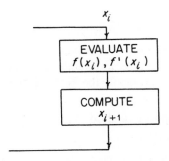

At each iteration, given the value x_i, one function and one derivative evaluation have to be performed, and the data obtained is 'combined' to compute the value x_{i+1}.

For the Secant method we have:

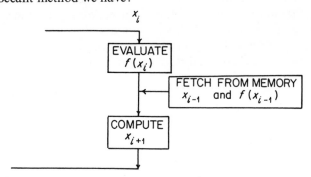

At each iteration, given two values, x_i and x_{i-1}, one function evaluation is performed and then the data obtained is combined to compute x_{i+1}.

In order to relate explicitly the iteration function ϕ to the 'cost' of an iteration for a given iterative method, we introduce the following parameters:

e the number of function evaluations per iteration (i.e. of f, f', etc.)

m the number of iterates, with the exception of the immediately preceding one, that occur in the iteration function (i.e. x_{i-1}, x_{i-2}, etc.).

We denote the iteration function by $\phi_{e,m}$.
Thus, for the Newton–Raphson method we have

$$\phi_{2,0} \text{ since } e = 2, m = 0$$

and for the Secant method we have

$$\phi_{1,1} \text{ since } e = 1, m = 1.$$

Well-known efficiency measures based on the order p and complexity parameter e are due to Traub (1964) and Ostrowski (1960):

$$E_1 = \frac{p}{e} \quad \text{(the Traub efficiency index)}$$

$$E_2 = p^{1/e} \quad \text{(the Ostrowski efficiency index)}.$$

These are asymptotic measures, that is they characterize a particular iterative method when the number of iterations tends to infinity. The higher the value of E_i, the more efficient asymptotically the iterative method.

Using these measures we find

for the Newton–Raphson method $\quad E_1 = 1, \qquad E_2 = 1.414,$

and for the Secant method $\qquad E_1 = 1.62, \quad E_2 = 1.62$

which shows that, in terms of these measures, the secant method is asymptotically more efficient than the Newton–Raphson. The efficiency indices E_1 and E_2 enable one to obtain a valuable overall picture of the relative efficiency of different iterative methods, e.g. an extensive comparison of the methods in terms of E_2 and the asymptotic error constant can be found in Jarratt (1973).

However, the indices E_1 and E_2, being asymptotic measures and so applicable only to the neighbourhood of the root α, are not always adequate for dealing with practical problems. Feldstein and Firestone (1969) have observed that algorithms of identical theoretical efficiency (say, in terms of E_2), do not perform equally well in practice. Two possible reasons for this may be indicated. First, e, the number of function evaluations, treats the function and its derivatives alike, which may be not appropriate in some problems. Second, a difference between methods is introduced by the very nature of the algorithm itself. For example, the Newton–Raphson method requires but 2 operations (1 division and 1 addition) per iteration, the Secant method needs 5 operations (1 division, 1 multiplication and 3 add/subt) per iteration, while the Traub method (see Section 3.9.4) needs 7 arithmetic + 1 shifting operations, etc. The effect of these differences however, is frequently masked by the deffering costs of evaluating f, f', etc. It should be possible, therefore, to develop practical measures of efficiency which would take this cost of computation into account.

In general, one would expect that such measures of efficiency would be obtained by defining efficiency as a function of

(i) the number of function and any needed derivative(s) evaluations per iteration and the cost of these evaluations (problem cost),

(ii) the cost of forming the iterative formula (combinatory cost),

(iii) the order of convergence, p,

(iv) size of the asymptotic error constant.

Feldstein and Firestone (1969) have suggested one such measure:

$$E_3 = p^{1/H}, \quad \text{where} \quad H = e\left(1 + \frac{A}{B}\right)$$

with p and e as before. Here, A is the number of arithmetic operations, exclusive of function evaluations, required to complete one step of the algorithm, B is the number of arithmetic operations required to evaluate f and any needed derivatives.

Another efficiency measure of similar type is due to Paterson (1972), namely

$$E_4 = \frac{1}{M}\log_2 p$$

where M is the number of divisions and multiplications required to calculate $\phi_{e,m}$.

Kung and Traub (1973) have proposed an efficiency measure which, as opposed to E_1, takes account of the number of arithmetic operations required. The measure is defined as follows:

$$E_5 = \sum_i \frac{\log_2 p}{n_i V(f^{(i)}) + c(\phi)}$$

where n_i is the number of evaluations of $f^{(i)}$ used in $\phi_{e,m}$, $V(f^{(i)})$ is the number of arithmetic operations for one evaluation of $f^{(i)}$, $c(\phi)$ is the minimum number of arithmetic operations required to combine the $f^{(i)}$ to form $\phi_{e,m}$ by any procedure.

We should note here, particularly in relation to the E_5 measure that various expressions of the iterative formula may not all be equally good numerically — there is some conflict here between efficiency and, for example, the numerical stability of an algorithm. The numerical stability of iterative methods is discussed in greater detail in Section 3.18. In relation to quantities E_3, E_4, and E_5 certain asymptotic estimates may be given in terms of the size of a problem. For example, for polynomial equations such estimates may be obtained in terms of the degree of a polynomial, n. Assuming that the polynomial and any needed derivatives are evaluated using the Horner algorithm which requires for a polynomial of degree n, n operations each of multiplication and addition, we get for large enough n, the (asymptotic) estimates for E_3, E_4, and E_5 as shown in Table 3.2.

Table 3.2 Asymptotic estimates of some efficiency measures of various iterative methods when applied to compute a simple real root of a polynomial of degree n

Efficiency measure Method	$E_3 = p^{1/H}$	$E_4 = \dfrac{1}{M}\log_2 p$	$E_5 = \sum_i \dfrac{\log_2 p}{n_i V(f^{(i)}) + c(\phi)}$
Secant	$\sqrt{1.618}$	$\dfrac{1}{1.44n}$	$\dfrac{1}{2.87n}$
Newton–Raphson's	$\sqrt{2}$	$\dfrac{1}{2n}$	$\dfrac{1}{4n}$
Traub's (see Section 3.14, formula (3))	$\sqrt[3]{4}$	$\dfrac{1}{1.5n}$	$\dfrac{1}{3n}$
Jarratt's (see Section 3.15, formula (6))	$\sqrt[3]{4}$	$\dfrac{1}{1.5n}$	$\dfrac{1}{3n}$
Fifth-order (see Section 3.15, formula (12))	$\sqrt[4]{5}$	$\dfrac{1}{1.7n}$	$\dfrac{1}{3.4n}$

3.11 Interpolation and One-Point Iterative Methods

Flow diagrams of the basic step of the Newton–Raphson and Secant methods show that at the $(i + 1)$st iteration, the function and its derivative(s) are evaluated at one point only, namely x_i. We refer to these evaluations as new data, other information, for example the values $f(x_{i-1})$, $f(x_{i-2})$, etc., being readily available in the computer memory for possible use in forming the next iterate, x_{i+1}. If the iterative method makes use of the 'old' information, then this information is simply 'passively' combined to form x_{i+1}. Iterative methods of this nature may be called one-point methods *with memory* (e.g. the Secant method) or *without memory* (e.g. the Newton–Raphson method).

The theory of these methods can be developed in a systematic and general form using the concept of direct and inverse interpolation.

3.11.1 Direct Polynomial Interpolation

Let $x_i, x_{i-1}, \ldots, x_{i-m}$ be $m + 1$ approximates to a zero of the function $f(x)$.

Let $Q_{e,m}$ be a polynomial of degree $e(m + 1) - 1$ whose first $e - 1$ derivatives agree with first $e - 1$ derivatives of $f(x)$ at these points, i.e.

$$Q_{e,m}^{(s)}(x_{i-j}) = f^{(s)}(x_{i-j}), \qquad j = 0, 1, \ldots, m; \; s = 0, \ldots, e - 1. \tag{1}$$

Define the next approximant, x_{i+1}, as the root of $Q_{e,m}$, i.e.

$$Q_{e,m}(x_{i+1}) = 0. \tag{2}$$

In this case, if $\phi_{e,m}$ is the iteration function that maps x_i into x_{i+1}, then it can be obtained from the polynomial $Q_{e,m}$.

The process is then repeated for the set $x_{i+1}, x_i, \ldots, x_{i-m+1}$ and so on.

The majority of computationally useful root-finding techniques fall into two groups: those whose iterative formulae require values of the function f only and those whose iterative formulae require values of both the function f and its derivative f' (processes which use higher derivatives than f' are relatively uncommon in practice, though some examples will be given later). We give below some examples of how iterative methods of either type can be derived using direct polynomial iterpolation.

1. Let $e = 1, m = 1$.
(The iteration function is generated using direct interpolation at two points.)

We have

$$Q_{1,1}(x_i) = f(x_i)$$
$$Q_{1,1}(x_{i-1}) = f(x_{i-1}),$$

and

$$Q_{1,1}(x) = f(x_i)\frac{x - x_{i-1}}{x_i - x_{i-1}} + f(x_{i-1})\frac{x - x_i}{x_{i-1} - x_i}. \tag{3}$$

Setting

$$Q_{1,1}(x_{i+1}) = 0, \text{ we get}$$

$$x_{i+1} = x_i - f(x_i)\frac{x_i - x_{i-1}}{f(x_i) - f(x_{i-1})} = \phi_{1,1},$$

which is the Secant method.

2. Let $e = 1, m = 2$.
(The iteration function is generated using direct interpolation at three points.)

We have

$$Q_{1,2}(x_i) = f(x_i)$$
$$Q_{1,2}(x_{i-1}) = f(x_{i-1})$$
$$Q_{1,2}(x_{i-2}) = f(x_{i-2})$$

and

$$
\begin{aligned}
Q_{1,2}(x) &= f(x_i)\frac{(x - x_{i-1})(x - x_{i-2})}{(x_i - x_{i-1})(x_i - x_{i-2})} \\
&\quad + f(x_{i-1})\frac{(x - x_i)(x - x_{i-2})}{(x_{i-1} - x_i)(x_{i-1} - x_{i-2})} \\
&\quad + f(x_{i-2})\frac{(x - x_i)(x - x_{i-1})}{(x_{i-2} - x_i)(x_{i-2} - x_{i-1})}.
\end{aligned}
\tag{4}
$$

Setting $Q_{1,2}(x_{i+1}) = 0$ we get an iterative method which is known as Muller's method:

$$x_{i+1} = x_i - \frac{2f(x_i)}{w + \{w^2 - 4f(x_i)f[x_i,x_{i-1},x_{i-2}]\}^{1/2}}$$

where $w = f[x_i,x_{i-1}] + (x_i - x_{i-1})f[x_i,x_{i-1},x_{i-2}]$.

Here $f[x_i,x_{i-1}]$ denotes the first divided difference

$$\frac{f(x_i) - f(x_{i-1})}{x_i - x_{i-1}}$$

and $f[x_i,x_{i-1},x_{i-2}]$, the second divided difference

$$\frac{f[x_i,x_{i-1}] - f[x_{i-1},x_{i-2}]}{x_i - x_{i-2}}.$$

The method was originally expounded by Muller in 1956 and is particularly suitable for solving high-order polynomial equations, although it can in principle be applied equally well to transcendental equations. It is used to find both real and complex roots. Muller's was the first published work in which the direct interpolation was proposed as a tool for developing iterative methods. The order of convergence of Muller's method is 1.839 and its discussion can be found in several good textbooks on numerical computation, for example, Conte and de Boor (1972).

3. Let $e = 2$, $m = 0$.
(The iteration function is generated using direct interpolation at one point.) Hence we have

$$Q_{2,0}(x_i) = f(x_i)$$
$$Q'_{2,0}(x_i) = f'(x_i)$$

and

$$Q_{2,0}(x) = f(x_i) + (x - x_i)f'(x_i). \tag{5}$$

Setting $Q_{2,0}(x_{i+1}) = 0$ we get

$$x_{i+1} = x_i - \frac{f(x_i)}{f'(x_i)} = \phi_{2,0},$$

which is the Newton–Raphson formula. We note that although $e(m + 1)$ values are necessary to start the process, thereafter, only one new function value per iteration is required.

The use of direct interpolating polynomials has a number of practical disadvantages. For example, to obtain any iterate, x_{i+1}, one has to solve a polynomial of degree $e(m + 1) - 1$. This makes the solution non-unique as we don't know which of the zeros of the polynomial we get. The rule (or rules) must be such that the point x_{i+1} is determined uniquely by the points x_i, $x_{i-1},\ldots,x_{i-m}$; that is, there must be a function which maps $x_i,\ldots,x_{i-m}$ into

x_{i+1}. Also, some or all the zeros of the polynomial may be complex, and although the fact that a new approximation may be predicted as being complex is acceptable in general for polynomial equations, it is a nuisance in the case where only real roots are of interest.

3.11.2 Inverse Polynomial Interpolation

To avoid the need for solving a polynomial equation, an interpolation of F, the inverse of f, at the points $y_i = f(x_i), y_{i-1} = f(x_{i-1}), \ldots, y_{i-m} = f(x_{i-m})$, can be carried out instead.

Let f' be non-zero (i.e. the root α is simple) and let $f^{(q)}$ be continuous in an interval J. Let f map J into K. Then f has an inverse F and $F^{(q)}$ is continuous on K. There exists a unique polynomial $R_{e,m}$ of degree $e(m+1) - 1$ such that

$$R_{e,m}^{(s)}(y_{i-j}) = F^{(s)}(y_{i-j}), \quad j = 0, \ldots, m; \quad s = 0, \ldots, e-1. \tag{1}$$

Then, evaluating the interpolatory polynomial at zero determines the point x_{i+1} uniquely, i.e. $x_{i+1} = R_{e,m}(0)$. The function that maps $x_i, x_{i-1}, \ldots, x_{i-m}$ into x_{i+1} is denoted by $\Psi_{e,m}$.

We now give examples of how iterative methods can be arrived at using inverse polynomial interpolation.

1. Let $e = 1, m = 1$, that is the iteration function is generated using inverse polynomial interpolation over two points. We write

$$R_{1,1}(y) = F(y_i)\frac{y - y_{i-1}}{y_i - y_{i-1}} + F(y_{i-1})\frac{y - y_i}{y_{i-1} - y_i} \tag{2}$$

where $y = f(x), \quad x = F(y)$.

An iterative method is obtained as

$$x_{i+1} = R_{1,1}(0) = -x_i\frac{f(x_{i-1})}{f(x_i) - f(x_{i-1})} - x_{i-1}\frac{f(x_i)}{f(x_{i-1}) - f(x_i)}$$

or

$$x_{i+1} = x_i - f(x_i)\frac{x_i - x_{i-1}}{f(x_i) - f(x_{i-1})},$$

which is the Secant method again.

2. Let $e = 1, m = 2$, that is the iteration function is generated using inverse polynomial interpolation over three points. Write

$$R_{1,2}(y) = F(y_i)\frac{(y - y_{i-1})(y - y_{i-2})}{(y_i - y_{i-1})(y_i - y_{i-2})} + F(y_{i-1})\frac{(y - y_i)(y - y_{i-2})}{(y_{i-1} - y_i)(y_{i-1} - y_{i-2})}$$

$$+ F(y_{i-2})\frac{(y - y_i)(y - y_{i-1})}{(y_{i-2} - y_i)(y_{i-2} - y_{i-1})}, \tag{3}$$

giving

$$x_{i+1} = R_{1,2}(0) = x_i \frac{f(x_{i-1})f(x_{i-2})}{(f(x_i)-f(x_{i-1}))(f(x_i)-f(x_{i-2}))}$$

$$+ x_{i-1} \frac{f(x_i)f(x_{i-2})}{(f(x_{i-1})-f(x_i))(f(x_{i-1})-f(x_{i-2}))}$$

$$+ x_{i-2} \frac{f(x_i)f(x_{i-1})}{(f(x_{i-2})-f(x_i))(f(x_{i-2})-f(x_{i-2}))}. \tag{4}$$

Expression (4) can be presented in the form:

$$x_{i+1} = x_i - \frac{f(x_i)}{f[x_i,x_{i-1}]} + \frac{f(x_i)f(x_{i-1})}{f(x_i)-f(x_{i-2})} \left[\frac{1}{f[x_i,x_{i-1}]} - \frac{1}{f[x_{i-1},x_{i-2}]} \right]. \tag{5}$$

The order of convergence of the method is 1.839.

3. We now deduce an algorithm for obtaining the iteration function $\Psi_{e,m}$ using inverse polynomial interpolation at one point. To do this we first observe that

$$\alpha = F(0),$$

since $x \equiv F(f(x))$, $x \in J$, and $y \equiv f(F(y))$, $y \in K$. (6)
By the Taylor formula:

$$\alpha = F(0) = F(y) + \sum_{k=1}^{r} (-1)^k \frac{y^k}{k!} F^{(k)}(y) + (-1)^{r+1} \frac{F^{(r+1)}(\eta)}{(r+1)!}, \quad 0 \leqslant \eta \leqslant y$$

we obtain

$$\alpha = x + \sum_{k=1}^{r} (-1)^k \frac{F^{(k)}(f(x))}{k!} [f(x)]^k + (-1)^{r+1} \frac{F^{(r+1)}(\eta)}{(r+1)!} [f(x)]^{r+1}. \tag{7}$$

For simplicity, we denote

$$F^{(k)}(f(x)) \equiv g_k(x) \text{ and } \Psi_{r,0}(x) = x + \sum_{k=1}^{r} (-1)^k \frac{g_k(x)}{k!} [f(x)]^k. \tag{8}$$

Equation $x = \Psi_{r,0}(x)$ has the root $x = \alpha$, since

$$\Psi_{r,0}(\alpha) = \alpha + \sum_{k=1}^{r} (-1)^k \frac{g_k(\alpha)}{k!} [f(a)]^k = \alpha.$$

Thus, setting

$$x_{i+1} = \Psi_{r,0}(x_i), \quad i = 0,1,\ldots, \quad x_0 \in J, \tag{9}$$

we obtain an iterative method.

If x_0 is chosen close enough to the root α, then the sequence $\{x_i\}$ converges to α, as there exists an interval containing the point α such that for any point x of this interval the following holds:

$$|\Psi'(x)| \leqslant C < 1. \tag{10}$$

It follows that for the convergence of the sequence $\{x_i\}$, it is only necessary for x_0 to belong to this interval.

The iteration function $\Psi_{r,0}(x)$ can be found in an explicit form in terms of $f(x)$ and its derivatives. By virtue of (6) we have:

$$F'(f(x))f'(x) = 1$$
$$F''(f(x))[f'(x)]^2 + F'(f(x))f''(x) = 0$$
$$F'''(f(x))[f'(x)]^3 + 3F''(f(x))f'(x)f''(x) + F'(f(x))f'''(x) = 0, \qquad (11)$$

giving

$$g_1(x)f'(x) = 1$$
$$g_2(x)[f'(x)]^2 + g_1(x)f''(x) = 0 \qquad (12)$$
$$g_3(x)[f'(x)]^3 + 3g_2(x)f'(x)f''(x) + g_1(x)f'''(x) = 0.$$

From (12), the functions $g_1(x), g_2(x), \ldots$ can be found successively.

We now give some examples of the use of inverse polynomial interpolation at one point.

3a. Let $e = 2$, $m = 0$.
From (9), (8), and (12), we have

$$x_{i+1} = \Psi_{2,0} = x_i - \frac{f(x_i)}{f'(x_i)},$$

and we have again the Newton–Raphson method.

3b. Let $e = 3$, $m = 0$.
We have

$$x_{i+1} = \Psi_{3,0} = x_i + (-1)g_1(x_i)f(x_i) + \frac{1}{2!}g_2(x_i)f^2(x_i),$$

where

$$g_1(x_i) = \frac{1}{f'(x_i)},$$

$$g_2(x_i) = -g_1(x_i)\frac{f''(x_i)}{[f'(x_i)]^2}.$$

It follows that

$$x_{i+1} = \Psi_{3,0} = x_i - \frac{f(x_i)}{f'(x_i)} - \left[\frac{f(x_i)}{f'(x_i)}\right]^2 \frac{f''(x_i)}{f'(x_i)}, \qquad (13)$$

which is known as the Chebyshev method. The order of the method is 3.

The orders of convergence of the corresponding direct and inverse methods are the same. For either of the two processes,—direct or inverse interpolation, $-x_0, x_1, \ldots, x_m$ must be available. One method for obtaining these starting values is to use $\phi_{e,1}$, followed successively by $\phi_{e,2}, \phi_{e,3}, \ldots, \phi_{e,m}$ at the beginning of a calculation.

3.11.3 Derivative Estimated Iterative Methods

The term identifying this useful group of methods is due to Traub (1964) who has extensively explored the possibilities for constructing new methods by taking an iterative formula involving the derivatives and replacing either all or the highest derivatives by estimates, generated in some suitable fashion.

Take, for example, the Newton–Raphson formula

$$x_{i+1} = x_i - \frac{f(x_i)}{f'(x_i)}.$$

Using direct polynomial interpolation over two approximations to the root, x_i and x_{i-1}, the polynomial of degree 1 is set up, giving

$$P_1(x) = f(x_i)\frac{x - x_{i-1}}{x_i - x_{i-1}} + f(x_{i-1})\frac{x - x_i}{x_{i-1} - x_i}.$$

The value of $P_1'(x_i)$ is now taken as an estimate of $f'(x_i)$ and used to replace the derivative in the Newton–Raphson formula, giving

$$x_{i+1} = x_i - f(x_i)\frac{x_i - x_{i-1}}{f(x_i) - f(x_{i-1})},$$

which, obviously, is the Secant method formula.

Similarly, using interpolation over three approximations, x_i, x_{i-1}, x_{i-2}, the polynomial of degree 2 is set up, for which $P_2(x_{i-j}) = f(x_{i-j})$, $j = 0, 1, 2$. The value of $P_1'(x_i)$ is now taken as an estimate of $f'(x_i)$ and used to replace the derivative in the Newton–Raphson formula. We get a new iterative method given by

$$x_{i+1} = x_i - \frac{f(x_i)}{f[x_i, x_{i-1}] + f[x_i, x_{i-2}] - f[x_{i-1}, x_{i-2}]} \tag{1}$$

where as before, $f[x_s, x_{s-1}]$ denotes the first divided difference.

This latter method (1) is one of the two new important methods which were explored by Traub as a result of utilizing the idea of derivative estimation. He has also shown that the order of convergence and the asymptotic error constant of the method are precisely the same as those of the method of direct interpolation using a quadratic fit; the order is 1.839.

Inverse polynomial intropolation can be employed in a similar fashion to estimate the derivative. Generally, the derivative estimates found using a direct and an inverse interpolating polynomial of the same degree are different. Thus, using inverse interpolation over three approximations, x_i, x_{i-1}, x_{i-2}, to estimate f' and then replacing the derivative in the Newton–Raphson formula by the estimate obtained lead to a new iterative formula given by

$$x_{i+1} = x_i - f(x_i)\left\{\frac{1}{f[x_i, x_{i-1}]} + \frac{1}{f[x_i, x_{i-2}]} - \frac{1}{f[x_{i-1}, x_{i-2}]}\right\} \tag{2}$$

which is the second of the two new methods obtained by Traub. The order of convergence of this method is again 1.839.

We now take the Chebyshev third order method:

$$x_{i+1} = x_i - \frac{f(x_i)}{f'(x_i)} - \left[\frac{f(x_i)}{f'(x_i)}\right]^2 \frac{f''(x_i)}{f'(x_i)}.$$

Using a two-point Hermite polynomial interpolation formula over two approximations, x_i and x_{i-1}, we get

$$H_3(x) = \frac{1}{(x_i - x_{i-1})^2}\left[\left(1 - 2\frac{x - x_i}{x_i - x_{i-1}}\right)(x - x_{i-1})^2 f(x_i)\right.$$

$$+ \left(1 - 2\frac{x - x_{i-1}}{x_{i-1} - x_i}\right)(x - x_i)^2 f(x_{i-1}) + (x - x_i)(x - x_{i-1})^2 f'(x_i)$$

$$\left. + (x - x_{i-1})(x - x_i)^2 f'(x_{i-1})\right] \tag{3}$$

Differentiating $H_3(x)$ twice gives an estimate of $f''(x_i)$ as

$$f''(x_i) = \frac{2}{x_i - x_{i-1}}\{2f'(x_i) + f'(x_{i-1}) - 3f[x_i, x_{i-1}]\} \tag{4}$$

while using inverse interpolation under similar conditions, one obtains

$$f''(x_i) = \frac{2}{f(x_i) - f(x_{i-1})}\left\{\frac{2}{f'(x_i)} + \frac{1}{f'(x_{i-1})} - \frac{3}{f[x_i, x_{i-1}]}\right\}. \tag{5}$$

Replacing the second derivative in Chebyshev's formula by the estimates given in (4) and (5) we obtain two different iterative formulae, each of order 2.73.

3.12 Order of Convergence and Optimality of One-Point Iterative Methods

We have noted earlier that the order of an iterative method may be considered a suitable measure of the improvement in accuracy of the numerical result gained at one iteration.

On the other hand, the order of a one-point iterative method is related to the number of 'new' function evaluations at one iteration step and this number includes evaluation of the function as well as its derivatives where appropriate. Traub was the first to observe that the efficiency of one-point iterative methods without memory is limited by the fact that such a method of order p depends on at least p-1 derivatives of a function f, and hence requires at least p 'new' evaluations of the function per iteration. In searching for ways to improve the efficiency of iterative methods, one possibility which suggests itself is an efficient use of the 'old' information. It is then natural to ask what is the most that information gathered at preceding iterations can add to the order of an iteration method. Again, Traub noted that the contribution of the "old" information to the order of convergence is less than unity when quantitatively measured in terms of the number of required function evaluations per iteration.

Traub (1964) conjectured subsequently that:

> an analytic one-point iterative method (with or without memory) which is based on p evaluations per iteration step is of order at most p.

This conjecture has since been proved by Brent, Winograd, and Wolfe (1973), and Kung and Traub (1974). In the next section we outline the proof, following Brent *et al.*

3.13 Order of Convergence of Polynomial Schemes

Formulation of the theory of one-point iterative methods using the concept of polynomial interpolation reveals that, in essence, all the iterative methods that have been discussed are based on the replacement at each iteration of a function f by a polynomial function Q which agrees with f in a specified way. Therefore, as always when one function is interpolated by another, a 'remainder' formula may be obtained for the error incurred. One such useful formula is due to Cauchy:

Let I be the smallest interval containing $(x_i, x_{i-1}, \ldots, x_{i-m})$.
For any x there is a ξ, in the smallest interval containing I and x, such that

$$f(x) - Q(x) = \frac{1}{n!}[f - Q]^{(n)}(\xi) \prod_{j=0}^{m} (x - x_{i-j})^e \tag{1}$$

where $n = e(m + 1)$.

This formula leads to the derivation of upper bounds on the order of convergence as a function of e for polynomial interpolation schemes with fixed m. Now, if Q_i is the interpolating polynomial agreeing with f at $x_i, \ldots, x_{i-m}$ for the first $(e - 1)$ derivatives, then Q_i is of degree $n - 1$, where $n = e(m + 1)$, so (1) gives

$$f(x) - Q_i(x) = \frac{1}{n!} f^{(n)}(\xi) \prod_{j=0}^{m} (x - x_{i-j})^e,$$

or

$$|Q_i(\alpha)| = \frac{1}{n!} |f^{(n)}(\xi)| \prod_{j=0}^{m} \varepsilon_{i-j}^e, \tag{2}$$

where $\varepsilon_k = |\alpha - x_k|$.
Since $Q_i(x_{i+1}) = 0$, then $|Q_i(\alpha)| = \varepsilon_{i+1} |Q_i'(\eta)|$ for some value of η in $[x_{i+1}, \alpha]$.
Assuming that in addition to $f'(\alpha) \neq 0$ (this assumption has been made at the beginning of the chapter), we have $f^{(n)}(\alpha) \neq 0$. Estimation of $|Q_i'(\eta)|$ and $|f^{(n)}(\xi)|$ reduces (2) to

$$\varepsilon_{i+1} = (C + \delta_i) \prod_{j=0}^{m} \varepsilon_{i-j}^e, \tag{3}$$

where

$$C = \left| \frac{f^{(n)}(\alpha)}{n! \, f'(\alpha)} \right| \quad \text{and } \delta_i \to 0.$$

Taking logarithms, we can rewrite (3) as

$$\log_2 \varepsilon_{i+1} = \log_2 (C + \delta_i) + e \sum_{j=0}^{m} \log_2 \varepsilon_{i-j}. \tag{4}$$

Since with i increasing, $\log_2 \varepsilon_{i+1} \to \infty$, the term $\log_2 (C + \delta_i)$ becomes unimportant, and we can consider the difference equation

$$\log_2 \varepsilon_{i+1} = e \sum_{j=0}^{m} \log_2 \varepsilon_{i-j}. \tag{5}$$

We solve this equation in a normal way by first letting

$\log_2 \varepsilon_s = \lambda^s$, and then considering

$$\lambda^{i+1} = e \sum_{j=0}^{m} \lambda^{i-j},$$

which gives, after multiplying both sides by λ^{m-i},

$$\lambda^{m+1} - e \sum_{j=0}^{m} \lambda^j = 0. \tag{6}$$

Let λ be the unique real root of largest modulus of equation (6). For $m = 0, 1, 2, 3$ and fixed integer value of e these roots are

$$\lambda = e, \quad \lambda = e + 0.732, \quad \lambda = e + 0.920, \quad \lambda = e + 0.974.$$

Moreover, $\lambda \to e + 1$ as $m \to \infty$.
Now, we have

$$\lim_{k \to \infty} \frac{\log \varepsilon_{k+1}}{\log \varepsilon_k} = \lambda. \tag{7}$$

We shall now show that the order of convergence of the one-point iterative method defined by $Q_i(x_{i+1}) = 0$, is equal to λ.

By definition we have

$$\lim_{k \to \infty} \frac{|\alpha - x_{k+1}|}{|\alpha - x_k|} = \lim_{k \to \infty} \frac{\varepsilon_{k+1}}{\varepsilon_k^p} = C \neq 0.$$

For large k, we can write

$$\frac{\varepsilon_{k+1}}{\varepsilon_k^p} \approx C.$$

Taking logarithms, we obtain

$$\log_2 \varepsilon_{k+1} - p \log_2 \varepsilon_k = \log_2 C,$$

which can be rewritten as

$$\frac{\log_2 \varepsilon_{k+1}}{\log_2 \varepsilon_k} = p + \frac{\log_2 C}{\log_2 \varepsilon_k},$$

giving

$$\lim_{k \to \infty} \frac{\log_2 \varepsilon_{k+1}}{\log_2 \varepsilon_k} = p, \tag{8}$$

since $\log_2 C$ is a constant.

Comparing (7) and (8), we get $p = \lambda = e + 1$.

We may thus conclude that if an iterative method uses only the values of the function f and its derivatives up to the $(e - 1)$st inclusive (evaluated at arbitrarily many previous values of the argument), then the order of convergence of the method is less or equal to $e + 1$.

This result is often referred to as the basic optimality theorem for one-point iterative methods.

3.14 Multi-Point Iterative Methods

The optimality theorem for one-point iterative methods establishes that the relation between the order of convergence of such methods and the number of function evaluations per iteration step is linear.

We may now pose a further question:

Is it possible to construct an iterative process which would not be subject to these restrictions?

The concept of the so called multi-point iterative method (without memory or with memory, in the same sense as used in distinguishing between two types of the one-point method) has been developed by Traub so as to give a positive answer to this question. A multi-point iterative method may be considered as one possible technique for increasing the efficiency of iterative methods.

The general idea of a multi-point method is based on the use of an appropriate combination of iterates. One simple example would be to use two successive Newton–Raphson iterates. The method obtained would be of fourth order and would require four function evaluations per iteration. We give below three further examples of multi-point iterative methods.

1. Suppose that the ith approximation to the root α, has been computed.

Now let $z_i = x_i - \dfrac{f(x_i)}{f'(x_i)}$ and $x_{i+1} = z_i - \dfrac{f(z_i)}{f'(x_i)}$.

Combining the two equations we get

$$x_{i+1} = x_i - u(x_i) - \frac{f(x_i - u(x_i))}{f'(x_i)}, \qquad \text{where } u = \frac{f}{f'}. \tag{1}$$

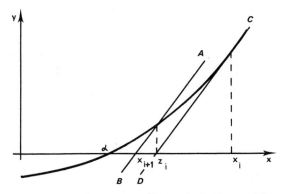

Fig. 3.9. An illustration of Example 1. The straight
line AB and CD are parallel

It may easily be verified that the iteration function

$$\phi_{3,0} = x_i - u(x_i) - \frac{f(x_i - u(x_i))}{f'(x_i)}$$

is of order 3.

A geometric illustration of the method is given in Fig. 3.9.

2. We now combine the Newton–Raphson method,

$$z_i = x_i - \frac{f(x_i)}{f'(x_i)},$$

and the Secant method,

$$x_{i+1} = z_i - f(z_i) \frac{z_i - x_i}{f(z_i) - f(x_i)},$$

to give (noting that $z_i + u_i = x_i$)

$$x_{i+1} = \phi_{3,0} = x_i - u(x_i) + f(x_i - u(x_i)) \frac{u(x_i)}{f(x_i - u(x_i)) - f(x_i)}$$

$$= x_i + \frac{u(x_i)f(x_i)}{f(x_i - u(x_i)) - f(x_i)}. \qquad (2)$$

The iteration function obtained is again of order 3. This method may be
visualized as the intersection with the x-axis of the secant line through the
points $(x, f(x))$ and $(x - u(x), f(x - u(x)))$. It is illustrated in Fig. 3.10.

3. Now consider the combination of the following two one-point iterative
methods

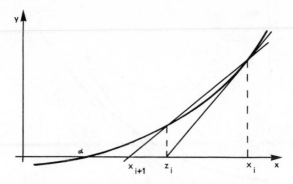

Fig. 3.10. An illustration of Example 2

$$z_i = x_i - \frac{f(x_i)}{f'(x_i)}, \qquad x_{i+1} = z_i - \frac{f(z_i)(z_i - x_i)}{2f(z_i) - f(x_i)}.$$

Here at every second step the derivative f' is replaced by a linear combination of $f(x)$ at two immediately preceding points. The method is illustrated in Fig. 3.11.

Combining the above two equations we obtain

$$x_{i+1} = \phi_{3,0} = x_i - u(x_i)\frac{f(x_i - u(x_i)) - f(x_i)}{2f(x_i - u(x_i)) - f(x_i)} \tag{3}$$

We shall call (3) the Traub iterative method.
The flow-diagram of the method's basic step is shown below:

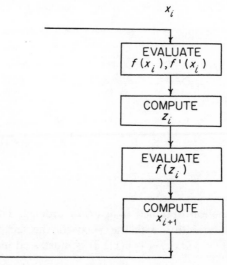

Here we have $e = 3$, $m = 0$, and, hence, $\phi = \phi_{3,0}$; however, as we shall now see, the method is of order 4.

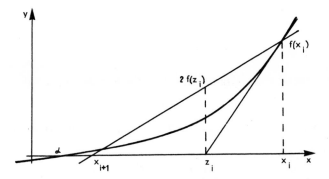

Fig. 3.11. The Traub method for a convex function

First re-write the formula (3) as follows:

$$x_{i+1} = x_i - (x_i - z_i)\frac{f(z_i) - f(x_i)}{2f(z_i) - f(x_i)}.$$

Introducing the error ε_i such that $x_i = \alpha + \varepsilon_i$, where α is the true value of the root, and noting that

$$z_i = \alpha + K\varepsilon_i^2 + M\varepsilon_i^3 + O(\varepsilon_i^4) \tag{4}$$

where $K = \frac{1}{2}\frac{f''}{f'}$ and $M = \frac{1}{2}\frac{f'''}{f'} - \frac{1}{3}\left(\frac{f''}{f'}\right)^2,$ (5)

we can write

$$\alpha + \varepsilon_{i+1} = \alpha + \varepsilon_i - (\alpha + \varepsilon_i - \alpha - K\varepsilon_i^2 - M\varepsilon_i^3)$$

$$\times \frac{f(\alpha + K\varepsilon_i^2 + M\varepsilon_i^3) - f(\alpha + \varepsilon_i)}{2f(\alpha + K\varepsilon_i^2 + M\varepsilon_i^3) - f(\alpha + \varepsilon_i)}. \tag{6}$$

Using the Taylor series expansion and some simple transformations, the relation (6) is brought to the form:

$$\varepsilon_{i+1} = \varepsilon_i - \varepsilon_i\frac{N}{D}, \tag{7}$$

where $N = -f' + \frac{1}{2}f''\varepsilon_i + (2Mf' - \frac{1}{6}f''')\varepsilon_i^2$
$$+ (\frac{1}{2}K^2f'' - \frac{1}{24}f^{iv} - KMf' + \frac{1}{6}Kf''')\varepsilon_i^2 + \dots,$$

and $D = -f' + \frac{1}{2}f''\varepsilon_i + (2Mf' - \frac{1}{6}f''')\varepsilon_i^2 + (K^2f'' - \frac{1}{24}f^{iv})\varepsilon_i^3 + \dots.$

which gives the leading term as

$$\varepsilon_{i+1} = \left[\frac{1}{8}\left(\frac{f''}{f'}\right)^3 - \frac{1}{12}\left(\frac{f''}{f'}\right)\left(\frac{f'''}{f'}\right)\right]\varepsilon_i^4 + O(\varepsilon_i^5). \tag{8}$$

It is customary to denote the ratio $\dfrac{f^{(r)}(\alpha)}{f'(\alpha)}$ by F_r. The relation (8) can then be rewritten as

$$\varepsilon_{i+1} = \left[\frac{1}{8}F_2^3 - \frac{1}{12}F_2F_3\right]\varepsilon_i^4 + O(\varepsilon_i^5),$$

where the expression in square brackets is the asymptotic constant of Traub's method. This completes the proof.

3.15 Further Examples of Multi-Point Iterative Methods

A particular feature of the Traub multi-point method given by 3.14—(3) lies in the fact that it is an iterative method whose order of convergence is higher than the number of function evaluations per iteration. The order of the method is $p = 4$ but it requires only 3 function evaluations per iteration. In this respect, however, even more dramatic examples of multi-point methods may be given. Below we illustrate some families of multi-point methods that possess specific useful features.

I. Suppose that the equation $f(x) = 0$ is such that the derivative f' can be rapidly evaluated compared with f itself (for example for a function defined as an integral). For problems of this kind a multi-point iterative method which minimizes the number of new function evaluations at the price of extra evaluations of its derivative(s) may be considered as more efficient. Traub (1964), Kogan (1967), and Jarratt (1966a, 1966c, 1969) have constructed multi-point iterative methods which utilize this feature.

Consider a family of methods defined by the general formula

$$\phi(x) = x - \sum_{j=1}^{s-1} \alpha_j^s \omega_j^s(x) \qquad (1)$$

where

$$\omega_j^s(x) = f(x)/f'\left[x + \sum_{i=1}^{j} \beta_{j,i}^s \omega_i^s(x)\right], \quad j > 1.$$

Expanding (1) about the root α and defining the ω_1^s in various ways, the free parameters, α_j^s and $\beta_{j,i}^s$, can be used to optimize, at least in the asymptotic sense, the order of convergence of a specific iterative method of form (1) for locating the simple roots of an arbitrary function f.

Traub considered in detail the simplified form

$$\phi(x) = x - \alpha_1 \omega_1(x) - \alpha_2 \omega_2(x) - \alpha_3 \omega_3(x), \qquad (2)$$

where

$$\omega_1 = u(x) = \frac{f(x)}{f'(x)}, \qquad (3)$$

$$\omega_2 = f/f'(x + \beta\omega_1), \qquad (4)$$

$$\omega_3 = f/f'(x + \gamma_1\omega_1 + \gamma_2\omega_2). \qquad (5)$$

He has shown that for $\alpha_3 = 0$ and appropriate choices of the parameters α_1, α_2

and β, it is possible to build a family of third order methods, each requiring one function and two derivative evaluations per iteration step. Furthermore, by including the term $-\alpha_3\omega_3(x)$ he has also shown that fourth order formulae exist that require function and three derivative evaluations per iteration step.

A further search for efficient iteration functions based on (1) resulted in a rather remarkable method due to Jarratt which is given by

$$x_{k+1} = \phi(x_k) = x_k - \frac{5}{8}\frac{f(x_k)}{f'(x_k)} - \frac{3}{8}f(x_k)\frac{f'(x_k)}{\left\{f'\left[x_k - \frac{2}{3}\frac{f(x_k)}{f'(x_k)}\right]\right\}^2}. \tag{6}$$

The method is of fourth order but requires one function and only two derivative evaluations.

Another simplified form of (1) is

$$\phi_r(x) = x - \frac{f(x)}{f'\left(x + \sum_{s=1}^{r}\beta_s(f(x))^s\right)}, \tag{7}$$

where $(f(x))^s$ denotes the sth power of $f(x)$.

Using this form leads to a third order iterative method due to Kogan and given by

$$x_{k+1} = \phi_1(x_k) = x_k - \frac{f(x_k)}{f'\left(x_k - \frac{f(x_k)}{2f'(x_k)}\right)}. \tag{8}$$

The process requires one function and two derivative evaluations per iteration.

Yet another group of interesting methods considered by Jarratt arises from the formula

$$\phi(x) = x - \frac{f(x)}{\alpha_1\omega_1(x) + \alpha_2\omega_2(x) + \alpha_3\omega_3(x)} \tag{9}$$

where

$$\omega_1 = f'(x), \qquad \omega_2 = f'(x + \alpha f(x)/\omega_1(x)), \tag{10}$$

$$\omega_3 = f'(x + \beta f(x)/\omega_1(x) + \gamma f(x)/\omega_2(x)). \tag{11}$$

Among the methods of this group, of chief importance is a fifth order method which has the form

$$x_{k+1} = x_k - \frac{6f(x_k)}{D}, \tag{12}$$

where $D = f'(x_k) + f'(x_k - f(x_k)/f'(x_k))$
$\qquad + 4f'(x_k - \frac{1}{8}f(x_k)/f'(x_k) - \frac{3}{8}f(x_k)/f'(x_k - f(x_k)/f'(x_k)))$.

It requires one function and three derivative evaluations per iteration.

II. Now suppose that the equation $f(x) = 0$ is such that the derivative f' is very difficult to evaluate.

Kung and Traub (1974) have developed a family of multi-point iterative methods without memory and of order 2^{n-1} that require just n evaluations of f per iteration. The family is based on the inverse interpolating polynomials, and n is the number of points over which the interpolation is carried out. An example of such a method for $n = 4$ may be given as

$$z_1 = x_k + \beta f(x_k),$$

$$z_2 = z_1 - \beta \frac{f(x_k) f(z_1)}{f(z_1) - f(x_k)},$$

$$z_3 = z_2 - \frac{f(x_k) f(z_1)}{f(z_2) - f(x_k)} \left\{ \frac{1}{f[z_1, x_k]} - \frac{1}{f[z_2, z_1]} \right\},$$

$$x_{k+1} = z_3 - \frac{f(x_k) f(z_1) f(z_2)}{f(z_3) - f(x_k)} \left\{ \frac{1}{f[z_1, x_k]} - \frac{1}{f[z_2, z_1]} - \frac{1}{f[z_3, z_2]} \right\}, \qquad (13)$$

where β is a suitably chosen non-zero constant and $f[x_i, x_{i-1}]$ denotes, as usual, the first divided difference.

3.16 Optimality of Multi-Point Iterative Methods

We can make the following general observation in relation to the multi-point iterative methods we have just discussed. If a multi-point method is composed of two or more one-point iterations directly, then such a method will retain the limitations on e and on p. For example, a multi-point method composed of n Newton–Raphson iterations is of order 2^n and requires n evaluations of f and n evaluations of f', that is, $2n$ evaluations per iteration.

On the other hand, Kung and Traub (1974) have shown that a multi-point iterative process without memory of order 2^{n-1} can be constructed from

either just n evaluations of f (see Section 3.15, Example (13))
or $n - 1$ evaluations of f and one evaluation of f'.

Their work confirms the view that the concept of multi-point iterative methods as a tool for constructing more efficient iterative methods as compared with presently existing methods is a potentially fruitful field of study.

Kung and Traub have conjectured that a multi-point iterative method without memory based on n evaluations has optimal order 2^{n-1}. Wozniakowski (1975a) has proved their conjecture for the so called Hermitian situation, that is under the assumption that if $f^{(j)}(x_k)$, j-th derivative of f at point x_k, is known, then $f^{(0)}(x_k), \ldots, f^{(j-1)}(x_k)$ are also known.

3.17 Conditioning of the Root-Finding Problem and Stopping Criteria for Computations

Our concern in this section is the conditioning of the root-finding problem and how it effects the computation of an accurate approximation to a simple root α of $f(x) = 0$. In what follows we shall use the approach of Dahlquist.

Let x_n be an approximation to α, $f(\alpha) \equiv 0$, $f'(\alpha) \neq 0$. By the mean value theorem we have

$$f(x_n) = (x_n - \alpha) f'(\xi), \quad \xi \in \text{int}(x_n, \alpha) = J \tag{1}$$

from which the method-independent estimate of the error is obtained as

$$|x_n - \alpha| \leqslant f(x_n) |/M, \quad |f'(x)| \geqslant M, \quad x \in J. \tag{2}$$

Denote by $\bar{f}(x_n)$ the computed value of the function and let

$$\bar{f}(x_n) = f(x_n) + \delta(x_n) \tag{3}$$

where $|\delta(x)| \leqslant \delta$ independently of x. We now show that the accuracy with which the root α can be determined is restricted by the size of δ. To do this, we first observe that the best one can hope for is to find x_n such that $\bar{f}(x_n) = 0$. The exact value of the function, then, will satisfy the condition

$$|f(x)| \leqslant \delta. \tag{4}$$

Assuming that $f'(x)$ does not change very rapidly in the vicinity of $x = \alpha$, from (2) we obtain

$$|x_n - \alpha| \leqslant \frac{\delta}{M} \approx E_\alpha \quad \text{where } E_\alpha = \frac{\delta}{|f'(\alpha)|}. \tag{5}$$

This means that E_α is the best method-independent error estimate available. We shall call it the limiting error of the root α. We further observe that if $|f'(\alpha)|$ is small, then E_α is large. In this case the problem of finding α is ill-conditioned. Two possible situations are illustrated in Fig. 3.12. Assume now that a root α is required to be computed with some prescribed tolerance ε. The error estimate, E_α, will not usually be known. If the prescribed tolerance happens to be such that $\varepsilon < E_\alpha$, then it is unlikely that the iterations will ever stop. Therefore, instead of specifying the tolerance ε it is often more convenient to determine the root with its limiting error E_α and then to estimate E. For iterative methods which converge faster than the first order, only a few extra iterations are normally needed to achieve an accuracy within the limiting error even when $\varepsilon \gg E_\alpha$.

We can therefore deduce a rule for a stopping criterion which is based on the following simple observation: Provided the sequence $\{x_n\}$ converges to the

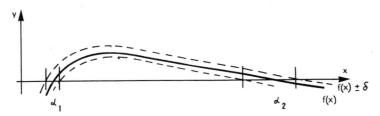

Fig. 3.12. Examples of a well-conditioned root (α_1) and an ill-condition-ed root (α_2)

root α, the differences $|x_n - x_{n-1}|$ will decrease until $|x_n - \alpha| \approx E_\alpha$, at least when n is greater than some fixed value, n_0. Thereafter, the error will be dominated by round-off effects and the differences will vary randomly. The iterations can therefore be stopped and the values of x_n accepted as a satisfactory value of the root as soon as the following two conditions are satisfied simultaneously

$$|x_{n+1} - x_n| \geqslant |x_n - x_{n-1}| \quad \text{and} \quad |x_n - x_{n-1}| < \delta.$$

Here δ is some coarse tolerance which is introduced with the purpose of preventing the iterations being stopped before x_n even gets into the neighbourhood of α.

The error in x_n is then adequately estimated by $|x_{n+1} - x_n|$. Using this stopping criterion considerably reduces the risk of the iterations continuing forever in the case of an ill-conditioned problem, e.g. for large E_α as well as taking care of the situation when the tolerance prescribed happens to be below the computational accuracy that can be achieved for specific well-conditioned roots.

3.18 The Numerical Stability of Iterative Algorithms

Until recently it was generally believed that there was less danger of serious error accumulation in iterative algorithms than in direct algorithms, i.e. algorithms that correspond to computing methods involving a finite number of operations. The argument was that at each iteration one works with the iteration function, the values of the parameters of which are preserved unchanged throughout the sequence of iterations. In consequence, the numerical stability problem for iterative algorithms did not seem important. In fact, for some specific groups of iterative methods this belief can be supported by an easily carried out analysis. For example, consider iterative methods of the general form

$$x_{n+1} = \phi(x_n) \tag{1}$$

where the iteration function depends on an immediately preceding approximation, x_n, only.

Denote by $\bar{x}_1, \bar{x}_2, \ldots$ the computed sequence of approximations to a root α and by δ_n the error in computing $\phi(\bar{x}_n)$.

We have

$$\bar{x}_{n+1} = \phi(\bar{x}_n) + \delta_n, \quad n = 0, 1, 2, \ldots, \tag{2}$$

wherefrom subtracting the equation $\alpha = \phi(\alpha)$ and using the mean value theorem on the result, we get

$$\bar{x}_{n+1} - \alpha = \phi'(\xi_n)(\bar{x}_n - \alpha) + \delta_n, \quad \xi_n \in \text{int}(\bar{x}_n, \alpha). \tag{3}$$

Rewrite expression (3) in the form

$$[1 - \phi'(\xi_n)](\bar{x}_{n+1} - \alpha) = \phi'(\xi_n)(\bar{x}_n - \bar{x}_{n+1}) + \delta_n. \tag{4}$$

Assuming further that

$$|\phi'(\xi_n)| \leqslant m < 1 \quad \text{and} \quad |\delta_n| < \delta$$

we obtain

$$|\bar{x}_{n+1} - \alpha| < \frac{m}{1-m}|\bar{x}_{n+1} - \bar{x}_n| + \frac{1}{1-m}\delta, \tag{5}$$

which gives a strict estimate of the error in $\bar{x}_{n+1}$ in terms of known quantities. The first term on the right hand side of (5) estimates the truncation error and the second: the round-off error.

We note that the final computing error depends only on the error, δ_n, occurring in the immediately preceding iteration, and $\bar{x}_n$ must be considered as an exact quantity when estimating δ_n. It follows that it is not necessary to work to full accuracy in the earlier iterations since round-off errors made at this stage do not effect the final accuracy. In this sense we can say that iterative methods are self-correcting.

In (5), for sufficiently large n, the round-off error becomes dominant. In this case further iterations do not give any improvement in accuracy while the error varies randomly with a constant size of order $\frac{\delta}{1-m}$. However, research carried out by Wozniakowski (1975b, 1975c) has shown that for wider groups of iterative methods the problem of numerical stability is of importance.

In his study, Wozniakowski defines the concepts of condition number, numerical stability and problem conditioning for solving single and sets of non-linear equations. He shows that only if a problem is well-conditioned and if the iterative algorithm used to solve the problem is numerically stable, can one expect to obtain a good approximation to the solution. He also demonstrates that under a natural assumption about the computed evaluation of f, the Newton–Raphson and the Secant methods are numerically stable.

Exercises

3.1 What is an iterative process?

3.2 Define the order of convergence of an iterative process.

3.3 Explain what is meant by (i) a one-point iterative method with or without memory, (ii) a multi-point iterative method with or without memory, (iii) a derivative estimated iterative method. Give examples for each group of the methods.

3.4 Discuss the optimality problem in relation to iterative processes for the root finding problem. Include in the discussion explanations of the terms: a measure of efficiency of an iterative process, the order of convergence, problem cost, direct and inverse polynomial interpolations; give examples.

3.5 Show that the multi-point iteration function

$$x_{i+1} = x_i - \frac{f(x_i)}{f'(x_i)}\frac{f(z_i)-f(x_i)}{2f(z_i)-f(x_i)} \quad \text{where} \quad z_i = x_i - \frac{f(x_i)}{f'(x_i)}$$

is of order 4.

3.6 Discuss the ways in which the efficiency of an iteration function can be measured.

3.7 A simple real zero α of a non-linear function $f(x)$ is computed using an iterative method of the general form

$$x_{i+1} = \phi(x_i, x_{i-1}, \ldots, x_{i-m}), \quad x_0 \text{ arbitrary.}$$

For $m = 0$ and $m = 1$, derive the conditions under which the sequence of values $x_0, x_1, \ldots$ converges to the root α.

3.8 Prove that the equation $x^{m+1} - e \sum_{j=0}^{m} x^j = 0$ has a unique largest real root x^* which for any $m \geq 0$, lies in the interval $e \leq x^* \leq e + 1$.

3.9 Prove that the largest attainable order of convergence of a one-point iterative method which uses no derivatives is 2.

3.10 Derive algorithms for computing the square root of a number using the iterative formulae of (a) Newton–Raphson's method, (b) the Secant method, (c) Traub's method, (d) Jarratt's method and (e) the fifth order method.

3.11 Use the Dekker algorithm to obtain the square root of number 5.

3.12 With reference to Exercise 3.10, deduce the corresponding formulae for Nth roots of a number.

3.13 The positive root of the equation $x = a - bx^2$, a and b positive, is computed using the iterative method $x_{i+1} = a - bx_i^2$. What is the condition for convergence?

3.14 Derive a multi-point iterative method by combining the Newton–Raphson and the method of false position. What is the order of convergence of this 'mixture' method?

3.15 Using each of the methods of Exercise 3.10, determine to an accuracy of 10^{-5} all the roots of the equations:

$$2^x - 2x^2 - 1 = 0,$$
$$2\log x - x/2 + 1 = 0,$$
$$e^{-x} - (x - 1)^2 = 0,$$
$$2 - xe^x = 0.$$

3.16 Describe the functioning of the Dekker algorithm in the worst case.

3.17 The order of convergence of an iterative method $x_{i+1} = \phi(x_i)$, $i = 0, 1, \ldots$ which solves $f(x) = 0$ for a simple root α, can be also defined as the number p such that $\phi'(\alpha) = \phi''(\alpha) = \ldots = \phi^{(p-1)}(\alpha) = 0$ but $\phi^{(p)}(\alpha) \neq 0$. Using this definition show that both the Traub and Jarratt methods are of order 4.

3.18 Using the inverse polynomial interpolation and setting $e = 1, m = 3$, derive a one-point iterative method of the form $x_{i+1} = \phi_{e,m}(x_i, \ldots, x_{i-m})$ which finds a simple root α of the equation $f(x) = 0$.

3.19 Show that the multi-point iterative method given by formula 3.14–(2) is of order 3.

3.20 Given a general multi-point formula

$$x_{i+1} = \phi(x_i) = x_i - \alpha_1 w_1(x_i) - \alpha_2 w_2(x_i),$$

where $w_1(x) = f(x)/f'(x),\quad w_2(x) = f(x)/f'(x + \beta w_1(x)),$

determine values for the parameters $\alpha_1, \alpha_2, \beta$ so as to obtain a third order multi-point method for finding a simple zero of $f(x)$, that requires one function and two derivative evaluations per iteration.

Chapter 4

Direct Methods for Solving Sets of Linear Equations

In this chapter we shall investigate the computational complexity and numerical accuracy of some of the better known computational algorithms for solving a set of linear algebraic equations. Inversion of matrices and evaluation of determinants are other operations which are related to the same general problem. As a rule, the operation of matrix multiplication forms one of the principal components of algorithms for solving these problems. Hence, a potential way of improving the efficiency of an algorithm for solving linear simultaneous equations or related problems is to explore the computational complexity of the matrix multiplication. In this respect we shall discuss some recently developed matrix multiplication algorithms which are capable of computing the product of two matrices using significantly fewer arithmetic operations as compared with the standard techniques. It should be pointed out, however, that these 'fast' multiplication algorithms can actually supersede the standard algorithms only when applied to solve problems of exceptionally large size. (In this context, the 'fast' algorithms are normally referred to as the 'asymptotically faster' multiplication algorithms.) Furthermore, the problem of numerical error control for these algorithms is not yet sufficiently well understood. Nevertheless, it is felt that the ideas of this chapter are worth considering because they draw attention to the fact that the obvious algorithms are not necessarily the best and they form a basis for the development of efficient and genuinely practical algorithms for this important family of problems.

We shall restrict ourselves to the so called direct methods for solving sets of linear equations as opposed to iterative methods, which are not the subject of this chapter.

A direct method of finding $\mathbf{x}$ to satisfy $\mathbf{A}\mathbf{x} = \mathbf{b}$ is one that defines $\mathbf{x}$ in terms of a finite number of operations on scalars. One such method is to invert $\mathbf{A}$ and then to calculate $\mathbf{x}$ from $\mathbf{x} = \mathbf{A}^{-1}\mathbf{b}$. However in practice the determination of the inverse, $\mathbf{A}^{-1}$, is a computationally difficult process in itself. Therefore, the method based on matrix inversion is never used for solving sets of linear equations. The inverse $\mathbf{A}^{-1}$ may however be important in contexts other than the solution of linear equations and, hence, techniques for the efficient computation

90

of A^{-1} are sought along with general methods for the solution of linear equations. Most direct methods are methods of factorisation in which the matrix A is transformed into the product of two easily invertible matrices; e.g. in triangular decomposition A is transformed into the product of a lower triangular matrix, L, and an upper triangular matrix, U. Thus

$$LUx = b,$$

giving

$$x = U^{-1}L^{-1}b.$$

We intent to study direct methods and to assess their relative efficiency basing our judgement of the methods upon two fundamental criteria, namely susceptibility to round-off errors and the number of arithmetic operations a particular method requires.

The analysis of round-off errors in direct methods has an interesting and educational history. The first analyses of computer round-off errors were attempted even before any modern computers had been constructed. Then, in 1947 von Neuman and Goldstine published a paper concerning the inversion of a positive definite matrix on a digital computer using an elimination method. They noted that computer arithmetic would not possess the fundamental properties of associativity and distributivity. However, because of the way they handled pivoting and because they looked at the error rather than the residual they were very pessimistic about the reliability of solutions to systems of order greater than about 20. This pessimism discouraged the use of direct methods in favour of iterative methods of solution because of the latter's numerical stability. The backward error analysis of Wilkinson has however led to a realization that the more stable direct methods are generally very accurate and the idea that more accurate results can usually be obtained by iterative methods has been abandoned.

4.1 Gaussian Elimination

The method first developed by Gauss early in the nineteenth century, has been studied by various authors and with various conclusions on its reliability but its complete analysis became possible only with the introduction of Wilkinson's backward error analysis. It has been established that the method is generally reliable and efficient.

We shall first introduce the method, examining in particular its most commonly used variant, known as triangular decomposition, and shall then proceed to analyse its computational complexity and numerical stability.

4.1.1 Ordinary Gaussian Elimination

Gaussian elimination is the process of subtracting suitable multiples of each row of the system of equations from all succeeding rows to produce a matrix with zeros below and to the left of the main diagonal. If the right-hand side, **b**, is carried along as an extra column these operations are known as forward elimination. The resulting augmented matrix represents a system of equations with the same solution as the original. This solution is easily found by a back substitution process, because the last equation involves only the last unknown, the next to last equation only the last two unknowns, and so on.

Consider the following system of equations:

$$a_{11}x_1 + a_{12}x_2 + \ldots\ldots + a_{1n}x_n = b_1 \tag{1}$$

$$a_{21}x_1 + a_{22}x_2 + \ldots\ldots + a_{2n}x_n = b_2 \tag{2}$$

$$\ldots\ldots\ldots\ldots\ldots\ldots\ldots\ldots\ldots\ldots\ldots\ldots\ldots\ldots\ldots\ldots\ldots\ldots\ldots\ldots\ldots\ldots\ldots\ldots$$

$$a_{n1}x_1 + a_{n2}x_2 + \ldots\ldots + a_{nn}x_n = b_n \tag{n}$$

The first step in Gaussian elimination involves the elimination of the variable x_1 from equations (2) to (n) as follows. For each $i = 2$ to n multiply the first equation by a_{i1}/a_{11} (assuming $a_{11} \neq 0$) and subtract it from the ith equation thereby obtaining a new set of equations (2) to (n). This transformation is also applied to **b**. We now have ($n-1$) equations in the ($n-1$) unknowns x_2 to x_n since the coefficients of x_1 have been made zero in the equations (2) to (n). The above process is then repeated on the new equations (2) to (n), again assuming that the leading coefficient of the first equation is non-zero. This second step produces ($n-2$) equations in ($n-2$) unknowns. Successive eliminations are made in this way until the process concludes with the nth equation involving only x_n. The matrix of the coefficients for the new system of equations is upper triangular. If at any point in the above process the leading coefficient (or pivot) of the first of the remaining equations is zero, then before the next elimination is performed the equations or the variables must be permuted so that one obtains a non-zero pivot. Schemes for doing this are known as pivoting. The 'partial pivoting', which is now used almost exclusively, involves searching the corresponding rth column for its largest element. Multiples of the row containing this element are then used to introduce the zeros in the other rows. The row containing the pivot is interchanged with the rth row. The other scheme known as 'complete pivoting' uses the largest element in the entire unreduced matrix. With either form of pivoting none of the multipliers is greater than unity in magnitude. Complete pivoting can produce more accurate results for some matrices but it is so much more expensive that its use may be justified only in very special cases. The transformation of **A** into upper triangular form is known as forward elimination. The remainder of the solution is obtained by back substitution in which one starts by substituting the value for x_n into the previous equation which involves x_n and x_{n-1}, to obtain the value of x_{n-1}. This process of substitution is continued until the value of x_1 is obtained. In ordinary Gaussian elimination the transformation of the vector **b** is done simultaneously with the processing of **A**. The multipliers created during forward elimination are normally saved along with the new coefficients. It is worth mentioning two important reasons for this. First, if there is more than one right-hand side, the forward elimination process need only be carried out once. A second reason for saving the multipliers occurs when carrying out the process of iterative improvement of the initial solution. Beginning with the initial solution, the multipliers along with the new coefficients are used to compute iteratively new and more accurate solutions.

4.1.2 Gaussian Elimination and Evaluation of a Matrix Determinant

Gaussian elimination developed in the preceding section for the solution of a linear system $\mathbf{Ax} = \mathbf{b}$, also may be applied to the computation of determinants. Evaluation of matrix determinants is of importance in many applications involving matrices. One straightforward example of necessity to evaluate the matrix determinant is in solution of a linear system $\mathbf{Ax} = \mathbf{b}$. Here, by virtue of the form of the solution given as $\mathbf{x} = \mathbf{A}^{-1}\mathbf{b}$, it follows that the system has a unique solution if and only if the matrix of the system is non-singular, that is inverse $\mathbf{A}^{-1}$ exists, which, in other words, means that the determinant of the matrix is non-zero, $\det \mathbf{A} \neq 0$. Thus, in every case, before embarking on numerical solution of the set $\mathbf{Ax} = \mathbf{b}$ one needs to check the value of $\det \mathbf{A}$ on the account of $\mathbf{A}$ being non-singular.

We shall now describe a computational scheme for evaluating a determinant using Gaussian elimination. Let the determinant of $\mathbf{A}$ be

$$\det \mathbf{A} = |\mathbf{A}| = \begin{vmatrix} a_{11} & a_{12} & \cdots & a_{1n} \\ a_{21} & a_{22} & \cdots & a_{2n} \\ \vdots & & & \\ a_{n1} & a_{n2} & \cdots & a_{nn} \end{vmatrix}$$

and let $a_{11} \neq 0$.

Removing the element a_{11} from the first row, we shall have

$$\det \mathbf{A} = |\mathbf{A}| = a_{11} \begin{vmatrix} 1 & b_{12} & \cdots & b_{1n} \\ a_{21} & a_{22} & \cdots & a_{2n} \\ \cdots & & & \\ a_{n1} & a_{n2} & \cdots & a_{nn} \end{vmatrix}, \qquad \text{where } b_{1j} = \frac{a_{1j}}{a_{11}}, \; j = 2, \ldots, n.$$

Next, subtracting from each row the first row multiplied by the first element of the particular row in hand, we obviously obtain

$$|\mathbf{A}| = a_{11} \begin{vmatrix} 1 & b_{12} & \cdots & b_{1n} \\ 0 & a_{22}^{(1)} & \cdots & a_{2n}^{(1)} \\ \cdots & & & \\ 0 & a_{n2}^{(1)} & \cdots & a_{nn}^{(1)} \end{vmatrix} = a_{11} \begin{vmatrix} a_{22}^{(1)} & \cdots & a_{2n}^{(1)} \\ \cdots & & \\ a_{n2}^{(1)} & \cdots & a_{nn}^{(1)} \end{vmatrix},$$

where $a_{ij}^{(1)} = a_{ij} - a_{i,1} b_{1j}, \qquad i,j = 2, \ldots, n.$

With the determinant of the $(n-1)$th order that remains, we shall now deal as before, provided $a_{22}^{(1)} \neq 0$. Carrying through the process, we shall find that the determinant is equal to the product of the leading elements:

$$\det \mathbf{A} = |\mathbf{A}| = a_{11} a_{22}^{(1)} \ldots a_{nn}^{(n-1)}.$$

If at any step it should turn out that $a_{ii}^{(i-1)} = 0$, pivoting can be performed so that a nonvanishing element appears in the upper left corner.

From an inspection of the process of computing the determinant we see that, with the exception of the last multiplication, it coincides with the forward course of Gaussian elimination applied to a system with the matrix for which the determinant is computed.

We are now ready to present a complete algorithm for solving a linear set $\mathbf{Ax} = \mathbf{b}$ using Gaussian elimination. The algorithm consists of two parts, where in the first part the matrix is transformed into an upper triangular form and its determinant is simultaneously evaluated and in the second part the solution vector is computed using back substitution.

Ordinary Gaussian Elimination Algorithm for Solving a Set of Linear Equations

The algorithm produces the solution to the set $\mathbf{Ax} = \mathbf{b}$. The input matrix $\mathbf{A}$ and the right-hand side vector $\mathbf{b}$ are stored in the two-dimensional array $\mathbf{A}(1:N, 1:N) = (A(I,J), \ I = 1,\ldots,N, \ J = 1,\ldots,N)$ and the one-dimensional array $\mathbf{B}(N) = (B(I), I = 1,\ldots,N)$, respectively. The output is stored in the one-dimensional array $\mathbf{X}(N) = (X(J), J = 1,\ldots,N)$ and DET contains the value of the matrix determinant. We also use an auxiliary array $\mathbf{P}(1:N)$ to store the reciprocals of the diagonal elements of input matrix $\mathbf{A}$.

(Triangularization of the matrix and evaluation of the determinant.)

1. Test $A(1, 1) = 0$?
   If true, go 'down' the first column and find the first non-zero element; let it be $A(P, 1)$. Exchange the first and the Pth rows.
2. Set $K \leftarrow 1$, DET $\leftarrow A(1, 1)$.
3. Compute $P(K) \leftarrow 1/A(K, K)$.
4. Set $I \leftarrow K + 1$.
5. Check whether the matrix element $A(K, K)$ is non-zero; if it is zero, continue 'down' the column K until a non-zero element, $A(Q, K)$, is encountered; exchange the two rows.
6. Compute $R \leftarrow A(I, K) * P(K)$, then set $A(I, K) \leftarrow 0$.
7. For $J = K + 1, K + 2, \ldots, N$ compute

$$A(I, J) \leftarrow A(I, J) - R * A(K, J)$$

8. Compute DET $\leftarrow$ DET $* A(I, K + 1)$
9. Compute $B(I) \leftarrow B(I) - R * B(K)$
10. $I \leftarrow I + 1$
11. Test for the end of the set, $I > N$?
    If false, go back to Step 6.

12. $K \leftarrow K + 1$
13. Test $K > N - 1$?
    If false, go back to Step 4.
14. Test the value of DET and if it is zero, terminate.
(Back Substitution.)
15. Compute $X(N) \leftarrow B(N)/A(N,N)$.
16. Set $K \leftarrow N - 1$.
17. Set SUM $\leftarrow 0$.
18. Set $J \leftarrow N$.
19. Compute SUM $\leftarrow$ SUM $+ A(K,J) * X(J)$
20. $J \leftarrow J - 1$
21. Test $J < K + 1$?
    If false, go back to Step 19.
22. Compute $X(K) \leftarrow (B(K) - \text{SUM})/A(K,K)$.
23. $K \leftarrow K - 1$
24. Test $K < 1$?
    If false, go back to Step 17.
25. Terminate.

Note that Steps 15 and 22 should take care of the cases when their divisors are zeros.

4.1.3 Triangular Decomposition

In triangular decomposition the matrix **A** is first decomposed into a lower triangular matrix with unit diagonal, **L**, and an upper triangular matrix, **U**, in the following manner:

$$\mathbf{A} = \mathbf{LU} \tag{1}$$

The equation $\mathbf{LUx} = \mathbf{b}$, then, corresponds to the solution of two equations:

$$\mathbf{Ly} = \mathbf{b}$$
$$\mathbf{Ux} = \mathbf{y} \tag{2}$$

If we denote the original set of equations by:

$$\mathbf{A}^{(1)}\mathbf{x} = \mathbf{b}^{(1)} \tag{3}$$

then each of the $(n - 1)$ steps of Gaussian elimination leads to a new set of equations which are equivalent to the original set. We denote the rth equivalent set by

$$\mathbf{A}^{(r)}\mathbf{x} = \mathbf{b}^{(r)} \tag{4}$$

$\mathbf{A}^{(r)}$ is already upper triangular as far as its first r rows are concerned and it has a square matrix of order $(n - r + 1)$ in the bottom right-hand corner. As an example take $\mathbf{A}^{(3)}$ when $n = 5$:

$$\mathbf{A}^{(3)} = \begin{bmatrix} a & a & a & a & a \\ 0 & a & a & a & a \\ 0 & 0 & a & a & a \\ 0 & 0 & a & a & a \\ 0 & 0 & a & a & a \end{bmatrix} \quad \text{where the } a\text{'s indicate non-zero elements}$$

The matrix $\mathbf{A}^{(r+1)}$ is derived from $\mathbf{A}^{(r)}$ by subtracting a multiplier m_{ir} of the rth row from the ith row for values of i from $r + 1$ to n. The multipliers m_{ir} are defined by:

$$m_{ir} = a_{ir}^{(r)} / a_{rr}^{(r)} \qquad i = r + 1, \ldots, n. \tag{5}$$

The rth row of $\mathbf{A}^{(r)}$ is called the rth pivotal row and $a_{rr}^{(r)}$ is called the rth pivot.

(i) Consider each element for which $i \leqslant j$:
The element is modified in each transformation until $\mathbf{A}^{(i)}$ is obtained, after which it remains constant. We may write

$$a_{ij}^{(i)} = a_{ij}^{(-1)} - m_{i,i-1} a_{i-1,j}^{(i-1)}$$

$$\text{or} \quad a_{ij}^{(i)} = a_{ij}^{(1)} - \sum_{l=1}^{i-1} m_{il} a_{lj}^{(l)} \tag{6}$$

where the $a_{ij}^{(1)}$ are the original elements of $\mathbf{A}$.

(ii) Consider the elements for which $i > j$:
The element is modified as in (i) until $A^{(j)}$ is obtained; $a_{jj}^{(j)}$ is then used to compute m_{ij}, and $a_{ij}^{(j+1)}$ to $a_{ij}^{(n)}$ are all taken to be exactly equal to zero. m_{ij} satisfies

$$m_{ij} = \frac{a_{ij}^{(j)}}{a_{jj}^{(j)}} \tag{7}$$

and $a_{ij}^{(j)} = a_{ij}^{(j-1)} - m_{i,j-1} a_{j-1,j}^{(j-1)}$

$$= a_{ij}^{(1)} - \sum_{l=1}^{j-1} m_{il} a_{lj}^{(l)} \tag{8}$$

If we take the terms involving m_{il} to the left hand side in equations (6) and (8) the set of n^2 equations is equivalent to the single matrix equation $\mathbf{LU} = \mathbf{A}$, where for $n = 4$ say we have:

$$\mathbf{F} = \begin{bmatrix} 1 & 0 & 0 & 0 \\ m_{21} & 1 & 0 & 0 \\ m_{31} & m_{32} & 1 & 0 \\ m_{41} & m_{42} & m_{43} & 1 \end{bmatrix} \tag{9}$$

$$\mathbf{U} = \begin{bmatrix} a_{11}^{(1)} & a_{12}^{(1)} & a_{13}^{(1)} & a_{14}^{(1)} \\ 0 & a_{22}^{(2)} & a_{23}^{(2)} & a_{24}^{(2)} \\ 0 & 0 & a_{33}^{(3)} & a_{34}^{(3)} \\ 0 & 0 & 0 & a_{44}^{(4)} \end{bmatrix}$$

It may happen that for some r, the $a_{rr}^{(r)}$ in (5) will be either zero or very small. In such a case, division by this particular $a_{rr}^{(r)}$ must be prevented.

4.1.4 Storage Requirements for Triangular Decomposition

The question of storage requirement is of significance in the problems involving manipulation of matrices. In particular, in the problem of solving $\mathbf{Ax} = \mathbf{b}$, we have, naturally, to assume that if the problem involves a matrix of dimension $n \times n$, then the input array of the same size is required. If, now, in addition, the intermediate results of the matrix manipulation are needed to be temporarily stored in an auxiliary array (or arrays), then the total volume of storage needed may become a restricting factor on the size of the problem, which may be solved on a particular computer, or on the speed of the solution. Accordingly a feature such as the fact that an algorithm for solving $\mathbf{Ax} = \mathbf{b}$ does not require any additional storage is usually explicitly noted. In the case of the triangular decomposition algorithm we note that no extra storage is required by the algorithm as the coefficients of the reduced equations do not need to be stored separately. Thus we can simply overwrite a previous equation by the new coefficients derived from the elimination process and record the multipliers in the spaces left by the coefficients which have been eliminated. In place of the pivots we can store their reciprocals, for further use in the formation of the multipliers and in the subsequent back substitution. For example the intermediate case $\mathbf{A}^{(3)}$ for $n = 5$, of what was originally the input array will look like

$$\mathbf{A}^{(3)} = \begin{bmatrix} [a_{11}^{(1)}]^{-1} & a_{12}^{(1)} & a_{13}^{(1)} & a_{14}^{(1)} & a_{15}^{(1)} \\ m_{21} & [a_{22}^{(2)}]^{-1} & a_{23}^{(2)} & a_{24}^{(2)} & a_{25}^{(2)} \\ m_{31} & m_{32} & [a_{33}^{(3)}]^{-1} & a_{34}^{(3)} & a_{35}^{(3)} \\ m_{41} & m_{42} & a_{43}^{(3)} & a_{44}^{(3)} & a_{45}^{(3)} \\ m_{51} & m_{52} & a_{53}^{(3)} & a_{44}^{(3)} & a_{55}^{(3)} \end{bmatrix}$$

We note, however, that if the solution computed is considered to be not sufficiently accurate an iterative technique (known as the iterative refinement method) can be used to improve the solution. In this case, the original matrix $\mathbf{A}$ has to be retained along with the triangular decomposition solution which implies the use of an auxiliary array of the same size as the data array.

LU Decomposition Algorithm

Given a non-singular square matrix $\mathbf{A}$ of dimension N, the algorithm decomposes it into the product of two matrices, $\mathbf{A} = \mathbf{LU}$, where $\mathbf{L}$ is a lower triangular matrix with unity entries along the main diagonal and $\mathbf{U}$ is an upper triangular matrix, respectively. The original matrix $\mathbf{A}$ is retained at the end of the process, while the results are stored in the auxiliary array $\mathbf{B}(1:N, 1:N)$, where the lower triangle, below the main diagonal, contains the elements of the lower triangular matrix $\mathbf{L}$ and the upper triangle including the main diagonal contains the elements of the upper triangular matrix $\mathbf{U}$.

1. Set $B(I,J) \leftarrow A(I,J)$ for $I = 1, \ldots, N$; $J = 1, \ldots, N$.

2. Set $K \leftarrow 1$.
3. (Partial pivoting.) Find the largest in absolute value element of $(B(L, K)$, $L = K, K + 1, \ldots, N)$. Let it be $B(P, K)$. If $P > K$, exchange the rows K and P.
4. For $J = K + 1, K + 2, \ldots, N$ compute

$$B \leftarrow B(J, K) - \sum_{L=1}^{K-1} B(J, L) * B(L, K),$$

$$B(J, K) \leftarrow B/B(K, K).$$

(At this step, Kth column of matrix L is obtained.)
5. Set $I \leftarrow K + 1$.
6. For $J = K + 1, K + 2, \ldots, N$ compute

$$B(I, J) \leftarrow B(I, J) - \sum_{L=1}^{K-1} B(I, L) * B(L, J)$$

(At this step, Ith row of matrix U is obtained.)
7. $K \leftarrow K + 1$.
8. Test for the end of matrix, $K > N - 1$?
   If false, go back to Step 3.
9. Terminate.

4.2 Algebraic Complexity: the Number of Arithmetic Operations Involved in Gaussian Elimination

We shall only consider triangular decomposition.
From 4.1.2–(6) if $i \leqslant j$ the ijth element of $\mathbf{A}$ is modified as

$$a_{ij}^{(i)} = a_{ij}^{(1)} - \sum_{l=1}^{i-1} m_{il} a_{lj}^{(l)}$$

Thus to calculate $a_{ij}^{(i)}$ requires $(i - 1)$ multiplications and $(i - 1)$ additions   (1)
(Note that as usual additions are taken to mean both additions and subtractions)
From 4.1.2–(8) if $i > j$ the ijth element of $\mathbf{A}$ is modified as

$$a_{ij}^{(j)} = a_{ij}^{(1)} - \sum_{l=1}^{j-1} m_{il} a_{lj}^{(l)}$$

Thus to calculate $a_{ij}^{(j)}$ requires $(j - 1)$ additions and $(j - 1)$ multiplications.   (2)
Also the calculation of the $n(n - 1)/2$ multipliers, m_{ij} requires:

$$n \text{ reciprocals and } \frac{n}{2}(n - 1) \text{ multiplications}$$ (3)

From (1), (2), and (3) the number of multiplications required to decompose $\mathbf{A}$ into $\mathbf{L}$ and $\mathbf{U}$ is given by:

$$\sum_{j=1}^{n} \left[\sum_{i=1}^{j} (i - 1) + \sum_{i=j+1}^{n} (j - 1) \right] + \frac{n}{2}(n - 1)$$

$$= \sum_{j=1}^{n} \left[\sum_{i=1}^{j} (i - j) + \sum_{i=1}^{n} (j - 1) \right] + \frac{n}{2}(n - 1) = \frac{n^3}{3} - \frac{n}{3}.$$

Similarly the number of additions is given by

$$\sum_{j=1}^{n}\left[\sum_{i=1}^{j}(i-1)+\sum_{i=j+1}^{n}(j-1)\right]=\frac{n^3}{3}-\frac{n^2}{2}+\frac{n}{6}.$$

In total, the decomposition of $\mathbf{A}$ into $\mathbf{L}$ and $\mathbf{U}$ requires

$$\frac{n^3}{3}-\frac{n}{3}\quad\text{multiplications,}$$

$$\frac{n^3}{3}-\frac{n^2}{2}+\frac{n}{6}\quad\text{additions,}\tag{4}$$

and n reciprocals

Back Substitution

To complete the solution of $\mathbf{Ax}=\mathbf{b}$, the following steps are performed:
(i) y is evaluated from $\mathbf{Ly}=\mathbf{b}$.
(ii) x is evaluated from $\mathbf{Ux}=\mathbf{y}$.
(i) can be written in the form:

$$l_{11}y_1 \qquad\qquad\qquad = b_1$$
$$l_{21}y_1+l_{22}y_2 \qquad\qquad = b_2$$
$$\dotfill$$
$$l_{n1}y_1+l_{n2}y_2+\ldots+l_{nn}y_n=b_n$$

The variables $y_1, y_2,\ldots y_n$ are determined in succession from the first, second, ..., nth equations, respectively,

i.e. $y_1=b_1/l_{11}$

$y_2=(b_2-l_{22}y_1)/l_{22}$, etc.

For a general step the evaluation of y_r requires one reciprocal (which has already been calculated), r multiplications and $r-1$ additions. Hence to solve the system of equations $\mathbf{Ly}=\mathbf{b}$ requires $n(n-1)/2$ additions and $n(n+1)/2$ multiplications. The solution of $\mathbf{Ux}=\mathbf{y}$ requires the same amount of arithmetic. However in our case all $l_{ii}=1$ and so the number of multiplications required is reduced by n.

To summarize, the amount of arithmetic required in the back substitution phase is

$$n^2\text{ multiplications}\tag{5}$$
and n^2-n additions,

and the total number of operations required to solve $\mathbf{Ax}=\mathbf{b}$ is

$$\frac{n^3}{3}+n^2-\frac{n}{3}\quad\text{multiplications}$$

$$\frac{n^3}{3}+\frac{n^2}{2}-\frac{5n}{6}\quad\text{additions}\tag{6}$$

n reciprocals

4.3 Matrix Inversion and Its Algebraic Complexity

An allied problem to the solution of a set of linear simultaneous equations is the inversion of a matrix. We shall give two methods for finding the inverse of a matrix $\mathbf{A}$.

Method 1. (a) $\mathbf{A}$ is decomposed into $\mathbf{L}$ and $\mathbf{U}$

        (b) $\mathbf{L}$ and $\mathbf{U}$ are inverted

        (c) $\mathbf{A}^{-1}$ is formed by premultiplying $\mathbf{L}^{-1}$ by $\mathbf{U}^{-1}$

Step (a) has already been discussed, so we shall turn to step (b).

If $\mathbf{U}$ has the elements a_{ij} and $\mathbf{U}^{-1}$ has the elements u_{ij} then by definition of the inverse, we have

$$\sum_{k=1}^{n} a_{ik}u_{kj} = \delta_{ij} = \begin{cases} 1 & \text{if } i=j, \\ 0 & \text{otherwise,} \end{cases}$$

e.g. if $n = 4$, we have to find the $\mathbf{U}_{ij}$ such that

$$\begin{bmatrix} a_{11} & a_{12} & a_{13} & a_{14} \\ 0 & a_{22} & a_{23} & a_{24} \\ 0 & 0 & a_{33} & a_{34} \\ 0 & 0 & 0 & a_{44} \end{bmatrix} \begin{bmatrix} u_{11} & u_{12} & u_{13} & u_{14} \\ 0 & u_{22} & u_{23} & u_{24} \\ 0 & 0 & u_{33} & u_{34} \\ 0 & 0 & 0 & u_{44} \end{bmatrix} = \begin{bmatrix} 1 & 0 & 0 & 0 \\ 0 & 1 & 0 & 0 \\ 0 & 0 & 1 & 0 \\ 0 & 0 & 0 & 1 \end{bmatrix}$$

To calculate the u_{ij}, we proceed in the following manner. Firstly for $i = 1,\ldots,n$ calculate $u_{ii} = 1/a_{ii}$. Then calculate u_{in} in the order $i = n,\ldots,1$. The next step is to calculate $u_{i,n-1}$ in the order $i = (n-1),\ldots,1$, etc. The elements of $\mathbf{U}^{-1}$ are, thus, given by

$$u_{ii} = \frac{1}{a_{ii}}, \tag{7}$$

$$u_{ij} = -u_{ii} \sum_{p=i+1}^{j} a_{ip}u_{pj}, \quad i \neq j. \tag{8}$$

The calculation of n reciprocals used in (7), is performed as part of the decomposition of $\mathbf{A}$ into $\mathbf{L}$ and $\mathbf{U}$. To compute u_{ij}, $i \neq j$, in (8), requires $(j - i + 1)$ multiplications and $(j - i - 1)$ additions which gives the number of multiplications as

$$\sum_{j=1}^{n}\sum_{i=1}^{j-1} (1 + j - i) = \frac{n(n-1)(n+4)}{6} \tag{9}$$

Similarly the number of additions is

$$\sum_{j=1}^{n}\sum_{i=1}^{j-1} (j - i - 1) = \frac{n(n-1)(n-2)}{6} \tag{10}$$

The same analysis is applicable to $\mathbf{L}^{-1}$ but as the diagonal elements of $\mathbf{L}$ are unity the number of multiplications is reduced by $n(n-1)$. To invert $\mathbf{L}$ requires

$$\frac{n(n-1)(n-2)}{6} \quad \text{multiplications}$$

$$\frac{n(n-1)(n-2)}{6} \quad \text{additions} \tag{11}$$

From (9), (10), and (11) the total number of operations required to invert $\mathbf{L}$ and $\mathbf{U}$ is

$$\frac{n(n^2 - 1)}{3} \qquad \text{multiplications} \tag{12}$$

and $\dfrac{n(n-1)(n-2)}{3}$ additions.

In a similar manner it may be shown that step (c), the formation of $\mathbf{A}^{-1} = \mathbf{U}^{-1}\mathbf{L}^{-1}$, requires the number of multiplications given as

$$\sum_{j=1}^{n}\left[\sum_{i=1}^{j}(n-j) + \sum_{i=j+1}^{n}(n-i+1)\right] = \frac{n(n^2-1)}{3}, \tag{13}$$

and the number of additions given as

$$\sum_{j=1}^{n}\left[\sum_{i=1}^{j}(n-j) + \sum_{i=j+1}^{n}(n-i)\right] = \frac{n^3}{3} - \frac{n^2}{2} + \frac{n}{6} \tag{14}$$

The total number of arithmetic operations to invert a matrix by this method is

$$n^3 - n \qquad \text{multiplications}$$
$$n^3 - 2n^2 + n \qquad \text{additions} \tag{15}$$

Method 2. (a) Decompose $\mathbf{A}$ into $\mathbf{L}$ and $\mathbf{U}$,
           (b) Invert $\mathbf{L}$,
           (c) Evaluate $\mathbf{A}^{-1}$ from $\mathbf{L}^{-1} = \mathbf{U}\mathbf{A}^{-1}$

Here, it is only step (c) which has not been considered before. It may be readily found that to calculate $\mathbf{A}^{-1}$ from $\mathbf{L}^{-1}$ and $\mathbf{U}$, the number of multiplications required is

$$\frac{n^3}{2} + \frac{n^2}{2} - 1 \tag{16}$$

and the number of additions is

$$\frac{n^3}{2} - n^2 + \frac{n}{2} \tag{17}$$

It follows then that to invert a matrix by Method 2 requires

$$n^3 - 1 \qquad \text{multiplications,}$$
$$\text{and } n^3 - 2n^2 + n \qquad \text{additions.} \tag{18}$$

Thus it appears that the explicit evaluation of $\mathbf{U}^{-1}$ required by Method 1 gives rise to the smaller number of multiplications. However, the gain of $(n-1)$ multiplications by using Method 1 as compared with Method 2 may be offset by the extra rounding error introduced in the process of computing $\mathbf{U}^{-1}$.

4.4 Error Analysis

We shall now analyse the numerical stability of the Gaussian elimination following Wilkinson's backward error analysis approach. For this purpose as well as for the numerical analysis of other methods later in the text we need some recollection of basic results on floating-point matrix multiplication and on the floating-point solution of sets of linear equations. A brief summary of these results can be found at the end of the book in Appendix A.

4.4.1 Conditioning of the Problem. Backward Error Analysis and Solution of Linear Simultaneous Equations

Consider the set

$$Ax = b. \tag{1}$$

Let $x = A^{-1}b$ be the exact solution of the set and $x^{(c)}$ be its computed solution. We shall distinguish between the difference

$$A^{-1}b - x^{(c)} = x - x^{(c)} \tag{2}$$

which is called the error and the difference

$$b - Ax^{(c)} \tag{3}$$

which is called the residual.

Studies on the conditioning of problem (1) show essentially that if both the error and the residual vectors for the given set are small then the problem is well-conditioned. On the other hand, it may happen that the residual is small while the error is very large. This situation indicates that the problem is ill-conditioned. The matrix of the ill-conditioned problem is usually 'nearly singular' or even exactly singular. (The matrix is called singular if its determinant is equal to zero.)

If problem (1) is ill-conditioned then no computational algorithm will solve such a problem accurately.

However it also follows from the above that for the well-conditioned problem, the numerical stability of a particular computational algorithm can be studied using either the concept of the error as given by (2) or the concept of the residual as given by (3). This means that if the solution, $x^{(c)}$, computed using the particular computational algorithm, is such that the error is small or that the residual is small then the algorithm is reliable (or it is called the numerically stable), and *vice versa*.

The classical (or forward) error analysis is based on handling the error, that is it attempts to bound the difference between the exact and the computed solutions at every step of computation. With the progress of computation, this process becomes too complicated, the bounds on the difference between the solutions becoming too 'loose', and this frequently leads to conclusions on the computational method's reliability being far too pessimistic. The backward error analysis, on the other hand, handles the residual instead of the error, and as it happens this approach is often simpler and gives reasonably sharp bound estimates on the residual which, in turn, gives more realistic conclusions about the reliability of the computational method as compared with the forward error analysis.

Since the introduction of backward error analysis many important results have been established, and the general reliability of the Gaussian elimination is one of them. Indeed the technique of backward error analysis is considered to be one of the most important ideas in mathematical error analysis.

4.4.2 Numerical Stability of Triangular Decomposition

In the error analysis which is to be carried out below we follow closely the work of Wilkinson (1963, 1965) and Forsythe and Moller (1967).

Wilkinson in his backward error analysis has essentially proved the following theorem.

Consider $Ax = b$ where A and b are a given matrix and vector with elements which are floating-point numbers. Let $x^{(c)}$ be the vector of floating-point numbers computed by Gaussian elimination. Then there is a matrix E (with real number elements) such that $x^{(c)}$ is the exact solution of

$$(A + E)x^{(c)} = b. \tag{1}$$

The matrix E is called the perturbation matrix.

For a well-conditioned problem the perturbation matrix is always small in comparison with A.

Equation (1) can be written as

$$b - Ax^{(c)} = Ex^{(c)} \tag{2}$$

which shows that the bounds on the residual $b - Ax^{(c)}$ are known as soon as the bounds on the perturbation E are known.

We shall now study bounds on the perturbation E under various conditions on the computed solution, $x^{(c)}$. We recall that the process of computing the $x^{(c)}$ using the triangular decomposition consists of two stages.

First, the factors L and U are computed such that $A = LU$. In fact, the computed factors L and U will satisfy the relation

$$LU = A + \partial A, \tag{3}$$

where ∂A is the error matrix due to the round-off errors accumulation.

Second, two sets of equations,

$$Ly = b \tag{4}$$

and $Ux = y,$ (5)

are solved. The computed solution $y^{(c)}$ of set (4) is the exact solution of the set

$$(L + \partial L)y^{(c)} = b \tag{6}$$

and similarly the computed solution $x^{(c)}$ of set (5) is the exact solution of the set

$$(U + \partial U)x^{(c)} = y, \tag{7}$$

where each of the matrices, ∂L and ∂U, is what we may call the local perturbation matrix.

Hence, the computed solution, $\mathbf{x}^{(c)}$, of the set (1) is the exact solution of the set

$$(\mathbf{L} + \partial \mathbf{L})(\mathbf{U} + \partial \mathbf{U})\mathbf{x}^{(c)} = \mathbf{b} \text{ where } \mathbf{LU} = \mathbf{A} + \partial \mathbf{A}. \tag{8}$$

We rewrite (8) as

$$(\mathbf{LU} + \partial \mathbf{A} + \mathbf{U}\partial \mathbf{L} + \mathbf{L}\partial \mathbf{U})\mathbf{x}^{(c)} = \mathbf{b},$$

and denoting the sum $\partial \mathbf{A} + \mathbf{U}\partial \mathbf{L} + \mathbf{L}\partial \mathbf{U}$ by $\mathbf{E}$, we get

$$(\mathbf{LU} + \mathbf{E})\mathbf{x}^{(c)} = \mathbf{b}. \tag{9}$$

In the analysis to follow we shall derive a bound on the perturbation matrix $\mathbf{E}$. To do that we consider successively the perturbations $\partial \mathbf{A}$, $\partial \mathbf{L}$ and $\partial \mathbf{U}$.

Bound Estimate on $\partial \mathbf{A}$

To estimate the bound on $\partial \mathbf{A}$, recall the process of triangular decomposition defined by equations 4.1.2–(3)–(9).

(a) Consider the case of the elements a_{ij} for which $i \leqslant j$. The element is modified in each transformation until $\mathbf{A}^{(i)}$ is obtained after which it will remain constant. We can write

$$a_{ij}^{(2)} \equiv a_{ij}^{(1)} - m_{i1}a_{1j}^{(1)} + d_{ij}^{(2)}$$
$$a_{ij}^{(3)} = a_{ij}^{(2)} - m_{i2}a_{2j}^{(2)} + d_{ij}^{(3)}$$
$$\dots \tag{10}$$
$$a_{ij}^{(i)} \equiv a_{ij}^{(i-1)} - m_{i,i-1}a_{i-1,j}^{(i-1)} + d_{ij}^{(i)}$$

$d_{ij}^{(k)}$ is the difference between the accepted $a_{ij}^{(k)}$ and the exact value which would be obtained using the computed $a_{ij}^{(k-1)}$, $m_{i,k-1}$ and $a_{k-1,j}^{(k-1)}$.

The summation of equations (10) gives

$$a_{ij}^{(i)} \equiv a_{ij}^{(1)} - \sum_{l=1}^{i-1} m_{il}a_{lj}^{(l)} + e_{ij}, \tag{11}$$

where

$$e_{ij} = \sum_{k=2}^{i} d_{ij}^{(k)} \tag{12}$$

(b) Consider the elements for which $i > j$. The element is modified as in (a) until $\mathbf{A}^{(j)}$ is obtained; $a_{ij}^{(j)}$ is then used to compute m_{ij}, and $a_{ij}^{(j+1)}$ to $a_{ij}^{(n)}$ are taken to be exactly equal to zero. The computed multiplier satisfies

$$m_{ij} \equiv a_{ij}^{(j)}/a_{jj}^{(j)} + q_{ij} \tag{13}$$

where q_{ij} is the rounding error involved in division. (13) can be written as

$$0 \equiv a_{ij}^{(i)} - m_{ij}a_{jj}^{(j)} + d_{ij}^{(j+1)}, \tag{14}$$

where

$$d_{ij}^{(j+1)} = a_{jj}^{(j)} q_{ij} \tag{15}$$

Thus we can write

$$a_{ij}^{(2)} \equiv a_{ij}^{(1)} - m_{i1} a_{1j}^{(1)} + d_{ij}^{(2)}$$
$$a_{ij}^{(3)} \equiv a_{ij}^{(2)} - m_{i2} a_{2j}^{(2)} + d_{ij}^{(3)}$$
$$\cdots \tag{16}$$
$$a_{jj}^{(j)} \equiv a_{jj}^{(j-1)} - m_{i,j-1} a_{j-1,j}^{(j-1)} + d_{ij}^{(j)}$$
$$0 \equiv a_{ij}^{(j)} - m_{ij} a_{jj}^{(j)} + d_{ij}^{(j+1)}$$

Summing (16) we have

$$0 \equiv a_{ij}^{(1)} - \sum_{l=1}^{j} m_{il} a_{lj} + e_{ij}, \tag{17}$$

where

$$e_{ij} = \sum_{k=2}^{j+1} d_{ij}^{(k)}. \tag{18}$$

We note that the set of n^2 equations defined by (11) and (17) is equivalent to the matrix equation

$$\mathbf{LU} \equiv \mathbf{A}^{(1)} + \partial \mathbf{A}, \tag{19}$$

where the elements ∂a_{ij} of $\partial \mathbf{A}$ are defined by (12) and (18).

The Bounds on $\partial \mathbf{A}$. Pivoting

The computed $a_{ij}^{(k)}$ is defined by,

$$a_{ij}^{(k)} \equiv \text{fl}(a_{ij}^{(k-1)} - m_{i,k-1} a_{k-1,j}^{(k-1)})$$
$$\equiv [a_{ij}^{(k-1)} - m_{i,k-1} a_{k-1,j}^{(k-1)} (1 + \delta_1)](1 + \delta_2) \tag{20}$$

Now $d_{ij}^{(k)}$ is defined by

$$d_{ij}^{(k)} = a_{ij}^{(k)} - [a_{ij}^{(k-1)} - m_{i,k-1} a_{k-1,j}^{(k-1)}]$$
$$= a_{ij}^{(k)} - \left[\frac{a_{ij}^{(k)}}{1 + \delta_2} + m_{i,k-1} a_{k-1,j}^{(k-1)} \delta_1 \right]$$
$$= \frac{a_{ij}^{(k)} \delta_2}{1 + \delta_2} + m_{i,k-1} a_{k-1,j}^{(k-1)} \delta_1 \tag{21}$$

The bounds on $d_{ij}^{(k)}$ obviously depend on the bounds on $a_{ij}^{(k)}$ and $m_{i,k-1}$. The expression for the multipliers,

$$m_{ij} = a_{ij}^{(j)}/a_{jj}^{(j)},$$

shows that on the one hand if $a_{jj}^{(j)} = 0$ the method of triangular decomposition will completely break down. On the other hand, if the method works then the

pivotal rows, i, and the pivots, $a_{ij}^{(j)}$, can at every stage be selected so as to ensure the smallest possible value for the m_{ij}.
Indeed, this value can always be made to satisfy

$$|m_{ij}| \leqslant 1 \quad \text{by} \tag{22}$$

either

(1) eliminating the columns in natural order but at the rth stage taking the pivotal row to be that one of the remaining $n - r + 1$ rows which has the element of largest modulus in column r; this new pivotal row and the rth row are then exchanged, i.e. by partial pivoting;

or

(2) selecting at the rth stage as pivot the element of largest modulus in the whole of the remaining $(n - r + 1)$ square array, i.e. by complete pivoting.

Of the two kinds of pivoting, partial pivoting has some advantages over complete pivoting. In particular, for matrices having a considerable number of zero elements organised in a special pattern, this pattern may be preserved by partial pivoting but destroyed by complete pivoting.

If partial pivoting is used and if all elements of $A^{(1)}$ satisfy $|a_{ij}^{(1)}| \leqslant a$, then all elements of $\mathbf{A}^{(r)}$ satisfy

$$|a_{ij}^{(r)}| \leqslant 2^{r-1} a. \tag{23}$$

It may be mentioned that this bound on the computed elements $a_{ij}^{(r)}$ is not very sharp, and in practice it is very uncommon for any element to attain a value even as large as $8a$.

If complete pivoting is used then it has been shown by Wilkinson that

$$|a_{rr}^{(r)}| < r^{1/2}(2^1 3^{1/2} 4^{1/3} \ldots r^{1/r-1})^{1/2} a \tag{24}$$

However Wilkinson again suggests that this is a severe overestimate. Assume that some form of pivoting has been used and that $|m_{ij}| \leqslant 1$ and $|a_{ij}^{(i)}| \leqslant 1$. Denote the maximum element in any $|A^{(r)}|$ by g. Thus from (21)

$$|d_{ij}^{(k)}| \leqslant \frac{g2^{-t}}{1 - 2^{-t}} + g2^{-t}$$

$$< (2.01)g2^{-t} \quad \text{(say)} \tag{25}$$

This applies to all $d_{ij}^{(k)}$ except $d_{ij}^{(j+1)}$ $(i > j)$. For these we use (13) and (15), and consider

$$m_{ij} \equiv \text{fl}(a_{ij}^{(j)}/a_{jj}^{(j)}) = [a_{ij}^{(j)}/a_{jj}^{(j)}](1 + \delta) \tag{26}$$

where

$$\delta < 2^{-1}$$

and

$$q_{ij} = a_{ij}^{(j)} \delta / a_{jj}^{(j)}, \tag{27}$$

so

$$|d_{ij}^{(j+1)}| = |a_{jj}^{(j)} a_{ij}^{(j)} \delta / a_{jj}^{(j)}| = |a_{ij}^{(j)} \sigma| \leqslant g2^{-1} \tag{28}$$

Noticing that $g2^{-1} < (2.01)g2^{-1}$ we can use the bound from (25) to cover all the d_{ij}. Combining (12), (18), and (25) we have the bounds on $\partial \mathbf{A}$ given by

$$|\partial \mathbf{A}| \leqslant (2.01)g2^{-1} \begin{bmatrix} 0 & 0 & 0 & \ldots & 0 & 0 \\ 1 & 1 & 1 & \ldots & 1 & 1 \\ 1 & 2 & 2 & \ldots & 2 & 2 \\ 1 & 2 & 3 & \ldots & 3 & 3 \\ & \ldots & & & & \\ & \ldots & & & & \\ 1 & 2 & 3 & & (n-1) & (n-1) \end{bmatrix} \tag{29}$$

Bound Estimates on $\partial \mathbf{L}$ and $\partial \mathbf{U}$

We assume now that the set of equations $\mathbf{Ax} = \mathbf{b}$ is expressed in the form

$$\mathbf{LUx} = \mathbf{b} \tag{1}$$

and to obtain its solution, we proceed by solving the two sets of equations,

$$\mathbf{Ly} = \mathbf{b} \tag{2}$$

and

$$\mathbf{Ux} = \mathbf{y} \tag{3}$$

Consider the solution of the lower triangular set, $\mathbf{Ly} = \mathbf{b}$,

$$\begin{aligned} l_{11}y_1 &= b_1 \\ l_{21}y_1 + l_{22}y_2 &= b_2 \\ \ldots \ldots \\ l_{n1}y_1 + l_{n2}y_2 + \ldots + l_{nn}y_n &= n_n \end{aligned} \tag{4}$$

At the time when y_r is computed from the rth equation, $y_1, y_2, \ldots, y_{r-1}$ have already been computed. In floating-point form we have

$$\begin{aligned} y_r &\equiv \mathrm{fl}\left(\frac{-l_{r1}y_1 - l_{r2}y_2 - \ldots - l_{r,r-1}y_{r-1} + b_r}{l_{rr}} \right) \\ &= \frac{\left(\sum_{j=1}^{r-1} -l_{rj}y_j(1 + \varepsilon_{rj}) \right) + b_r(1 + \delta_r)}{l_{rr}(1 + \beta_r)}, \end{aligned} \tag{5}$$

where

$$\begin{aligned} &|\beta_r|, |\delta_r| \leqslant 2^{-t_1}, \\ &|\varepsilon_{rj}| \leqslant (r + 2 - j)2^{-t_1} \end{aligned} \tag{6}$$

and

$$2^{-t_1} = (1.06)2^{-t}.$$

In (5), dividing the numerator and the denominator of the right-hand side expression by $(1 + \delta_r)$ and denoting $(1 + \varepsilon_{rj})/(1 + \delta_r)$ and $(1 + \beta_r)/(1 + \delta_r)$ by $1 + \alpha_{ri}$ and $1 + \alpha_{rr}$, respectively, we get

$$y_r \equiv \frac{\sum\limits_{j=1}^{r-1} (-l_{rj}y_j)(1 + \alpha_{rj}) + b_r}{l_{rr}(1 + \alpha_{rr})}, \tag{7}$$

where

$$|\alpha_{rj}| \leqslant (r + 1 - i)2^{-t}$$

and

$$|\alpha_{rr}| \leqslant 2^{-t+1}. \tag{8}$$

Expression (7) can be written in the form

$$\sum_{j=1}^{r} l_{rj}y_j(1 + \alpha_{rj}) \equiv b_r, \tag{9}$$

which shows that the computed vector-solution is the exact solution of

$$(\mathbf{L} + \partial\mathbf{L})\mathbf{y} = \mathbf{b}, \tag{10}$$

where say for $n = 6$, we have

$$|\partial\mathbf{L}| \leqslant 2^{-t} \begin{bmatrix} 2|l_{11}| & 0 & 0 & 0 & 0 & 0 \\ 2|l_{21}| & 2|l_{22}| & 0 & 0 & 0 & 0 \\ 3|l_{31}| & 2|l_{32}| & 2|l_{33}| & 0 & 0 & 0 \\ 4|l_{41}| & 3|l_{42}| & 2|l_{43}| & 2|l_{44}| & 0 & 0 \\ 5|l_{51}| & 4|l_{52}| & 3|l_{53}| & 2|l_{54}| & 2|l_{55}| & \\ 6|l_{61}| & 5|l_{62}| & 4|l_{63}| & 3|l_{64}| & 2|l_{65}| & 2|l_{66}| \end{bmatrix} \tag{11}$$

If the maximum element in any $|\mathbf{A}^{(r)}|$ is $\leqslant g$, i.e. $|l_{ij}| \leqslant g$, then

$$\|\partial\mathbf{L}\|_1 = \|\partial\mathbf{L}\|_\infty \leqslant \tfrac{1}{2}(n^2 + n + 2)g\,2^{-t}. \tag{12}$$

Note that for these matrix norms, we have

$$\|\partial\mathbf{L}\| \leqslant n\|\mathbf{L}\|2^{-t}. \tag{13}$$

Now, rewrite (10) as

$$\mathbf{b} - \mathbf{Ly} = \partial\mathbf{Ly}. \tag{14}$$

If we assume that $\mathbf{L}$ and $\mathbf{b}$ are scaled so that all elements are of modulus < 1, then we can take $g = 1$ and so from (12) we get the following bound for the residual

$$\|\mathbf{b} - \mathbf{Ly}\|_\infty \leqslant \tfrac{1}{2}(n^2 + n + 2)\|\mathbf{y}\|_\infty 2^{-t} \tag{15}$$

Note that the bound for the residual is directly dependent on $\| \mathbf{y} \|_\infty$.

If $\| \mathbf{y} \|_\infty$ is of order unity, then since in practice the factor $(n^2 + n + 2)$ is very unlikely to be attained, the residual (15) is of necessity very small whether or not $\mathbf{y}$ is an accurate solution.

This completes the analysis on the local perturbation matrix, $\partial \mathbf{L}$. A similar analysis is readily carried out for the local perturbation matrix of the second set, $\partial \mathbf{U}$.

We are now in a position to give a bound estimate for the total perturbation matrix, $\mathbf{E}$, given by

$$\mathbf{E} = \partial \mathbf{A} + \mathbf{U} \partial \mathbf{L} + \mathbf{L} \partial \mathbf{U}.$$

Indeed, we have shown that assuming pivoting has been used and all $|l_{ij}| \leqslant 1$, the following bounds hold

$$\| \partial \mathbf{A} \|_\infty \leqslant (2.01)g \left(\frac{n}{2} + 1 \right)(n - 1)2^{-t} \tag{16}$$

$$\| \partial \mathbf{L} \|_\infty \leqslant \tfrac{1}{2}(n^2 + n + 2)2^{-t}, \tag{17}$$

$$\| \partial \mathbf{U} \|_\infty \leqslant \tfrac{1}{2}g(n^2 + n + 2)2^{-t}, \tag{18}$$

$$\| \mathbf{L} \|_\infty \leqslant n, \tag{19}$$

$$\| \mathbf{U} \|_\infty \leqslant gn, \tag{20}$$

where g is the maximum element of any $|\mathbf{A}^{(r)}|, r = 1, \ldots, n$. Combining equations (16) to (20) we get a bound for $\| \mathbf{E} \|_\infty$ which for sufficiently large n is effectively given by

$$\| \mathbf{E} \|_\infty \leqslant g(n^3 + 2.005n^2)2^{-t}, \tag{21}$$

and the bound on the residual is

$$\| \mathbf{b} - \mathbf{A}\mathbf{x} \|_\infty \leqslant g(n^3 + 2.005n^2) \| \mathbf{x} \|_\infty 2^{-t}. \tag{22}$$

We see that in (22) the dominant term involves n^3. This term comes from the errors in the solution of the triangular sets of equations, but as was pointed out by Wilkinson, the bounds for $\| \partial \mathbf{L} \|$ and $\| \partial \mathbf{U} \|$ may only be attained in exceptional circumstances, and, in fact, he suggests a more realistic bound for the residual as

$$\| \mathbf{b} - \mathbf{A}\mathbf{x} \|_\infty \leqslant gn \| \mathbf{x} \|_\infty 2^{-t}. \tag{23}$$

4.5 Iterative Refinement of the Solution

If the matrix $\mathbf{A}$ is somewhat ill-conditioned then the computed solution $\mathbf{x}^{(1)}$ of $\mathbf{A}\mathbf{x} = \mathbf{b}$ may not be sufficiently accurate. The solution $\mathbf{x}^{(1)}$ can be improved by a process of iterative refinement in which a sequence of vectors $\mathbf{x}^{(1)}, \mathbf{x}^{(2)}, \ldots,$ $\mathbf{x}^{(r)}$ is calculated which in certain circumstances converges to the true solution $\mathbf{x}$.

We already have an $\mathbf{L}$ and $\mathbf{U}$ such that,

$$\mathbf{L}\mathbf{U} = \mathbf{A} + \mathbf{E} \tag{1}$$

We define the residual vectors $\mathbf{r}^{(s)}$ by

$$\mathbf{r}^{(s)} = \mathbf{b} - \mathbf{A}\mathbf{x}^{(s)} \tag{2}$$

where

$$\mathbf{x}^{(s+1)} = \mathbf{x}^{(s)} + (\mathbf{LU})^{-1}\mathbf{r}^{(s)} \tag{3}$$

The sth iteration of the process involves three steps.

(1) Calculate the residuals $\mathbf{r}^{(s)} = \mathbf{b} - \mathbf{A}\mathbf{x}^{(s)}$

(2) Solve the system $(\mathbf{LU})^{-1}\mathbf{r}^{(s)} = \mathbf{d}^{(s)}$

(3) Add the correction, $\mathbf{x}^{(s+1)} = \mathbf{x}^{(s)} + \mathbf{d}^{(s)}$

Each system $(\mathbf{LU})\mathbf{d}^{(s)} = \mathbf{r}^{(s)}$ has the same matrix $\mathbf{LU}$ and so we can make use of the multipliers already stored from the first solution. This is why iterative improvement of a first solution adds only a moderate amount (about 25%) to the computational time of the algorithm. It is absolutely essential that the residuals $\mathbf{r}^{(s)}$ be computed with a higher precision than the rest of the computation because the residual is the critical computation.

If the sequence of residuals and vectors can be solved without further rounding error, then we have from (2) and (3)

$$\mathbf{x}^{(s+1)} = \mathbf{x}^{(s)} + (\mathbf{LU})^{-1}(\mathbf{A}\mathbf{x} - \mathbf{A}\mathbf{x}^{(s)}) \tag{4}$$

or $\quad \mathbf{x}^{(s+1)} - \mathbf{x} = [\mathbf{I} - (\mathbf{LU})^{-1}\mathbf{A}]^s(\mathbf{x}^{(1)} - \mathbf{x}) \tag{5}$

Similarly we have

$$\mathbf{r}^{(s+1)} = \mathbf{A}(\mathbf{x} - \mathbf{x}^{(s+1)}) = [\mathbf{I} - \mathbf{A}(\mathbf{LU})^{-1}]\mathbf{r}^{(s)} \tag{6}$$

From (5) and (6) we can guarantee convergence if

$$[\mathbf{I} - (\mathbf{A} + \mathbf{E})^{-1}\mathbf{A}]^s \text{ tends to 0 as } s \text{ tends to } \infty \tag{7}$$

i.e. $\quad \| \mathbf{I} - (\mathbf{A} + \mathbf{E})^{-1}\mathbf{A} \| < 1 \tag{8}$

which is satisfied if

$$\| \mathbf{A}^{-1} \| \, \| \mathbf{E} \| < \tfrac{1}{2} \tag{9}$$

If $\mathbf{A}$ has been triangularized by Gaussian elimination with partial pivoting using floating-point with accumulation, then we have in general

$$\| \mathbf{E} \| \leqslant n2^{-t} \tag{10}$$

and hence the exact iterative process will converge if

$$\| \mathbf{A}^{-1} \| \leqslant n2^{t-1} \tag{11}$$

Now if

$$n2^{-t}\| \mathbf{A}^{-1} \| < 2^{-p}, \quad \text{for } p > 1 \tag{12}$$

it then follows from (5) and (6) that

$$\| \mathbf{x}^{(s+1)} - \mathbf{x} \| \leqslant [2^{-p}/(1 - 2^{-p})] \| \mathbf{x}^{(s)} - \mathbf{x} \| \tag{13}$$

and $\|\mathbf{r}^{(s+1)}\| \leqslant [2^{-p}/(1-2^{-p})]\|\mathbf{r}^{(s)}\|$ (14)

Both the error and the residual decrease by at least the factor $2^{-p}/(1-2^{-p})$ with each iteration so that if p is appreciably greater than 2 we effectively gain at least p binary digits per iteration.

In practice the iterative process cannot be performed exactly and thus the effect of rounding errors could nullify the whole process. If the iteration is performed using t-digit arithmetic, the accuracy of $\mathbf{x}^{(s)}$ cannot increase indefinitely with s since we are using only t digits to represent the components. If we use floating-point arithmetic with accumulation, then the residual corresponding to the first solution is almost certain to be of the same order of magnitude as that corresponding to the correctly rounded solution. We can therefore scarcely expect the residual to diminish by the factor 2^{-p} with each iteration and it appears improbable that the accuracy of the successive solutions could steadily improve if the residuals remain roughly constant. Nevertheless this is precisely what does happen and the performance of the practical process using any mode of arithmetic does not fall far short of that corresponding to exact iteration provided only that inner products are accumulated in the computation of the residuals.

For further details of iterative improvement see Forsyth and Moler (1967), Moler (1967), and Wilkinson (1963).

4.6 Cholesky Decomposition of Symmetric Matrices

A special case, important in practical problems, is given by the set $\mathbf{Ax} = \mathbf{b}$, where $\mathbf{A}$ is so-called positive definite. A real symmetric matrix $\mathbf{A}$ is called positive definite if the quadratic form $\mathbf{x}^T\mathbf{Ax} > 0$ for all real $\mathbf{x} \neq \mathbf{0}$. (A necessary and sufficient condition for the matrix to be positive definite is that all its eigenvalues should be positive.)

When the matrix $\mathbf{A}$ of the set $\mathbf{Ax} = \mathbf{b}$ is symmetric and positive definite, the diagonal elements of the lower triangular matrix $\mathbf{L}$ can be chosen in such a way that $\mathbf{L}$ is real and the upper triangular matrix $\mathbf{U}$ is the transpose of $\mathbf{L}$. Thus we have,

$$\mathbf{A} = \mathbf{LL}^T$$ (1)

This is known as the Cholesky decomposition.

4.6.1 The Cholesky Method

The computation of $\mathbf{L}$ is illustrated by the following example. Consider the case of a 3×3 matrix. We have to find the elements l_{ij} corresponding to known elements a_{ij} for which the following matrix equation holds.

$$\begin{bmatrix} l_{11} & 0 & 0 \\ l_{21} & l_{22} & 0 \\ l_{31} & l_{32} & l_{33} \end{bmatrix} \begin{bmatrix} l_{11} & l_{21} & l_{31} \\ 0 & l_{22} & l_{32} \\ 0 & 0 & l_{33} \end{bmatrix} = \begin{bmatrix} a_{11} & a_{12} & a_{13} \\ a_{12} & a_{22} & a_{23} \\ a_{13} & a_{23} & a_{33} \end{bmatrix}$$ (2)

From (2) we have 6 independent equations.

$$l_{11}^2 = a_{11},$$
$$l_{11}l_{21} = l_{21}l_{11} = a_{12}, \tag{3}$$
$$l_{11}l_{31} = l_{31}l_{11} = a_{13},$$

giving successively l_{11}, l_{21} and l_{31};

$$l_{21}^2 + l_{22}^2 = a_{22},$$
$$l_{21}l_{31} + l_{22}l_{32} = l_{31}l_{21} + l_{32}l_{22} = a_{23}, \tag{4}$$

giving successively l_{22} and l_{32};

$$l_{31}^2 + l_{32}^2 + l_{33}^2 = a_{33}, \tag{5}$$

giving l_{33}.

We can see that only the matrix **L** needs to be stored, and not $\mathbf{L}^T$. It is unnecessary to compute the inverse of **L** to solve the set of equations $\mathbf{Ax} = \mathbf{b}$. From (1) we have

$$\mathbf{LL}^T\mathbf{x} = \mathbf{b} \tag{6}$$

Writing $\mathbf{L}^T\mathbf{x} = \mathbf{y}$ we have two sets of triangular equations

$$\mathbf{Ly} = \mathbf{b} \tag{7}$$
$$\mathbf{L}^T\mathbf{x} = \mathbf{y} \tag{8}$$

and two processes of back substitution give the solution **x**.

The proof of $\mathbf{LL}^T = \mathbf{A}$, due to Cholesky, is by induction and may be found in Fox (1954) or Wilkinson (1965). A practically useful property of Cholesky's decomposition lies in the fact that if the matrix **A** is scaled originally so that all elements of $|\mathbf{A}|$ are bounded by unity then so, too, are all elements of **L**. This fact can be easily observed when equating the diagonal elements in the equation (1), i.e. we have

$$l_{i1}^2 + l_{i2}^2 + \ldots + l_{ii}^2 = a_{ii}, \qquad i = 1, \ldots, n. \tag{9}$$

It follows that no pivoting is required in Cholesky's decomposition.

If **A** is symmetric but not positive definite then Cholesky's method is not possible without complex arithmetic as in general $\mathbf{A} = \mathbf{LL}^T$ does not hold. In fact the symmetric decomposition of a matrix which is not positive definite enjoys none of the stability of the positive definite case. To be sure of stability we must use interchanges and this will destroy the symmetry.

The Cholesky decomposition algorithm

We assume that input matrix **A** symmetric and positive definite. The matrix is also scaled so that none of its elements, in modulus, exceeds unity. The matrix is stored in the array $\mathbf{A}(1:N, 1:N) = (A(I,J), I = 1, \ldots, N, J = 1, \ldots, N)$. The algorithm computes the elements of a lower triangular matrix **L** such that $LL^T = A$. The computed lower triangular matrix L at the end of the process is stored in the lower triangle of the array $A(1:N, 1:N)$, except for the reciprocals of the diagonal elements of L which are stored in the array $(P(I), I = 1, \ldots, N)$. The upper triangle of the array contains the elements of the input matrix A.

1. Set $I \leftarrow 1$.
2. Set $J \leftarrow I$.
3. Set SUM $\leftarrow A(I, J)$.
4. $I > 1$?
    If false, go to Step 6.
5. Compute SUM $\leftarrow$ SUM $- \sum_{K = I-1}^{1} A(J, K) * A(I, K)$
6. $J = I$?
    If false, go to Step 8.
7. $P(I) \leftarrow 1/\text{SQRT(SUM)}$ then go to Step 9.
8. $A(J, I) \leftarrow \text{SUM} * P(I)$.
9. $J \leftarrow J + 1$.
10. $J > N$?
    If false, go back to Step 3.
11. $I \leftarrow I + 1$.
12. $I > N$?
    If false, go back to Step 2.
13. Terminate.

Warning: If due to rounding errors, the matrix A ceases to be positive definite, the Cholesky decomposition algorithm becomes invalid.

4.6.2 Algebraic Complexity: the Number of Arithmetic Operations Required to Solve a Set of Linear Equations

As the example of Section 4.6.1 shows, the basic steps in the Cholesky decomposition are

$$l_{ii} = \left(a_{ii} - \sum_{j=1}^{i-1} l_{ij}^2 \right)^{1/2}, \qquad i = 1, \ldots, n, \tag{1}$$

and

$$l_{ij} = \frac{a_{ij} - \sum_{k=1}^{j-1} l_{ik} l_{jk}}{l_{jj}} \qquad \begin{aligned} i &= 1, \ldots, n. \\ j &= 1, \ldots, i-1. \end{aligned} \tag{2}$$

The division in (2) is usually replaced by computing instead the reciprocal of the square root in (1) and then multiplying the numerator in (2) by the corresponding reciprocal. In fact, the computed reciprocals are stored in the diagonal positions of L and used directly at later steps of computation.

The total number of multiplications required by the process is readily given by

$$\sum_{j=1}^{n} \left[(j-1) + \sum_{i=j+1}^{n} j \right] = \frac{n^3}{6} + \frac{n^2}{2} - \frac{2}{3}n \tag{3}$$

and the total number of additions by

$$\sum_{j=1}^{n} \left[(j-1) + \sum_{i=j+1}^{n} (j-1) \right] = \frac{n(n^2-1)}{6}. \tag{4}$$

For solving linear equations from the formulae $Ly = b$, $L^Tx = y$, we require a further $n(n-1)$ additions and $n(n+1)$ multiplications. The n divisions required for each triangular system do not have to be carried out as the reciprocal of the diagonal elements $l_{ii}(i = 1, \dots n)$ are normally already stored in the diagonal positions of L. Thus the total number of arithmetic operations required by Cholesky decomposition to solve a set of linear equations is

$$\frac{n^3}{6} + \frac{3}{2}n^2 + \frac{n}{3} \quad \text{multiplications,}$$

$$\frac{n^3}{6} + n^2 - \frac{7}{6}n \quad \text{additions,} \tag{5}$$

$$n \quad \text{square root reciprocal evaluations.}$$

A variant of Cholesky decomposition which avoids calculation of square roots altogether is suggested in Martin, Peters and Wilkinson (1965).

4.6.3 Matrix Inversion

To invert a symmetric positive matrix A, we use

$$A = LL^T$$
$$A^{-1} = (L^T)^{-1}L^{-1} \tag{1}$$
$$L^TA^{-1} = L^{-1}$$

The use of (1) avoids the explicit inversion of L. Consider a 3×3 system.

Let l_{ij} be the elements of L^T,
$\quad d_{ij}$ be the elements of L^{-1},
and a_{ij} be the elements of A^{-1} (which is symmetric)

The (1) can be expanded as

$$\begin{bmatrix} l_{11} & l_{21} & l_{31} \\ 0 & l_{22} & l_{23} \\ 0 & 0 & l_{33} \end{bmatrix} \begin{bmatrix} a_{11} & a_{12} & a_{13} \\ a_{12} & a_{22} & a_{23} \\ a_{13} & a_{23} & a_{33} \end{bmatrix} = \begin{bmatrix} d_{11} & 0 & 0 \\ d_{21} & d_{22} & 0 \\ d_{31} & d_{32} & d_{33} \end{bmatrix}$$

Thus $a_{33} = d_{33}/l_{33}$

$$a_{23} = -l_{23}a_{33}/l_{22}$$

$$a_{22} = (d_{22} - l_{23}a_{23})/l_{22}$$

$$a_{13} = -(l_{21}a_{13} + l_{31}a_{33})/l_{11}$$

$$a_{12} = -(l_{21}a_{22} + l_{31}a_{23})/l_{11}$$

$$a_{11} = (d_{11} - l_{21}a_{12} - l_{31}a_{13})/l_{11}$$

Now the only elements of $\mathbf{L}^{-1}$ required are the diagonal elements but these are just the reciprocals of the diagonal elements of $\mathbf{L}$ which are already stored in $\mathbf{L}$. Thus we do not need to compute $\mathbf{L}^{-1}$.

The elements of $\mathbf{A}^{-1}$ are given by

$$a_{ii} = \frac{1}{l_{ii}}\left[\frac{1}{l_{ii}} - \sum_{k=i+1}^{n} l_{ik}a_{ik}\right] \tag{2}$$

which requires $(n - i + 1)$ multiplications and $(n - i)$ additions,

and $a_{ij} = -\dfrac{1}{l_{ii}} \displaystyle\sum_{k=i+1}^{n} l_{ki}a_{kj}, \qquad i < j,$ $\hfill(3)$

which requires $(n - i + 1)$ multiplications and $(n - i - 1)$ additions. From (2) and (3) the number of multiplications is

$$\sum_{j=1}^{n}\sum_{i=1}^{j}(n + 1 - i) = \frac{n^3}{3} + \frac{n^2}{2} + \frac{n}{6} \tag{4}$$

and the number of additions is

$$\sum_{i=1}^{n}(n - i) + \sum_{j=1}^{n}\sum_{i=1}^{j-1}(n - 1 - j) = \frac{n^3}{3} - \frac{n^2}{2} + \frac{n}{6} \tag{5}$$

From 4.5.2 $-$ (3), (4) and from (4) and (5) the total number of arithmetic operations required to invert a symmetric positive definite matrix is

$\dfrac{n^3}{2} + n^2 - \dfrac{n}{2}$ multiplications,

$\dfrac{n^3}{2} - \dfrac{n^2}{2}$ additions, $\hfill(6)$

$\qquad n$ square root reciprocals.

4.6.4 Numerical Stability

The error analysis of the Cholesky decomposition may be carried out in a way similar to the analysis of the general triangularisation method. In fact, Wilkinson (1961) has shown that the Cholesky decomposition has a guaranteed stability even without pivoting, provided the original matrix $\mathbf{A}$ is scaled so that all its elements are bounded by unity. For the computation in floating point with accumulation, the bounds on the error matrix $\mathbf{E}$, where $\mathbf{LL}^T = \mathbf{A} + \mathbf{E}$, are given as

$$e_{ij} = \begin{cases} l_{ij}l_{jj}2^{-t}, & i < j. \\ l_{ij}l_{ii}2^{-t}, & i > j, \\ l_{jj}^2 2^{-t}, & i = j. \end{cases} \tag{1}$$

4.7 The Orthogonal Reduction Methods

Another group of methods for solving sets of linear equations and related problems is known as the group of orthogonal reduction methods. In methods of this type a triangular form of the original matrix is achieved by premultiplication with orthogonal matrices.

The original matrix is transformed successively to $\mathbf{A}_1, \mathbf{A}_2, \ldots, \mathbf{A}_s$ where each member of the sequence has at least one more zero below the diagonal than the previous matrix. These zeroes may be produced one at a time as in, say, the Given's method or a whole subcolumn at a time as in the Householder method. In Gaussian elimination the main obstacle to finding error bounds is the possibility of a progressive increase in size of the elements of the reduced matrices, since in general the current rounding error is proportional to the current size of the elements. With orthogonal or nearly orthogonal matrices this problem is no longer present. The Euclidean norm of each column of $\mathbf{A}_1$ is preserved by exact orthogonal transformations. Hence, if originally the sum of the squares in each column is less than unity, it will remain true for all subsequent matrices.

The methods of Givens (1957, 1958) and of Householder (1958) are the best known and most efficient methods of the group.

We shall only briefly mention the Givens method and concentrate our attention on the Householder method. Strictly speaking the methods achieve the triangularisation of the matrix of the set $\mathbf{Ax} = \mathbf{b}$, after which any standard technique may be employed to complete the solution of the set. In the Givens method the triangular form is brought about by means of a sequence of $n(n-1)/2$ plane rotations, whose product is an orthogonal matrix. Each rotation requires the extraction of a square root. An important advantage in using the method as compared to the Gaussian elimination method lies in the fact that an orthogonal matrix is perfectly conditioned and so the condition of the matrix cannot deteriorate through successive transformations.

4.7.1 The Householder Reduction

The method developed by Householder may be considered as a variant of the Givens method in the sense that it preserves all the advantages of the former method but requires fewer arithmetic operations. We may say that in terms of computational complexity Householder's method is the most efficient of the orthogonal reduction methods.

To illustrate how the method works let us consider an intermediate step in the Householder reduction. We have A_r which is upper triangular in its first r columns and we may write

$$A_r = \left[\begin{array}{c|c} B_r & C_r \\ \hline 0 & W_{n-r} \end{array}\right] \tag{1}$$

where B_r is an upper triangular matrix of order r. In the $(r+1)$th step we use a matrix P_r of the form

$$P_r = \left[\begin{array}{c|c} I & 0 \\ \hline 0 & I - 2uu^* \end{array}\right], \qquad \|u\|_2 = 1 \tag{2}$$

We then compute

$$A_{r+1} = P_r A_r = \left[\begin{array}{c|c} B_r & C_r \\ \hline 0 & (I - 2uu^*)W_{n-r} \end{array}\right] \tag{3}$$

so that only W_{n-r} is modified. We must choose u so that the first column of $(I - 2uu^*)W_{n-r}$ is null apart from its first element.

This choice is based on the following

Lemma (Householder, 1958).

For any vector $a \neq 0$ and any unit vector v, a unit vector u exists such that

$$(I - 2uu^*)a = \|a\|_2 v, \tag{4}$$

where

$$\|a\|_2 = (a^*a)^{1/2} = \left[\sum_{i=1}^{n} |a_i|^2\right]^{1/2}.$$

(By a unit vector one means the vector whose norm equals unity.)

The computation of the vector u requires two square root and one reciprocal evaluations. The calculation of $\|a\|_2$ requires one of the necessary square root evaluations. Then, from (4) it is required that

$$a - 2uu^*a = \|a\|_2 v,$$

which, for convenience we rewrite as

$$a - 2(u^*a)u = \|a\|_2 v. \tag{5}$$

Now, let

$$q = 2u^*a, \tag{6}$$

then

$$qu = a - \|a\|_2 v \tag{7}$$

and

$$(qu)^2 = (a - \|a\|_2 v)^2$$

gives

$$q^2 = 2 \| \mathbf{a} \|_2 (\| \mathbf{a} \|_2 - \mathbf{v}^*\mathbf{a}), \tag{8}$$

since

$$\| \mathbf{u} \|_2 = \| \mathbf{v} \|_2 1.$$

The evaluation of q from (8) accounts for the other square root evaluation.
  Further, from (8) it follows that q is real, since

$$\| \mathbf{a} \|_2 - \mathbf{v}^*\mathbf{a} \geqslant 0, \tag{9}$$

as $\| \mathbf{a} \|_2$ is the length of vector $\mathbf{a}$ and $\mathbf{v}^*\mathbf{a}$ is the projection of $\mathbf{a}$ upon the unit vector $\mathbf{v}$. Hence, q can be taken to be non-negative. Now, if $\| \mathbf{a} \|_2 = \mathbf{v}^*\mathbf{a}$ then Lemma (4) is verified with $\mathbf{u} = \mathbf{0}$. Otherwise we can take $q > 0$ defined by (8), and then the vector $\mathbf{u}$ can always be computed from (7). As we see, this computation requires evaluation of a reciprocal, q^{-1}.
  Now let $\mathbf{a}$ be the first column of A and take $\mathbf{v} = \mathbf{e}_1$, the first column of the identity matrix. Application of the lemma provides a unitary matrix

$$\mathbf{P}_1 = (\mathbf{I} - 2\mathbf{u}_1 \mathbf{u}_1^*) \tag{10}$$

such that the first column of $\mathbf{P}_1 A$ is null except in the first element.
(Note that a matrix $\mathbf{B}$ is called unitary if $\mathbf{B}^{-1} = \mathbf{B}^*$.)
  One continues after suppressing the first row and first column of the transformed matrix. After $(n-2)$ steps, at most, the matrix $\mathbf{A}$ is triangularized. We then have

$$\mathbf{P}\mathbf{A} = \mathbf{R} \tag{11}$$

where $\mathbf{R}$ is an upper triangular matrix and

$$\mathbf{P} = \mathbf{P}_{n-1}\mathbf{P}_{n-2}\cdots\mathbf{P}_2\mathbf{P}_1.$$

We note that as described, the computation process requires $2(n-2)$ square root evaluations. However, half of the square root evaluations can be evaded using an elegant device developed by Wilkinson (1965).
  Consider the following problem. Given a vector $\mathbf{x}$ of length n, construct an elementary Hermitian matrix $\mathbf{P} = \mathbf{I} - 2\mathbf{w}\mathbf{w}^*$, such that

$$\mathbf{P}\mathbf{x} = k\mathbf{e}_1 \tag{12}$$

Since the l_2-norm is invariant, if we write

$$S^2 = x_1^2 + x_2^2 + \ldots + x_n^2 \tag{13}$$

then

$$k = \pm S. \tag{14}$$

From (12) we have

$$x_1 - 2w_1(\mathbf{w}^T\mathbf{x}) = \pm S, \tag{15}$$

$$x_i - 2w_i(\mathbf{w}^T\mathbf{x}) = 0, \qquad i = 2, \ldots, n, \tag{16}$$

and, hence,

$$2Kw_1 = x_1 \mp S, \tag{17}$$

$$2Kw_i = x_i, \qquad i = 2, \ldots, n, \tag{18}$$

where

$$K = \mathbf{w}^T\mathbf{x} \tag{19}$$

Squaring the equations (17) and (18), adding, and dividing by factor 2 we have

$$2K^2 = S^2 \mp x_1 S \tag{20}$$

(In computations the sign in (20) is decided upon so as to avoid K becoming very small, that is

if $x_1 > 0$ then $2K^2 = S^2 + x_1 S$ is used, and

if $x_1 < 0$ then $2K^2 = S^2 - x_1 S$ is used.)

If we write

$$\mathbf{u}^* = (x_1 \mp S, x_2, x_3, \ldots, x_n), \tag{21}$$

then

$$\mathbf{w} = \mathbf{u}/2K \tag{22}$$

and

$$\mathbf{P} = \mathbf{I} - \mathbf{u}\mathbf{u}^*/2K^2. \tag{23}$$

Hence, only one square root evaluation is required in the process, namely that of S^2 to give K.

We shall now sketch the Householder triangularization process for a matrix of dimension $n = 5$.

Given matrix

$$\mathbf{A} = \mathbf{A}^{(1)} = \begin{bmatrix} a_{11} & a_{12} & a_{13} & a_{14} & a_{15} \\ \hline a_{21} & a_{22} & a_{23} & a_{24} & a_{25} \\ a_{31} & a_{32} & a_{33} & a_{34} & a_{35} \\ a_{41} & a_{42} & a_{43} & a_{44} & a_{45} \\ a_{51} & a_{52} & a_{53} & a_{54} & a_{55} \end{bmatrix} = \begin{bmatrix} \mathbf{B}_1 & \mathbf{C}_1 \\ \hline a_{21} & \\ a_{31} & \mathbf{W}_{n-1} \\ a_{41} & \\ a_{51} & \end{bmatrix}$$

1. Form a symmetric orthogonal matrix

$$\mathbf{P}^{(1)} = \begin{bmatrix} 1 & 0 & 0 & 0 & 0 \\ \hline 0 & & & & \\ 0 & & \mathbf{I} - 2\mathbf{u}_1\mathbf{u}_1^* & & \\ 0 & & & & \\ 0 & & & & \end{bmatrix},$$

where

$$\|\mathbf{u}_1\|_2 = 1$$

and

$$\mathbf{u}_1 \mathbf{u}_1^* = (\mathbf{v}_1 \mathbf{v}_1^*)/4K^2$$

with

$$
\begin{aligned}
S^2 &= a_{21}^2 + a_{31}^2 + a_{41}^2 + a_{51}^2, \\
\mathbf{v}_1^{\mathrm{T}} &= (a_{21} \mp S, a_{31}, a_{41}, a_{51}), \\
2K^2 &= S^2 \mp a_{21} S
\end{aligned}
$$

Then obtain

$$
\mathbf{A}^{(2)} = \mathbf{P}^{(1)}\mathbf{A}^{(1)} =
\left[\begin{array}{c|ccccc}
x & x & x & x & x \\
\hline
0 & x & x & x & x \\
0 & 0 & x & x & x \\
0 & 0 & x & x & x \\
0 & 0 & x & x & x
\end{array}\right]
=
\left[\begin{array}{cc|ccc}
x & x & x & x & x \\
0 & x & x & x & x \\
\hline
0 & 0 & & & \\
0 & 0 & & \mathbf{W}_{n-2} & \\
0 & 0 & & &
\end{array}\right]
=
\left[\begin{array}{c|c}
\mathbf{B}_2 & \mathbf{C}_2 \\
\hline
\mathbf{0} & \mathbf{W}_{n-2}
\end{array}\right]
$$

$$\underbrace{\qquad\qquad\qquad\qquad}_{\text{this is matrix}}$$

$$(\mathbf{I} - 2\mathbf{u}_1\mathbf{u}_1^*)\mathbf{W}_{n-1}$$

2. Form a symmetric orthogonal matrix

$$
\mathbf{P}^{(2)} =
\left[\begin{array}{cc|ccc}
1 & 0 & 0 & 0 & 0 \\
0 & 1 & 0 & 0 & 0 \\
\hline
0 & 0 & & & \\
0 & 0 & & \mathbf{I} - 2\mathbf{u}_2\mathbf{u}_2^* & \\
0 & 0 & & &
\end{array}\right],
$$

where

$$\|\mathbf{u}_2\|_2 = 1$$

and

$$\mathbf{u}_2 \mathbf{u}_2^* = (\mathbf{v}_2 \mathbf{v}_2^*)/4K^2$$

with

$$
\begin{aligned}
S^2 &= a_{32}'^2 + a_{42}'^2 + a_{52}'^2 \\
\mathbf{v}_2^{\mathrm{T}} &= (a_{32}' \mp S, a_{42}', a_{52}') \\
2K^2 &= S^2 \mp a_{32}' S.
\end{aligned}
$$

Then obtain

$$\mathbf{A}^{(3)} = \mathbf{P}^{(2)}\mathbf{A}^{(2)} = \begin{bmatrix} x & x & x & x & x \\ 0 & x & x & x & x \\ \hline 0 & 0 & x & x & x \\ 0 & 0 & 0 & x & x \\ 0 & 0 & 0 & x & x \end{bmatrix} = \begin{bmatrix} x & x & x & x & x \\ 0 & x & x & x & x \\ 0 & 0 & x & x & x \\ \hline 0 & 0 & 0 & x & x \\ 0 & 0 & 0 & x & x \end{bmatrix} = \begin{bmatrix} \mathbf{B}_3 & \vdots & \mathbf{C}_3 \\ \hline \mathbf{0} & \vdots & \mathbf{W}_{n-3} \end{bmatrix}$$

$$\underbrace{\qquad\qquad\qquad}_{\text{this is matrix}}$$

$$(\mathbf{I} - 2\mathbf{u}_2\mathbf{u}_2^*)\mathbf{W}_{n-2}$$

3. Form a symmetric orthogonal matrix

$$\mathbf{P}^{(3)} = \begin{bmatrix} 1 & 0 & 0 & 0 & 0 \\ 0 & 1 & 0 & 0 & 0 \\ 0 & 0 & 1 & 0 & 0 \\ \hline 0 & 0 & 0 & & \\ 0 & 0 & 0 & \mathbf{I} - 2\mathbf{u}_3\mathbf{u}_3^* \\ 0 & 0 & 0 & & \end{bmatrix},$$

where

$$\|\mathbf{u}_3\|_2 = 1$$

and

$$\mathbf{u}_3\mathbf{u}_3^* = (\mathbf{v}_3\mathbf{v}_3^*)/4K^2$$

with

$$S^2 = a_{43}''^2 + a_{53}''^2$$
$$\mathbf{v}_2^{\mathrm{T}} = (a_{43}'' \mp S, a_{53}'')$$
$$2K^2 = S^2 \mp a_{43}' S,$$

Then obtain

$$\mathbf{A}^{(4)} = \mathbf{P}^{(3)}\mathbf{A}^{(3)} = \begin{bmatrix} x & x & x & x & x \\ 0 & x & x & x & x \\ 0 & 0 & x & x & x \\ \hline 0 & 0 & 0 & x & x \\ 0 & 0 & 0 & 0 & x \end{bmatrix} = \begin{bmatrix} x & x & x & x & x \\ 0 & x & x & x & x \\ 0 & 0 & x & x & x \\ 0 & 0 & 0 & x & x \\ \hline 0 & 0 & 0 & 0 & x \end{bmatrix} = \begin{bmatrix} \mathbf{B}_4 & \vdots & \mathbf{C}_4 \\ \hline \mathbf{0} & \vdots & \mathbf{W}_{n-4} \end{bmatrix}$$

$$\underbrace{\qquad\qquad\qquad}_{\text{this is matrix}}$$

$$(\mathbf{I} - 2\mathbf{u}_3\mathbf{u}_3^*)\mathbf{W}_{n-3}$$

The matrix $\mathbf{A}^{(4)}$ is upper triangular.
The Householder process is complete.

Following a lucid presentation of the algorithmic structure of the process in Businger and Golub (1971), we write for a general $n \times n$ matrix $\mathbf{A}$:

Let $\mathbf{A} = \mathbf{A}^{(1)} = \{a_{i,j}^{(1)}\}$ and let $\mathbf{A}^{(2)}, \mathbf{A}^{(3)}, \ldots, \mathbf{A}^{(n+1)}$ be defined as follows:

$$\mathbf{A}^{(k+1)} = \mathbf{P}^{(k)} \mathbf{A}^{(k)} \qquad \text{for } k = 1, 2, \ldots, n.$$

$\mathbf{P}^{(k)}$ is a unitary (or symmetric orthogonal in the case of real $\mathbf{A}$) matrix of the form

$$\mathbf{P}^{(k)} = \mathbf{I} - \beta_k \mathbf{u}^{(k)} \mathbf{u}^{(k)*},$$

where the elements of $\mathbf{P}^{(k)}$ are derived so that

$$a_{i,k}^{(k+1)} = 0, \qquad i = k+1, \ldots, n.$$

Algorithmically, $\mathbf{P}^{(k)}$ is generated as follows:

$$S_k = \left[\sum_{i=k}^{n} (a_{i,k}^{(k)})^2 \right]^{1/2},$$

$$\beta_k = [S_k(S_k + |a_{k,k}^{(k)}|)]^{-1},$$

$$u_i^{(k)} = 0 \quad \text{for } i < k,$$

$$u_k^{(k)} = \text{sgn}(a_{k,k}^{(k)})(S_k + |a_{k,k}^{(k)}|),$$

$$u_i^{(k)} = a_{i,k}^{(k)} \quad \text{for } i > k.$$

The matrix $\mathbf{P}^{(k)}$ is not computed explicitly. Rather it is noted that

$$\mathbf{A}^{(k+1)} = (\mathbf{I} - \beta_k \mathbf{u}^{(k)} \mathbf{u}^{(k)*}) \mathbf{A}^{(k)}$$
$$= \mathbf{A}^{(k)} - \mathbf{u}^{(k)} \mathbf{w}_k^{\mathsf{T}},$$

where

$$\mathbf{w}_K^{\mathsf{T}} = \beta_k \mathbf{u}^{(k)*} \mathbf{A}^{(k)}.$$

In computing the vector $\mathbf{w}_k$ and $\mathbf{A}^{(k+1)}$, one takes advantage of the fact that the first $(k-1)$ components of $\mathbf{u}^{(k)}$ are equal to zero.

At the kth step the column of $\mathbf{A}^{(k)}$ is chosen which will maximize $|a_{k,k}^{(k+1)}|$. Let

$$s_j^{(k)} = \sum_{i=k}^{n} (a_{i,j}^{(k)})^2, \qquad j = k, k+1, \ldots, n.$$

Then since $|a_{k,k}^{(k+1)}| = S_k$, one should choose that column for which $s_j^{(k)}$ is maximized. After $\mathbf{A}^{(k+1)}$ has been computed, one can compute $s_j^{(k+1)}$ as follows:

$$s_j^{(k+1)} = s_j^{(k)} - (a_{k,j}^{(k+1)})^2$$

since the orthogonal transformations leave the column lengths invariant. We shall next estimate the computational power of the Householder process.

4.7.2 Algebraic Complexity: the Number of Arithmetic Operations Required to Solve a Set of Linear Equations Using the Householder Reduction

First we shall count the number of arithmetic operations required to triangulize the matrix $\mathbf{A}$ of the set $\mathbf{Ax} = \mathbf{b}$ using the Householder method. Recall the sequence of operations performed at an intermediate $(r + 1)$st step, at which the partially triangular matrix

$$\mathbf{A}_r = \left[\begin{array}{c|c} \mathbf{B}_r & \mathbf{C}_r \\ \hline \mathbf{0} & \mathbf{W}_{n-r} \end{array} \right],$$

obtained at the preceding rth step is transformed into the matrix

$$\mathbf{A}_{r+1} = \left[\begin{array}{c|c} \mathbf{B}_r & \mathbf{C}_r \\ \hline \mathbf{0} & (1 - 2\mathbf{u}\mathbf{u}^*)\mathbf{W}_{n-r} \end{array} \right],$$

with one subcolumn of zeroes more than $\mathbf{A}_r$.

To compute $\mathbf{A}_{r+1}$ we have to calculate

(1) $\|\mathbf{a}\|_2 = \left(\sum_{i=1}^{r} |a_{ir}|^2 \right)^{1/2}$, where a_{ir} are the elements of $\mathbf{W}_{n-r}$.

To calculate $\|\mathbf{a}\|_2$ will require

> r multiplications
> $r - 1$ additions   (1)

and one square root evaluation.

(2) $q = [2\|\mathbf{a}\|(\|\mathbf{a}\| - \mathbf{v}_r^*\mathbf{a}_r)]^{1/2}$

where $\mathbf{a}_r$ is the rth column of $\mathbf{W}_{n-r}$. $\mathbf{v}_r$ is the rth column of identity matrix and so $\mathbf{v}_r^*\mathbf{a}_r$ does not require any multiplications or additions. Also multiplication by factor 2 can be regarded as a shifting operation and not a multiplication. To calculate q requires

> one multiplication,
> one addition,   (2)
> one square root evaluation

and one shifting operation.

(3) $\mathbf{u}_r = (\mathbf{a}_r - \|\mathbf{a}\|_2\mathbf{v}_r)/q,$

where $\mathbf{u}_r$ is the column vector $\mathbf{u}$ at the step $r + 1$. To calculate $\mathbf{u}_r$ requires

> r multiplications,
> one reciprocal   (3)

and one addition.

(4) Next, to calculate $\mathbf{u}_r^*\mathbf{a}_i$ requires r multiplications and $(r - 1)$ additions for a particular a_i, $i = 2, \ldots, r$, which gives in total

> $r(r - 1)$ multiplications
> and $(r - 1)^2$ additions.   (4)

(5) To calculate $2u_r u_r^* a_i$ requires r multiplications for a particular a_i and so we have in total,

$$r(r-1) \text{ multiplications.} \tag{5}$$

(6) Finally, to calculate $a_i - 2uu^* a_i$ requires r additions for a particular a_i and in total

$$r(r-1) \text{ additions.} \tag{6}$$

Thus, from (1) to (6), the number of multiplications required to triangularize an $n \times n$ matrix is

$$\sum_{r=2}^{n} [2r(r-1) + 2r + 1] = \frac{(n-1)(2n^2 + 5n + 9)}{3} \tag{7}$$

and the number of additions is

$$\sum_{r=2}^{n} [2r^2 - 2r + 2] = \frac{2n^3}{3} + \frac{4n}{3} - 2; \tag{8}$$

the number of reciprocal evaluations $= (n-1)$, $\qquad$ (9)

the number of square root evaluations $= 2(n-1)$, $\qquad$ (10)

and the number of shifting operations $= n - 1$. $\qquad$ (11)

Note that if the Wilkinson technique given by $4.7.1 - (12) - (23)$, is used in computation then the number of square root evaluations is halved.

At the same time as we are triangularizing $\mathbf{A}$ we must also process $\mathbf{b}$, the right hand side of the set $\mathbf{Ax} = \mathbf{b}$. If we regard $\mathbf{b}$ as an extra column, $\mathbf{a}_{r+1}$, in steps (4), (5), and (6) we can see that the number of multiplications required to transform $\mathbf{b}$ is

$$\sum_{r=2}^{n} 2r = n^2 + n - 2 \tag{12}$$

and the number of additions required is

$$\sum_{r=2}^{n} (2r - 1) = n^2 - 1 \tag{13}$$

To complete the solution of $\mathbf{Ax} = \mathbf{b}$ we must solve a triangular set of equations which will require

$$\begin{array}{ll} n(n+1)/2 & \text{multiplications} \\ n(n-1)/2 & \text{additions} \\ n & \text{reciprocals} \end{array} \tag{14}$$

The total number of arithmetic operations required to solve an $n \times n$ system using the Householder reduction is

$$\frac{2}{3}n^3 + \frac{5}{2}n^2 + \frac{17}{6}n - 5 \text{ multiplications}$$

$$\frac{3}{3}n^3 + \frac{3}{2}n^2 + \frac{5}{6}n - 3 \quad \text{additions}$$

$$2n - 1 \qquad\qquad\qquad \text{reciprocals}$$

or $\left.\begin{array}{c} 2(n-1) \\[12pt] (n-1) \end{array}\right\}$ square root evaluations

4.7.3 Matrix Inversion by the Householder Method and Its Computational Complexity

The Householder triangularization gives

$$\mathbf{PA} = \mathbf{R} \tag{1}$$

where $\mathbf{R}$ is upper triangular matrix. To invert $\mathbf{A}$ we use

$$\mathbf{A}^{-1}\mathbf{P}^{-1} = \mathbf{R}^{-1}$$

which gives

$$\mathbf{A}^{-1} = \mathbf{R}^{-1}\mathbf{P} \tag{2}$$

(1) The Householder reduction requires

$$\frac{2}{3}n^3 + n^2 + \frac{4}{3}n - 3 \qquad \text{multiplications} \tag{3}$$

(2) The inversion of $\mathbf{R}$ requires

$$\frac{n(n-1)(n+4)}{6} \text{ multiplications} \tag{4}$$

(3) $\mathbf{P} = \mathbf{P}_{n-1}\mathbf{P}_{n-2}\cdot\cdot\mathbf{P}_2\mathbf{P}_1$, where $\mathbf{P}_r = \mathbf{I} - 2\mathbf{u}_r\mathbf{u}_r^*$
To evaluate all the $\mathbf{P}_r$'s requires

$$\sum_{r=2}^{n} r^2 = \frac{(2n^3 + 3n^2 + n)}{6} - 1 \text{ multiplications}$$

To evaluate $\mathbf{P}$ by multiplying the $\mathbf{P}_r$'s requires a further

$$\sum_{r=2}^{n-1} r^2 = \frac{(2n^3 - 3n^2 + n)}{6} - 1 \text{ multiplications}$$

Thus, to calculate $\mathbf{P}$ requires

$$\frac{2n^3}{3} + \frac{1}{3}n - 2 \text{ multiplications} \tag{5}$$

(4) To multiply $\mathbf{R}^{-1}$ by $\mathbf{P}$ requires

$$\frac{n(n+1)}{2} \text{ multiplications} \tag{6}$$

From (3) to (6) the number of multiplications required to invert $\mathbf{A}$ is given by

$$\frac{3n^3}{2} + 2n^2 + \frac{3}{2}n - 5 \tag{7}$$

4.7.4 Error Analysis of Householder Reduction

We shall now study the numerical stability of the Wilkinson variant of Householder's triangularization, that is we assume that the unitary matrices $\mathbf{P}_r$, $r = 1, \ldots, n - 1$, required for the process of triangularization of matrix $\mathbf{A}$, are determined using the algorithm given by $4.7.1 - (12) - (23)$. The error analysis of the method is due to Wilkinson (1965) and we shall highlight the main points of this analysis.

The first basic step in the process of triangularization consists in computing the current unitary matrix $\mathbf{P}_r$. We wish to obtain some estimates, in terms of the matrix norms, on the error accumulation in the computed matrix, $\mathbf{P}_r^{(c)}$, i.e. we wish to estimate $\| \mathbf{P}_r^{(c)} - \mathbf{P}_r \|$.

To obtain this estimate we consider the accumulation of errors in the sequence of computational steps which calculates $\mathbf{P}_r^{(c)}$. So, suppose we have a vector $\mathbf{x}$ of length n, and we wish to construct an elementary Hermitian matrix $\mathbf{P}$, such that

$$\mathbf{P}\mathbf{x} = k\mathbf{e}_1$$

where $\mathbf{e}_1$ is the first column of the identity matrix. Suppose that $\mathbf{x}$ has floating-point components and we use fl_2 computation, i.e. floating-point with accumulation, to derive the computed matrix $\mathbf{P}$. As usual the double-length floating-point representation of a number, y, will be denoted by either $\bar{y}$ or $\mathrm{fl}_2(y)$, i.e. $\bar{y} \equiv \mathrm{fl}_2(y)$.

The steps in the computation are

(1) Calculate $\bar{a}$, where

$$\bar{a} \equiv \mathrm{fl}_2(\bar{x}_1^2 + \bar{x}_2^2 + \ldots + \bar{x}_n^2). \tag{1}$$

Assuming that the 2t-digit mantissa is retained as it will be used in later computation steps, from $2.6.1 - (16)$, we have

$$\bar{a} \equiv (x_1^2 + x_2^2 + \ldots + x_n^2)(1 + \varepsilon) = S^2(1 + \varepsilon), \tag{2}$$

where

$$|\varepsilon| < \tfrac{3}{2}n2^{-2t_2}, \qquad 2^{-t_2} = (1.06)2^{-2t}. \tag{3}$$

(2) Calculate $\bar{S}$.

$$\bar{S} \equiv \mathrm{fl}_2((\bar{a})^{1/2}) = [S^2(1 + \varepsilon)]^{1/2}(1 + \alpha), \tag{4}$$

where it may be assumed that

$$|\alpha| < (1.00001)2^{-t}, \tag{5}$$

From (4), it follows that

$$\bar{S} = S(1 + \beta),\tag{6}$$

where

$$|\beta| < (1.00002)2^{-t}.\tag{7}$$

(3) Calculate $2\bar{K}^2$

$$2\bar{K}^2 = \text{fl}_2(\bar{x}_1^2 + \bar{x}_2^2 + \ldots + \bar{x}_n^2 + \bar{x}_1\bar{S})$$
$$\equiv [x_1^2(1 + \varepsilon_1) + x_2^2(1 + \varepsilon_2) + \ldots + x_n^2(1 + \varepsilon_n) + x_1\bar{S}(1 + \varepsilon_{n+1})](1 + e)\tag{9}$$

where again from 2.6.1—(16),

$$|\varepsilon_i| < \frac{3}{2}(n + 1)2^{-2t_2}, \quad |e| \leqslant 2^{-t}\tag{10}$$

In equation 4.6.1—(20) there is a choice of sign. To ensure the numerical stability of the method, the correct choice of the sign should always give the larger of the two possible values for $2K^2$ because this quantity is used as a denominator in an operation of division, and division by small numbers, as a rule, leads to a disastrous loss of significant digits.

We re-write (9) as

$$2\bar{K}^2 = (x_1^2 + x_2^2 + \ldots + x_n^2 + x_1\bar{S})(1 + \varepsilon)(1 + e),\tag{11}$$

where

$$\varepsilon = \max\{|\varepsilon_i|\}, \quad \varepsilon < \frac{3}{2}(n + 1)2^{-2t_2}, \quad |\varepsilon| \leqslant 2^{-t}.$$

Now, assume that x_1 is positive and take

$$2K^2 = S^2 + x_1 S\tag{12}$$

(No generality is lost in taking x_1 to be positive, provided we take the stable choice). From (9) and (10) and remembering that x_1 is positive, we have

$$2\bar{K}^2 \equiv (x_1^2 + x_2^2 + \ldots + x_n^2 + x_1\bar{S})(1 + \varepsilon)(1 + e)$$
$$\equiv [S^2 + x_1 S(1 + \gamma)](1 + \varepsilon)(1 + e)\tag{13}$$

where

$$|\varepsilon| < \frac{3}{2}(n + 1)2^{-2t_2}.$$

Now, if we write

$$S^2 + x_1 S(1 + \gamma) \equiv (S^2 + x_1 S)(1 + \kappa),\tag{14}$$

then

$$\kappa = \left(\frac{x_1 S}{S^2 + x_1 S}\right)\gamma.\tag{15}$$

Since $x_1 S$ and S^2 are positive and $x_1 \leqslant S$ this gives

$$|\kappa| \leqslant |\gamma|/2 \tag{16}$$

Substituting (14) into (13) we obtain

$$2\bar{K}^2 \equiv [S^2 + x_1 S](1 + \kappa)(1 + \varepsilon)(1 + e), \tag{17}$$

giving

$$2\bar{K}^2 \equiv 2K^2(1 + \theta), \text{ where } |\theta| < (1.501)2^{-t} \tag{18}$$

(4) Finally, compute u where $u^* = (x_1 + S, x_2, x_3, \ldots, x_n)$. As we can see only the first element of u requires evaluation and we have

$$\bar{u} = \mathrm{fl}(x_1 + \bar{S}) \equiv (x_1 + \bar{S})(1 + \phi), \qquad |\phi| \leqslant 2^{-t}$$
$$\equiv [x_1 + S(1 + \gamma)](1 + \phi) \tag{19}$$

Here again the choice of sign ensures a low relative error and we have

$$\bar{u}_1 \equiv (x_1 + S)(1 + \psi)$$
$$\equiv u_1(1 + \psi), \qquad |\psi| < (1.501)2^{-t} \tag{20}$$

We can write

$$\bar{\mathbf{u}}^* \equiv (\bar{u}_1, u_2, u_3, \ldots, u_n) \equiv \mathbf{u}^* + \partial \mathbf{u}^* \tag{21}$$

where

$$\|\partial \mathbf{u}\|_2 = |u_1|\,|\psi| < (1.501)2^{-t}\|\mathbf{u}\|_2 \tag{22}$$

and thus

$$\|\bar{\mathbf{u}}\|_2 = \|\mathbf{u} + \partial \mathbf{u}\|_2 < [1 + (1.501)2^{-t}]\|\mathbf{u}\|_2 \tag{23}$$

Now

$$\mathbf{P}^{(c)} - \mathbf{P} = \frac{\bar{\mathbf{u}}\bar{\mathbf{u}}^*}{2\bar{K}^2} - \frac{\mathbf{u}\mathbf{u}^*}{2K}.$$

In order to estimate the norm $\|\mathbf{P}^{(c)} - \mathbf{P}\|_2$, consider first the norm $\left\|\dfrac{\mathbf{u}\mathbf{u}^*}{2K^2}\right\|_2$.

We have

$$\|\mathbf{u}\|_2 = \|\mathbf{u}^*\|_2 = ((x_1 + S)^2 + x_2^2 + \ldots + x_n^2)^{1/2} = (2S(S + x_1))^{1/2}, \tag{24}$$

where

$$S = \|\mathbf{x}\|_2.$$

Using (24) and (12) we find

$$\left\|\frac{\mathbf{u}\mathbf{u}^*}{2K^2}\right\| = \frac{\|\mathbf{u}\|_2\|\mathbf{u}^*\|_2}{2K^2} = \frac{2S(S + x_1)}{S(S + x_1)} = 2 \tag{25}$$

Now consider

$$\| \mathbf{P}^{(c)} - \mathbf{P} \|_2 = \left\| \frac{\bar{\mathbf{u}}\bar{\mathbf{u}}^*}{2\bar{K}^2} - \frac{\mathbf{u}\mathbf{u}^*}{2K^2} \right\|_2.$$

From (17), (21), (22) and (24) we have

$$\frac{\bar{\mathbf{u}}\bar{\mathbf{u}}^*}{2\bar{K}^2} = \frac{(\mathbf{u} + \mathrm{d}\mathbf{u})(\mathbf{u}^* + \mathrm{d}\mathbf{u}^*)}{2K^2(1 + \theta)} = \frac{\mathbf{u}\mathbf{u}^* + (\mathbf{u} + \mathbf{u}^*)\mathrm{d}\mathbf{u} + \mathrm{d}\mathbf{u}\mathrm{d}\mathbf{u}^*}{2K^2(1 + \theta)},$$

giving for the norm

$$\left\| \frac{\bar{\mathbf{u}}\bar{\mathbf{u}}^*}{2\bar{K}^2} - \frac{\mathbf{u}\mathbf{u}^*}{2K^2} \right\|_2 = \left\| \frac{\mathbf{u}\mathbf{u}^*}{2K^2}\left(1 - \frac{1}{1+\theta}\right) + \frac{(\mathbf{u}+\mathbf{u}^*)\mathrm{d}\mathbf{u} + \mathrm{d}\mathbf{u}\mathrm{d}\mathbf{u}^*}{2K^2(1+\theta)} \right\|_2$$

$$\leqslant \left\| \frac{\mathbf{u}\mathbf{u}^*}{2K^2} \right\|_2\left(1 - \frac{1}{1+\theta}\right) + 2\left\| \frac{\mathbf{u}\mathrm{d}\mathbf{u}}{2K^2} \right\|_2^{1/1+\theta} + \left\| \frac{\mathrm{d}\mathbf{u}\mathrm{d}\mathbf{u}^*}{2K^2} \right\|_2\frac{1}{1+\theta}$$

$$\leqslant 2\left(1 - \frac{1}{1+\theta}\right) + 2\frac{2\theta}{1+\theta} + \frac{\theta^2}{1+\theta}$$

$$= \frac{2\theta + 4\theta + \theta^2}{1 + \theta} < 6\theta < (9.01)2^{-t}.$$

Thus,

$$\| \mathbf{P}^{(c)} - \mathbf{P} \|_2 < (9.01)2^{-t}. \tag{26}$$

In (26) the matrix $\mathbf{P}$ may be any one of the matrices $\mathbf{P}_r$, $r = 1, \ldots, n - 1$, which are computed for the purpose of triangularization of $\mathbf{A}$.

The second basic step in the triangularization of matrix $\mathbf{A}$ is premultiplication of $\mathbf{A}$ by the approximate elementary Hermitian matrix, $\bar{\mathbf{P}}_r$, $r = 1, \ldots, n - 1$.

We again consider a general case for the matrix $\bar{\mathbf{P}}$, where $\bar{\mathbf{P}}$ stands for any of the matrices $\bar{\mathbf{P}}_r$, $r = 1, \ldots, n - 1$. The matrix $\bar{\mathbf{P}}$ has, in general, the following structure

$$\bar{\mathbf{P}} = \left[\begin{array}{c|c} \mathbf{I} & \mathbf{O} \\ \hline \mathbf{O} & \mathbf{R}^{(c)} \end{array} \right] \tag{27}$$

with appropriate dimension $(n - r) \times (n - r)$ for the matrix $\mathbf{R}^{(c)}$.

It follows that the premultiplications, $\bar{\mathbf{P}}\mathbf{A}$, leave the first r rows of $\mathbf{A}$ unaltered and the last $(n - r)$ rows are premultiplied by $\mathbf{R}^{(c)}$, and so, the error matrix

$$\mathrm{fl}_2(\bar{\mathbf{P}}\mathbf{A}) - \bar{\mathbf{P}}\mathbf{A} \tag{28}$$

has nulls in its first r rows.

The bounds for the error matrix (28) can be obtained in terms of the Euclidean norm of $\mathbf{A}$ to give

$$\| \mathrm{fl}_2(\bar{\mathbf{P}}\mathbf{A}) - \bar{\mathbf{P}}\mathbf{A} \|_E \leqslant (3.55)\| \mathbf{A} \|_E 2^{-t}. \tag{29}$$

Further, if $\bar{\mathbf{A}}_n$ is the computed triangular matrix, $\mathbf{A}_1$ the original matrix and $\mathbf{G}$

the error matrix, then for

$$\bar{\mathbf{A}}_n = \bar{\mathbf{P}}_{n-1}\bar{\mathbf{P}}_{n-2}\ldots\bar{\mathbf{P}}_1(\mathbf{A}_1 + \mathbf{G}) \tag{30}$$

the bounds on $\mathbf{G}$ can readily be verified to yield

$$\|\mathbf{G}\|_E \leqslant 3.55(n-1)[1 + (9.01)2^{-t}]^{n-2}2^{-t}\|\mathbf{A}_1\|_E. \tag{31}$$

Similarly, if

$$\bar{\mathbf{b}}_n = \bar{\mathbf{P}}_{n-1}\bar{\mathbf{P}}_{n-3}\ldots\bar{\mathbf{P}}_1(\mathbf{b}_1 + k) \tag{32}$$

where $\bar{\mathbf{b}}_n$ is the final computed right-hand side vector, then the bounds on k are given as

$$\|k\|_2 < 3.55(n-1)[1 + (9.01)2^{-t}]^{n-1}2^{-t}\|\mathbf{b}_1\|_2. \tag{33}$$

Comparing formulae (31) and (33) with the bounds on the relative error in the solution $\mathbf{x}$, of the set $\mathbf{Ax} = \mathbf{b}$, due to perturbations in either $\mathbf{A}$ or $\mathbf{b}$, (cf. Appendix A) we conclude that the Householder reduction is a numerically stable method.

4.8 How the Efficiency of a Direct Method May be Increased

We have seen that various direct methods for solving sets of linear equations all require the number of multiplications and additions of $O(n^3)$ but the different coefficients in the leading term of the corresponding expressions for this number, make the actual number differ significantly. For instance, the Gaussian elimination requires in total fewer arithmetic operations than the other methods, except for the Cholesky decomposition.

However, the latter is not as general as the other methods. On the basis of the number of arithmetic operations required, one may speak of a 'fast' method or a 'slow' method. So, it is only natural to raise a general question: What is the optimal number of arithmetic operations which is required by an algorithm to solve a general set of n linear equations in n unknowns? The answer to this question depends to a great extent on what kind of operations such an algorithm is allowed to use. Klyuev and Kokovkin–Shcherbak (1965) have proved that $(n^3/3) + n^2 - (n/3)$ multiplications and $(n^3/3) + (n^2/2) - (5n/6)$ additions (the same as the number of operations in the Gaussian elimination) are the minimal number of these operations that are required to solve an $n \times n$ system of linear equations, when using methods which employ linear combinations of rows and columns only. However, Winograd (1967, 1968, 1970) and Strassen (1969) have put forward methods which do not employ just linear combinations of rows and columns. Winograd's method takes only $(n^3/6)$ multiplications but $(n^3/2)$ additions. Since we assume that multiplication takes longer than addition (which is the case for most modern computers), this method should be faster than Gaussian elimination. Strassen's method requires less than $6n^{\log_2 7}$ multiplications and additions. This method would be even faster than Winograd's for large n. We shall consider both methods. We shall see that for problems of very large size both the Winograd and Strassen methods are more 'efficient' than Gaussian elimination, in terms of computational complexity.

This being the case however, for practical usefulness of the methods, further ideas are needed for enhancement of their computational algorithms. However, the methods are of principal importance in the development of the theory of computational algorithms because they demonstrate conceptually new ways of 'trading-off' more 'expensive' arithmetic operations for 'cheaper' mathematical manipulation of the problem data. This general concept of trading-in one kind of operation for another commands at present, wide support as one of the ways to improve the efficiency of computational algorithms.

4.9 The Winograd Method

4.9.1 The Winograd Identity

Consider the identity

$$x_1 y_1 + x_2 y_2 = (x_1 + y_2)(x_2 + y_1) - x_1 x_2 - y_1 y_2. \tag{1}$$

Now, the Winograd identity is an expansion of (1) for the even number, $n = 2k$, of pairwise products

$$\sum_{i=1}^{2k} x_i y_i = \sum_{u=1}^{k} (x_{2u-1} + y_{2u})(x_{2u} + y_{2u-1}) - \sum_{u=1}^{k} x_{2u-1} x_{2u} - \sum_{u=1}^{k} y_{2u-1} y_{2u}. \tag{2}$$

Consider the computation of $c = Ab$ where A is an $m \times n$ matrix and b is an n-vector. Assume that n is even, i.e. $n = 2k$. (If n is odd, simply apply the algorithm to the matrix A' and vector b' of dimensions $m \times (n+1)$ and $(n+1)$, respectively, where $a'_{i,n+1} = 0$ for all i and $b'_{n+1} = 0$.)

Using the Winograd identity, we can write

$$c_i = \sum_{j=1}^{n} a_{ij} b_j$$

$$= \sum_{u=1}^{k} (a_{i,2u-1} + b_{2u})(a_{i,2u} + b_{2u-1})$$

$$- \sum_{u=1}^{k} a_{i,2u-1} a_{i,2u} - \sum_{u=1}^{k} b_{2u-1} b_{2u}, \qquad i = 1, \dots, m. \tag{3}$$

From (3) it follows that to compute all c_i requires

$$\frac{3}{2} nm \qquad \text{multiplications}$$

and $\left(\frac{5}{2} n - 1\right) m$ additions. $\tag{4}$

In comparison the standard method requires only nm multiplications and $(n-1)m$ additions. However, method (3) becomes advantageous over the standard method of computing c_i, when the same matrix has to be multiplied by many vectors as is, for example, the case in computing the product of two matrices. The savings in the number of arithmetic operations comes from the fact that the product $f_i = \sum_{u=1}^{k} a_{i,2u-1} a_{i,2u}$ has to be calculated only once.

4.9.2 The Winograd Matrix Multiplication Algorithm

Let $\mathbf{A}$ and $\mathbf{B}$ be two matrices of dimension $m \times n$ and $n \times p$, respectively. The algorithm for computing $\mathbf{C} = \mathbf{AB}$ using the Winograd identity is given as

(assuming $n = 2k$)

1. Compute $\quad f_i = \sum_{u=1}^{k} a_{i,2u-1} a_{i,2u}, \qquad i = 1, \ldots, m.$ $\qquad$ (1)

2. Compute $\quad g_j = \sum_{u=1}^{k} b_{2u-1,j} b_{2u,j}, \qquad j = 1, \ldots, p.$ $\qquad$ (2)

3. Compute $\quad c_{ij} = \sum_{u=1}^{k} (a_{i,2u-1} + b_{2u,j})(a_{i,2u} + b_{2u-1,j}) - f_i - g_j,$ $\qquad$ (3)

$$i = 1, \ldots, m,$$
$$j = 1, \ldots, p.$$

The total number of operations required by the process (1)–(3) is

$$\frac{nmp}{2} + \frac{n}{2}(m + p) \qquad \text{multiplications}$$

$$\text{and } \frac{3}{2}nmp + mp + \left(\frac{n}{2} - 1\right)(m + p) \qquad \text{additions.} \qquad (4)$$

This number may be compared with nmp multiplications and $(n - 1)mp$ additions required by the standard method. We note that with the use of Winograd's identity, the number of multiplications is about halved while some price for this reduction is paid in terms of an increased number of additions required.

4.9.3 Optimality of Winograd's Formula

We have seen that the Winograd method of matrix multiplication is founded on what we may call the basic formula, (1), that is, it uses the identity involving the sum of two pairwise products.

Now, we raise the question of whether it is possible to use as a basic formula for deriving some matrix multiplication algorithm similar to Winograd's, the identity involving, say, three pairwise products, i.e. $a_1 b_1 + a_2 b_2 + a_3 b_3$, or more. Would the algorithms based on such identities allow even greater savings in the number of operations than the Winograd method? In fact, it has been shown by Harter (1972) that the answer to this question is no. The Winograd formula is the best of its kind.

The Harter theorem states:

Suppose that $\mathbf{p}, \mathbf{q}, \mathbf{r},$ and $\mathbf{s}$ are vectors of length k and f and g are functions such that for all vectors $\mathbf{a}$ and $\mathbf{b}$ of length k the following identity holds

$$\sum_{i=1}^{k} a_i b_i = \sum_{i=1}^{k} (p_i a_i + q_i b_i) \sum_{j=1}^{k} (r_j a_j + s_j b_j) + f(a_i) + g(b_i), \tag{1}$$

then $k \leqslant 2$.

4.9.4 Algebraic Complexity: the Number of Arithmetic Operations Required by Winograd's Method

We first consider the matrix inversion.

Matrix Inversion
Suppose that an $n \times n$ matrix $\mathbf{A}$ is to be inverted. We can view $\mathbf{A}$ as an $m \times m$ matrix whose entries are $k \times k$ matrices. Gaussian elimination is then performed on this $m \times m$ matrix and whenever a multiplication of two $k \times k$ matrices is required use of Winograd's identity is made.

From 4.3—(18) the number of multiplications of $k \times k$ matrices required to invert $\mathbf{A}$ is

$$m^3 - 1$$

To multiply two $k \times k$ matrices using Winograd's identity requires, from 4.9.2—(8) $k^3/2 + k^2$ multiplications.

(We assume that the vectors f_i and g_j of 4.9.2—(1) and (2), respectively, are generated as required and are not stored even though some of the $k \times k$ entries are used more than once in computation.)

Further, m reciprocals are required to invert $\mathbf{A}$. These reciprocals are treated as inversions of $k \times k$ matrices. The inversions are carried out by ordinary Gaussian elimination and this will require $k^3 - 1$ multiplications plus k reciprocals giving $k^3 + k - 1$ multiplicative operations. Thus the number of multiplications required, to invert $\mathbf{A}$ is

$$(m^3 - 1)\left(\frac{k^3}{2} + k^2\right) + m(k^3 + k - 1)$$

which is approximately equal to

$$\frac{n^3}{2} + \frac{n^3}{k} + nk^2 + n \tag{1}$$

We wish to minimise this function with respect to k. Write

$$F = \frac{n^3}{2} + \frac{n^3}{k} + nk^2 + n,$$

and the minimum of the function will occur when

$$\frac{dF}{dk} = -\frac{n^3}{k^2} + 2nk = 0,$$

i.e. when $k = \left(\dfrac{n^2}{2}\right)^{1/3}$, $\qquad$ (2)

giving $F = \dfrac{n^3}{2} + 1.89 n^{7/3} + n$ $\qquad$ (3)

We now turn to the number of additions required to invert **A**.

(1) From 4.3—(18) $(m^3 - 2m^2 + m)$ additions of $k \times k$ matrices are required. Each addition of two $k \times k$ matrices requires k^2 additions. Thus the number of additions required by this part of the inversion is

$$(m^3 - 2m^2 + m)k^2 \qquad (4)$$

(2) From 4.9.2—(8) each multiplication of two $k \times k$ matrices requires $(3k^2/2) + 2k^2 - 2k$ additions. Thus the number of additions contributed by the matrix multiplications is

$$(m^3 - 1)\left(\frac{3k}{2} + 2k^2 - 2k\right) \qquad (5)$$

(3) Further, each inversion of a $k \times k$ matrix requires $(k^3 - 2k^2 + k)$ additions and so the m inversions required give

$$m(k^3 - 2k^2 + k) \text{ additions} \qquad (6)$$

Thus from (4), (5), and (6) the total number of additions required to invert A is

$$(m^3 - 2m^2 + m)k^2 + (m^3 - 1)\left(\frac{3k^3}{2} + 2k^2 - 2k\right) + m(k^3 - 2k^2 + k)$$

which is approximately (i.e. ignoring the 1 of $m^3 - 1$)

$$\frac{3n^3}{2} + \frac{3n^3}{k} - 2n^2 - \frac{2n^3}{k^2} + nk^2 - nk + n. \qquad (7)$$

From (2), we chose the optimum k to reduce the number of multiplications to be $(n^2/2)^{1/3}$. Substituting this value of k in (7) we obtain

$$\frac{3n^3}{2} + \left(\frac{7}{2}\right)^{2/3} n^{7/3} - 2n^2 + O(n^{5/3}) \quad \text{additions} \qquad (8)$$

Solution of a Set of Linear Equations

Next, we consider the computational complexity of an algorithm which solves the set $\mathbf{Ax = b}$ using the Winograd matrix multiplication method.

We assume that the set $\mathbf{Ax = b}$ is solved by first decomposing $\mathbf{A}$ into the product $\mathbf{LU}$ and then applying back-substitution.

(1) The decomposition of **A** into **LU** requires $m(m^2 - 1)/3$ multiplications and m reciprocals of $k \times k$ matrices. Thus the decomposition phase requires

$$\frac{1}{3}m(m^2 - 1)\left(\frac{k^3}{2} + k^2\right) + m(m^3 + k - 1) \quad \text{multiplications} \tag{9}$$

(2) The back-substitution process requires m^2 multiplications of $k \times k$ matrices with k–vectors. Each of these multiplications requires $k^2 + (k/2)$ actual multiplications. Thus the back-substitution phase requires

$$m^2\left(k^2 + \frac{k}{2}\right) \quad \text{multiplications} \tag{10}$$

In total, the number of multiplications required to solve a set of n linear equations is

$$\frac{1}{3}(m^3 - m)\left(\frac{k^3}{2} + k^2\right) + m^2\left(k^2 + \frac{k}{2}\right) + m(k^3 + k - 1)$$

$$= \frac{n^3}{6} + \frac{n^3}{3k} + n^2 + \frac{5nk^2}{6} + \frac{n^2}{2k} - \frac{nk}{3} + n - m. \tag{11}$$

Simplifying somewhat by preserving only the first four terms of (11), we write

$$F = \frac{n^3}{6} + \frac{n^3}{3k} + n^2 + \frac{5nk^2}{6}, \tag{12}$$

and the minimum of this function is

$$\frac{\mathrm{d}F}{\mathrm{d}k} = -\frac{n^3}{3k^2} + \frac{5nk}{3} = 0 \tag{13}$$

giving

$$k = \left(\frac{n^2}{5}\right)^{1/3} \tag{14}$$

If we substitute this value of k into (11) we obtain

$$F = \frac{n^3}{6} + 0.86\,n^{7/3} + n^2 + 0(n^{5/3}) \tag{15}$$

We now turn to the number of additions required.

(1) The decomposition of **A** into **LU** requires $\frac{1}{3}m(m^2 - 1)$ multiplications of $k \times k$ matrices each of which will require $(3k^3/2 + 2k^2 - 2k)$ additions. Also $[(m^3/3) - (m^2/2) - m/6]$ additions of $k \times k$ matrices are required and each addition will require k^2 actual additions. The m inversions will each require $k^3 - 2k^2 + k$ additions. Thus the total number of additions required to triangularize **A** is

$$\frac{1}{3}m(m^2 - 1)\left(\frac{3k^2}{2} + 2k^2 - 2k\right) + \left(\frac{m^3}{3} - \frac{m^2}{2} + \frac{m}{6}\right)k^2 + m(k^3 - 2k^2 + k) \tag{16}$$

(2) In the back-substitution process m^2 multiplications of $k \times k$ matrices with k-column vectors are required. Each matrix–vector multiplication requires $\left(2k^2 + \dfrac{k}{2} - 1\right)$ additions. The $m(m-1)$ additions of k-column vectors each require k additions. The back-substitution phase thus requires the following number of additions:

$$m^2\left(2k^2 + \frac{k}{2} - 1\right) + m(m-1)k \tag{17}$$

Combining (16) and (17), the total number of additions is essentially given by

$$\frac{n^3}{2} + \frac{n^3}{k} + \frac{nk^2}{2} + \frac{3n^2}{2} \tag{18}$$

Substituting the optimum value of $k = (n^2/5)^{1/3}$ from (14), we obtain

$$\frac{n^3}{2} + 1.88\, n^{7/3} + \frac{3}{2}n^2 + O(n^{5/3}) \tag{19}$$

From (3), (8), (15), and (9) we can see that using Winograd's method, it is possible to reduce by one half the number of multiplications required to invert a matrix and to solve a set of equations. However, the number of additions required by both of the above problems is increased by a factor of $\frac{3}{2}$.

4.9.5 Error Analysis of Winograd's Identity

In this analysis we follow closely the method of Brent (1970). To find the inner product $\sum\limits_{i=1}^{n} x_i y_i$ using the Winograd method, for n even, we compute

$$\bar{w} = \mathrm{fl}(\bar{p} - \bar{r}), \tag{1}$$

where

$$\bar{p} = \mathrm{fl}\left(\sum_{i=1}^{n/2} (x_{2i-1} + y_{2i})(x_{2i} + y_{2i-1})\right), \tag{2}$$

$$\bar{r} = \mathrm{fl}(\bar{f} + \bar{g}), \tag{3}$$

$$\bar{f} = \mathrm{fl}\left(\sum_{i=1}^{n/2} x_{2i-1} x_{2i}\right) \tag{4}$$

and

$$\bar{g} = \mathrm{fl}\left(\sum_{i=1}^{n/2} y_{2i-1} y_{2i}\right) \tag{5}$$

Now, we denote the error in f by e_f, so that

$$e_f = f - \mathrm{fl}\left(\sum_{i=1}^{n/2} x_{2i-1} x_{2i}\right), \tag{6}$$

and define the errors e_g, e_r, e_p, e_w in a similar way. Next, we assume that the computations are carried out under conditions of normalized t-digit rounded floating-point arithmetic, and further assume that n is such that $n2^{-t} \leqslant 1$, which for our purposes may be more conveniently written as

$$\frac{n2^{1-t}}{2} \leqslant 1. \tag{7}$$

Now, let

$$t_1 = t - \frac{n2^{1-t}}{\log 2}. \tag{8}$$

This choice of t_1 ensures the inequality

$$2^{t-t_1} \geqslant 1 + n2^{1-t}. \tag{9}$$

In practice condition (7) will usually be satisfied very easily and the distinction between t and t_1 may be ignored.

Let $a = \max|x_i|$ (10)

and $b = \max|y_i|$ (11)

The following inequalities can now be proved. The error in the computed inner product $S = \sum\limits_{i=1}^{n} x_i y_i$ for $n \geqslant 1$ is bounded by

(1) $\quad |e_s| \leqslant \frac{1}{4}ab(n^2 + 3n)2^{1-t_1}$, (12)

where S stands for 'standard method' if S is computed in the usual way.

(2) $\quad |e_w| \leqslant \frac{1}{8}(a + b)^2(n^2 + 16n)2^{1-t_1}$, (13)

where W stands for 'Winograd's method', if S is computed by Winograd's method.

(3) $\quad |e_w| \leqslant \frac{3}{8}ab(n^2 + 17n)2^{1-t_1}$ (14)

if S is computed by Winograd's method and $a = b$.

These three bounds can be proven as follows

(1) Due to Wilkinson we have

$$e_n = \sum_{i=1}^{n} x_i y_i e_i, \tag{15}$$

where

$$|e_1| \leqslant (1 + \tfrac{1}{2}2^{1-t})^n - 1$$

and

$$|e_i| \leqslant (1 + \tfrac{1}{2}2^{1-t})^{n+2-i} - 1, \qquad \text{for } 2 \leqslant i \leqslant n. \tag{16}$$

It follows that

$$|e_g| = |S - \bar{S}| \leq ab \left[\left(\sum_{k=2}^{n} (1 + \tfrac{1}{2}2^{1-t})^k - 1 \right) + (1 + \tfrac{1}{2}2^{1-t})^n - 1 \right] \tag{17}$$

Now consider $(1 + z)^k$ and expand it as a binomial series to obtain

$$(1 + z)^k - 1 = kz + \frac{k(k-1)}{2}z^2 \left(1 + \frac{(k-2)z}{3} \right) + O(z^4)$$

$$\leq kz + \frac{k(k-1)z^2}{2} \left[1 - \frac{(k-2)z}{3} \right]^{-1}. \tag{18}$$

The maximum value of $\left[1 - \dfrac{(k-2)z}{3} \right]^{-1}$ will occur when k and z are at their maximum.

If we take the bounds on k and z as

$$k \leq n + 2 \quad \text{and} \quad z \leq \frac{1}{n} = \frac{1}{2}2^{1-t}, \tag{19}$$

then

$$\left[1 - \frac{(k-2)z}{3} \right]^{-1} \leq \frac{3}{2}. \tag{20}$$

Substituting (18) into (17) and using (20) we obtain

$$|e_S| \leq abz \frac{n^2 + 3n - 2}{2} \left(1 + \frac{nz}{2} \right). \tag{21}$$

From (9)

$$\left(1 + \frac{nz}{2} \right) \leq \left(1 + \frac{n^{\frac{1}{2}-t}}{4} \right) \leq 2^{t-t_1}$$

Thus

$$|e_S| \leq \tfrac{1}{4}ab(n^2 + 3n - 2)2^{1-t}2^{t-t_1}$$

$$\leq \tfrac{1}{4}ab(n^2 + 3n)2^{1-t_1},$$

which proves the bound (12).

To prove (13), we first suppose that $n = 2m$ is even, so from (7),

$$\tfrac{1}{2}m2^{1-t} \leq \tfrac{1}{2} \tag{22}$$

Now we have

$$|e_f| \leq a^2 \left[\left(\sum_{k=2}^{m} (1 + z)^k - 1 \right) + (1 + z)^m - 1 \right], \tag{23}$$

where the same bounds as given by (19), are valid for k and z. Using arguments analogous to those of (18)–(20), we obtain

$$|e_f| \leq \tfrac{1}{2}a^2z(m^2 + 3m - 2)(1 + \tfrac{2}{5}mz)$$

$$\leq \tfrac{6}{5}a^2m^2z, \tag{24}$$

which is obvious when $m = 1$ and follows from (22) when $m > 1$. Similarly, we get

$$|e_g| \leqslant \tfrac{6}{5}b^2 m^2 z. \tag{25}$$

From (6) it is also easy to see that

$$|f| \leqslant a^2 m + |e_f|,$$

and so

$$|f| \leqslant a^2 m(1 + \tfrac{6}{5}mz). \tag{26}$$

Again, a similar estimate, with a replaced by b, is valid for g. Hence, we can write

$$|f| + |g| \leqslant (a^2 + b^2)m(1 + \tfrac{6}{5}mz). \tag{27}$$

Further, the error in r is

$$\begin{aligned}|e_r| &\leqslant |e_f| + |e_g| + z(|f| + |g|) \\ &\leqslant \tfrac{1}{2}z(a^2 + b^2)(m^2 + 5m - 2)(1 + 2mz).\end{aligned} \tag{28}$$

Similarly it can be shown that the error in p is given by

$$|e_p| \leqslant \tfrac{1}{2}z(a + b)^2(m^2 + 7m - 2)(1 + \tfrac{7}{6}mz) \tag{29}$$

Now,

$$|e_w| \leqslant |e_p| + |e_r| + z(|p| + |r|),$$

where

$$z \leqslant \tfrac{1}{2}2^{1-t},$$

and with the use of (28) and (29), we obtain

$$\begin{aligned}|e_{\mathrm{w}}| &\leqslant z(a + b)^2(m^2 + 8m - 2)(1 + 3mz) \\ &\leqslant \frac{1}{2}(a + b)^2\left(\frac{n^2}{4} + 4n - 2\right)\left(1 + \frac{3}{4}n2^{1-t}\right)2^{1-t},\end{aligned} \tag{30}$$

where we used $m = \dfrac{n}{2}$ and $z \leqslant \dfrac{1}{2}2^{1-t}$.

Finally, since from (9) we have

$$1 + \tfrac{3}{4}n2^{1-t} \leqslant 2^{t-t_1},$$

it follows that

$$|e_{\mathrm{W}}| = |e_w| \leqslant \tfrac{1}{8}(a + b)^2(n^2 + 16n)2^{1-t_1}.$$

This completes the proof of the bound given in (13).
For $a = b$ a somewhat stronger result can be derived, namely

$$|e_w| \leqslant \tfrac{3}{8}ab(n^2 + 17n)2^{1-t_1}. \tag{31}$$

(13) shows that Winograd's method is very bad numerically if a/b is either large

or small compared to unity, for then $(a + b)^2 \gg ab$. By comparison the bound for the standard method is not influenced by a/b. For Winograd's method to be effective scaling must be employed. Ignoring the simple cases of $a = 0$ and $b = 0$, there is an integer k such that:

$$2^{-1/2} \leqslant \frac{a}{b} 2^k < 2^{1/2} \tag{32}$$

If we replace x by $2^k x$, computed without rounding errors, then using the Winograd identity and multiplying the final result by 2^{-k}, we get

$$|e_W| \leqslant \tfrac{1}{8}(2 + 2^{1/2} + 2^{-1/2})ab(n^2 + 16n)2^{1-t_1} \tag{33}$$

and this expression lacks the term $(a + b)^2$.

If we require the matrix product $\mathbf{C} = \mathbf{AB}$, then even a crude scaling, say, $\mathbf{A}$ into $2^k \mathbf{A}$ and $\mathbf{B}$ into $2^{-k} \mathbf{B}$, such that

$$2^{-1} \leqslant \frac{2^k a}{2^{-k} b} < 2,$$

followed by the application of the Winograd algorithm, will give a result $\mathbf{AB} + \mathbf{E}$ where

$$\max|e_{ij}| \leqslant \tfrac{9}{8} ab(n^2 + 16n)2^{-t_1}, \tag{34}$$

$$a = \max|a_{ij}|, \qquad b = \max|b_{ij}|.$$

We conclude that Winograd's method with scaling is nearly as accurate as the usual method (without accumulation of inner products) and the scaling takes time of $O(n^2)$ compared with the total time of $O(n^3)$.

4.9.6 Numerical Stability of LU Decomposition Obtained Using Winograd's Matrix Multiplication Algorithm

We have seen that numerical stability of the Winograd matrix multiplication algorithm may only be guaranteed if an a priori scaling by the appropriate factor 2^k is carried out on the matrices-multipliers.

Winograd (1968) suggests that his algorithm can be used to solve the set $\mathbf{Ax} = \mathbf{b}$ by partitioning the $n \times n$ matrix $\mathbf{A}$ into $m \times m$ submatrices of dimension $k \times k$ and, then, applying the algorithm where appropriate. Brent (1970) has, however, indicated that for practical implementation of this idea clarification is needed on what, if any, form of pivoting can be used. He suggests one algorithm which uses scaling and partial pivoting to obtain an error bound which is satisfactory unless the so called growth factor, g, is too large. The growth factor, g, is defined by

$$g = \max_{i,j}|u_{ij}|, \tag{1}$$

where u_{ij} are elements of matrix $\mathbf{U}$ in $\mathbf{A} = \mathbf{LU}$. We may point out that g

also affects the error bound of Gaussian elimination with partial pivoting, and in the same way.

LU *Decomposition Algorithm which uses the Winograd Identity, and Scaling of* **L** *and* **U** *as proposed by Brent*

Given a non-singular matrix **A** of dimension N, the algorithm decomposes it into the product of two matrices, $\mathbf{A} = \mathbf{LU}$, where **L** is a lower triangular and **U** is an upper triangular matrices, respectively. The matrix **A** is retained at the end of the process, and the results are stored in the auxiliary array $\mathbf{B}(1:N, 1:N)$, where the lower triangle below the main diagonal, contains the elements of **L** and the upper triangle including the main diagonal contains the elements of **U**.

1. Set $B(I, J) \leftarrow A(I, J)$ for $I = 1, \ldots, N$; $J = 1, \ldots, N$.
2. Set $K \leftarrow 1$.
3. (Partial Pivoting.) Find the largest in absolute value element of $(B(L, K), L = K, K + 1, \ldots, N)$. Let it be $B(P, K)$. If $P > K$, exchange the rows K and P.
4. (Scaling Kth row of **U**.) Find the largest in absolute value element of $(B(K, L), L = 1, \ldots, N)$. Let it be $B(K, P)$.
   Then compute $T \leftarrow \text{SQRT}(\text{ABS}(B(K, P)))$.
   For $L = 1, 2, \ldots, N$ compute $B(K, L) \leftarrow B(K, L)/T$
5. (Computing Kth column of **L**.) For $J = K + 1, K + 2, \ldots, N$ compute

$$B \leftarrow B(J, K) - \sum_{L=1}^{K-1} B_{,}(J, L) \ B(L, K),$$

   where the sum is computed using the Winograd identity 4.9.1—(2).

   Compute $B(J, K) \leftarrow B/B(K, K)$.

6. (Scaling Kth column of **L**.)
   For $L = K + 1, \ldots, N$ compute

$$B(L, K) \leftarrow T * B(L, K)$$

7. Set $I \leftarrow K + 1$.
8. (Computing Ith row of **U**.)
   For $J = K + 1, K + 2, \ldots, N$ compute

$$B(I, J) \leftarrow B(I, J) - \sum_{L=1}^{K-1} B(I, L) * B(L, J)$$

   where the sum is computed with the use of the Winograd identity 4.9.1—(2).
9. $K \leftarrow K + 1$.
10. Test for the end of matrix, $K > N - 1$?
    If false, go back to Step 3.

11. Unscale matrices $\mathbf{L}$ and $\mathbf{U}$.

12. Terminate.

Below is given

The Error Analysis of the Brent Algorithm

We seek to decompose the matrix $\mathbf{A}$ so that

$$\mathbf{PA} + \mathbf{E} = \mathbf{LU} \tag{2}$$

where $\mathbf{P}$ is a permutation matrix, $\mathbf{L}$ is lower triangular with unit diagonal, $\mathbf{U}$ is upper triangular and $\mathbf{E}$ is the error matrix.

$$\text{Let} \qquad u'_{ij} = \begin{cases} 0, & \text{if } d_i = 0, \\ \dfrac{u_{ij}}{d_i}, & \text{if } d_i \neq 0, \end{cases} \tag{3}$$

$$m'_{ij} = m_{ij} d_j, \tag{4}$$

$$a'_{ij} = (\mathbf{PA})_{ij}, \tag{5}$$

where d_i is the scaling factor which is defined by Brent as

$$d_i = \left[2 \max_j |u_{ij}| \right]^{1/2} \tag{6}$$

and where u'_{ij}, m'_{ij}, a'_{ij} are the elements of matrices $\mathbf{U}'$, $\mathbf{L}'$, $\mathbf{A}'$ satisfying

$$\mathbf{L}'\mathbf{U}' = \mathbf{A}' + \mathbf{E}.$$

We have

$$m'_{ik} u'_{kj} = m_{ik} u_{kj} \tag{7}$$

which holds even if d_k vanishes, as then both sides of (7) vanish.

By definition

$$a'_{ij} + e_{ij} = \sum_{k=1}^{\min(i,j)} m'_{ik} u'_{kj}, \tag{8}$$

but for $i \leq j$ we have

$$u_{ij} = \mathrm{fl}\left(a'_{ij} - \mathrm{fl_W} \sum_{k=1}^{i-1} m'_{ik} u'_{kj} \right) \tag{9}$$

where $\mathrm{fl_W}$ indicates that the calculation of the inner product is performed using Winograd's identity.

Pivoting ensures that $|m_{ik}| < 1$ and scaling ensures that $|d_k| < 2^{1/2} g^{1/2}$. Thus

$$|m'_{ik}| < 2^{1/2} g^{1/2}. \tag{10}$$

Also if $d_k \neq 0$, then

$$|u_{kj}| = \left|\frac{u_{kj}}{d_k}\right| = \left|\frac{u_{kj}}{d_k^2}\right|^{1/2} |u_{kj}|^{1/2} = \left[\frac{|u_{kj}|}{2\max_j |u_{kj}|}\right]^{1/2} |u_{kj}|^{1/2},$$

so $\quad |u_{kj}| \leqslant 2^{-1/2} g^{1/2}.$ (11)

The relation (11) holds if $d_k = 0$, for then $u_{kj}' = u_{kj} = 0$. Using 4.9.5—(14) and setting $a = b = \sqrt{2g}$, $n = i - 1$, we obtain

$$\left| \text{fl}_\text{W} \sum_{k=1}^{i-1} m_{ik}' u_{kj}' - \sum_{k=1}^{i-1} m_{ik}' u_{kj}' \right| < \frac{3}{8} g(i^2 + 15i - 16)2^{2-t_1},$$ (12)

but from (9) we have

$$u_{ij}(1 + e) = a_{ij}' - \text{fl}_\text{W} \sum_{k=1}^{i-1} m_{ik}' u_{kj}',$$ (13)

where

$$|e| \leqslant \tfrac{1}{2} 2^{1-t}.$$

Hence, using $|u_{ij}| \leqslant g \leqslant \frac{3}{4}(2g)$, we obtain

$$|a_{ij}' - u_{ij} - \sum_{k=1}^{i-1} m_{ik}' u_{kj}'| \leqslant |u_{ij}| \, |e| + \frac{3}{8} g(i^2 + 15i - 16)2^{2-t_1}$$

$$\leqslant \frac{3}{8} g(i^2 + 15i - 16)2^{2-t_1},$$ (14)

for $2 \leqslant i \leqslant j \leqslant n$.

Observing that $u_{ij}' = m_{ii} u_{ij}$, from (7), (8) and (14) we get

$$e_{ij} = -a_{ij}' + \sum_{k=1}^{\min(i,j)} m_{ik}' u_{kj}' = -a_{ij}' + \sum_{k=1}^{i} m_{ik}' u_{kj}',$$

since $i \leqslant j$.

Further, since $u_{ij} = m_{ii} u_{ij} = m_{ii}' u_{ij}'$, we get

$$e_{ij} = -a_{ij}' + u_{ij} + \sum_{k=1}^{i-1} m_{ik}' u_{kj}',$$ (15)

which, finally, gives

$$|e_{ij}| \leqslant \tfrac{3}{8} g(i^2 + 15i - 15)2^{2-t_1} \qquad \text{for } 2 \leqslant i \leqslant j \leqslant n.$$

In a similar fashion we can bound e_{ij} for $i > j$.
We have

$$m_{ij} = \begin{cases} 0, & \text{if } u_{jj} = 0, \\[2mm] \text{fl}\dfrac{\left(a_{ij} - \text{fl}_\text{W} \sum_{k=1}^{j-1} m_{ik}' u_{kj}'\right)}{u_{jj}}, & \text{otherwise,} \end{cases}$$ (16)

for $1 \leqslant j < i \leqslant n$.

Thus

$$m_{ij}u_{jj}(1 + e) = a_{ij} - \text{fl}_W \sum_{k=1}^{j-1} m'_{ik}u'_{kj} \tag{17}$$

where $|e| \leqslant 2^{1-t_1}$.

If $u_{jj} = 0$ then both sides of (17) vanish.
Now, proceeding as before we obtain

$$\left| a_{ij} - m_{ij}u_{jj} - \sum_{k=1}^{j-1} m'_{ik}u'_{kj} \right| \leqslant \frac{3}{8}g(j^2 + 15j - 14)2^{2-t_1} \tag{18}$$

for $1 \leqslant j < i \leqslant n$.
From (18), (7) and (8), the error bound is given by

$$|e_{ij}| \leqslant \frac{3}{8}g(j^2 + 15j - 14)2^{2-t_1} \tag{19}$$

for $1 \leqslant j < i \leqslant n$.
From (15) and (19) we obtain the uniform bound

$$|e_{ij}| \leqslant \frac{3}{8}g(n^2 + 15n)2^{2-t_1} \tag{20}$$

For the usual triangular decomposition, cf. Wilkinson (1963), we have

$$|e_{ij}| \leqslant \frac{1}{8}g(n^2 + n)2^{2-t_1} \tag{21}$$

The error bound for Winograd's method is, thus, worse by a factor of 3. Without scaling it can only be guaranteed that $|m_{ij}| < 1$ and $|u_{ij}| \leqslant g$. This would give a factor of $(1 + g)^2$ in the error bound, i.e.

$$|e_{ij}| \leqslant \frac{1}{8}(1 + g)^2(n^2 + 14n)2^{1-t_1} \tag{22}$$

so for any hope of success $\mathbf{A}$ has to be scaled so that max $|a_{ij}|$ is approximately unity, for otherwise g could be arbitrarily large or small.

4.9.7 Cholesky Decomposition Using Winograd's Algorithm

Winograd's matrix multiplication algorithm can be used to obtain the Cholesky decomposition $\mathbf{A} = \mathbf{L}\mathbf{L}^T$ with $n^3/12$ multiplications and $n^3/4$ additions instead of the usual $n^3/6$ of each. There is no need for pivoting and, if $|a_{ij}| \leqslant 1$ and the square roots are computed with a relative error of 2^{1-t}, then

$$\mathbf{L}\mathbf{L}^T = \mathbf{A} + \mathbf{E} \tag{1}$$

where

$$|e_{ij}| < \frac{3}{8}(n^2 + 15n)2^{1-t_1}. \tag{2}$$

4.10 The Strassen Method

In 1969 Strassen suggested an algorithm for computing the product of two square matrices $\mathbf{A}$ and $\mathbf{B}$ of order n, which requires less than $4.7\,n^{\log_2 7}$ arithmetic operations. The algorithm can be used to induce algorithms for inverting a matrix of order n, solving a system of n linear equations in n unknowns, computing a determinant of order n, etc., all requiring the total number of arithmetic operations proportional to $n^{\log_2 7}$. We shall first consider the basic Strassen algorithm which finds the product of two matrices.

4.10.1 Matrix Multiplication

Let $\mathbf{A}$ and $\mathbf{B}$ be two matrices of order $n = m2^{k+1}$ and their product be denoted by $\mathbf{C}$.

We write

$$\mathbf{A} = \begin{bmatrix} \mathbf{A}_{11} & \mathbf{A}_{12} \\ \mathbf{A}_{21} & \mathbf{A}_{22} \end{bmatrix}, \quad \mathbf{B} = \begin{bmatrix} \mathbf{B}_{11} & \mathbf{B}_{12} \\ \mathbf{B}_{21} & \mathbf{B}_{22} \end{bmatrix}, \quad \mathbf{C} = \begin{bmatrix} \mathbf{C}_{11} & \mathbf{C}_{12} \\ \mathbf{C}_{21} & \mathbf{C}_{22} \end{bmatrix},$$

where $\mathbf{A}_{ij}, \mathbf{B}_{ij}, \mathbf{C}_{ij}$ are matrices of order $m2^k$. To evaluate $\mathbf{C} = \mathbf{AB}$ we compute

$$
\begin{aligned}
(Q1) &= (\mathbf{A}_{11} + \mathbf{A}_{22})(\mathbf{B}_{11} + \mathbf{B}_{22}), \\
(Q2) &= (\mathbf{A}_{21} + \mathbf{A}_{22})\mathbf{B}_{11}, \\
(Q3) &= \mathbf{A}_{11}(\mathbf{B}_{12} - \mathbf{B}_{22}), \\
(Q4) &= \mathbf{A}_{22}(-\mathbf{B}_{11} + \mathbf{B}_{21}), \\
(Q5) &= (\mathbf{A}_{11} + \mathbf{A}_{12})\mathbf{B}_{22}, \\
(Q6) &= (-\mathbf{A}_{11} + \mathbf{A}_{21})(\mathbf{B}_{11} + \mathbf{B}_{12}), \\
(Q7) &= (\mathbf{A}_{12} - \mathbf{A}_{22})(\mathbf{B}_{21} + \mathbf{B}_{22}).
\end{aligned}
\tag{1}
$$

Then

$$
\begin{aligned}
\mathbf{C}_{11} &= (Q1) + (Q4) - (Q5) + (Q7), \\
\mathbf{C}_{21} &= (Q2) + (Q4), \\
\mathbf{C}_{12} &= (Q3) + (Q5), \\
\mathbf{C}_{22} &= (Q1) + (Q3) - (Q2) + (Q6).
\end{aligned}
\tag{2}
$$

We now define $a(m, k)$ as algorithms which multiply matrices of order $m2^k$, where $a(m, 0)$ is the usual algorithm for matrix multiplication and requires m^3 multiplications and $m^2(m - 1)$ additions. From (1) and (2) it follows that algorithm $a(m, k)$ requires 7 multiplications of matrices of order $m2^{k-1}$. Denoting by $Ma(m, k)$ the number of multiplications required by algorithm $a(m, k)$, we get

$$
\begin{aligned}
Ma(m, k) &= 7Ma(m, k - 1) \\
&= 7^k Ma(m, 0) \\
&= 7^k m^3.
\end{aligned}
\tag{3}
$$

Define, further, by $Sa(m, k)$ the number of additions or subtractions required by the algorithm $a(m, k)$. Now, from (1) and (2) it follows that algorithm $a(m, k)$ requires 18 additions of matrices of order $m2^{k-1}$. Also the 7 multiplications of matrices of order $m2^{k-1}$, each require $Sa(m,k-1)$ additions. Thus

$$Sa(m,k) = 18(m2^{k-1})^2 + 7Sa(m,k-1) \tag{4}$$

$$= (5+m)m^2 7^k - 6(m2^k)^2 \tag{5}$$

Relation (5) can be proved by induction in the following way. For $k = 0$ we have

$$Sa(m,0) = (5+m)m^2 - 6m^2 = m^2(m-1) \text{ which is correct.}$$

Now, assuming that (5) is true for some k, we consider the relation for $k+1$:

$$Sa(m,k+1) = (5+m)m7^{k+1} - 6(m2^{k+1})^2$$

$$= 7(5+m)m^2 7^k - 24(m2^k)^2$$

$$= 7[(5+m)m^2 7^k - 6(m2^k)^2] + 18(m2^k)^2$$

$$= 18(m2^k)^2 + 7\,Sa(m,k)$$

This completes the proof.
From (3) and (5) the total number of arithmetic operations is:

$$T = 7^k m^3 + (5+m)m^2 7^k - 6(m2^k)^2$$

$$= (5+2m)m^2 7^k - 6(m2^k)^2. \tag{6}$$

Now, set

$k = \lfloor \log_2 n - 4 \rfloor,$ with n assumed to be $\geqslant 16$, and introduced to include

$m = \lfloor n2^{-k} \rfloor + 1,$ all n in the range $m2^{k-1} < n \leqslant m2^k$.

We get

$$T < [5 + 2(n2^{-k} + 1](n2^{-k} + 1)^2 7^k$$

$$= 7^k(2n^3 2^{-3k} + 11n^2 2^{-2k} + 16n2^{-k} + 2)$$

$$< 7^k(2n^3 2^{-3k} + 12.03n^2 2^{-2k}),$$

where we make use of the above assumption that $16(2^k) \leqslant n, \quad k \geqslant 0$,

$$= 2(\tfrac{7}{8})^k n^3 + 12.03(\tfrac{7}{4})^k n^2$$

$$= [2(\tfrac{8}{7})^{\log_2 n - k} + 12.03(\tfrac{4}{7})^{\log_2 n - k}]n^{\log_2 7}$$

$$\leqslant \max_{4 \leqslant t \leqslant 5} [2(\tfrac{8}{7})^t + 12.03(\tfrac{4}{7})^t]n^{\log_2 7}$$

$$\leqslant 4.7n^{\log_2 7}, \quad \text{by virtue of convexity of the function in square brackets.} \tag{7}$$

We have thus shown that using the matrix multiplication algorithm defined by (1) and (2), the product of two matrices of order n can be computed using no more than $4.7n^{\log_2 7}$ arithmetic operations.

    The crucial importance of the Strassen algorithm for matrix multiplication lies in the fact that discovery of such a method to multiply two matrices has broken an unwritten common belief that multiplication of two matrices of order n is inherently a process proportional to n^3.

Next, we shall consider how the Strassen algorithm can be used to compute the inverse of a matrix.

4.10.2 Matrix Inversion

We define by $b(m,k)$ the algorithms which invert matrices of order $m2^k$ by induction on k, where $b(m,0)$ is the usual Gaussian elimination algorithm. If A is a matrix of order $m2^{k+1}$, to be inverted, then write

$$A = \begin{bmatrix} A_{11} & A_{12} \\ A_{21} & A_{22} \end{bmatrix} \qquad A^{-1} = \begin{bmatrix} C_{11} & C_{12} \\ C_{21} & C_{22} \end{bmatrix}$$

To evaluate A^{-1} we must compute

$$
\begin{aligned}
(Q1) &= A_{11}^{-1}, \\
(Q2) &= A_{21}(Q1), \\
(Q3) &= (Q1)A_{12} \\
(Q4) &= A_{21}(Q3), \\
(Q5) &= (Q4) - A_{22}, \\
(Q6) &= (Q5)^{-1}, \\
C_{12} &= (Q3)(Q6), \\
C_{21} &= (Q6)(Q2), \\
(Q7) &= (Q3)C_{21}, \\
C_{11} &= (Q1) - (Q7), \\
C_{22} &= -(Q6).
\end{aligned}
\tag{1}
$$

Now, let $Mb(m,k)$ be the number of multiplications required to invert a matrix of order $m2^k$, $Sb(m,k)$ be the number of additions required to invert a matrix of order $m2^k$. In particular, we have $Mb(m,0) = m^3$; $Sb(m,0) = m^3 - 2m^2 + m$. The inversion of a matrix of order $m2^k$ requires

2 inversions of matrices of order $m2^{k-1}$
6 multiplications of matrices of order $m2^{k-1}$
2 additions of matrices of order $m2^{k-1}$

Thus

$$Mb(m,k) = 6Ma(m,k-1) + 2Mb(m,k-1) \tag{2}$$

$$\leqslant 6m^3 7^k/5, \tag{3}$$

and

$$Sb(m,k) = 2Sb(m,k-1) + 6Sa(m,k-1) + 2(m2^{k-1})^2 \tag{4}$$

$$\leqslant 6(5+m)m^2 7^k - 7(m2^k)^2 \tag{5}$$

Combining (3) and (5) we have that the inverse of a matrix of order n can be calculated with less than

$$5.64\, n^{\log_2 7} \text{ arithmetic operations} \tag{6}$$

148

Applying results (3) and (5), we can give an indication of the number of arithmetic operations which may be required to solve a set of linear equations using the Strassen algorithm for matrix multiplication.

To solve the set of equations $\mathbf{Ax} = \mathbf{b}$, we form

$$\mathbf{x} = \mathbf{A}^{-1}\mathbf{b}. \tag{7}$$

To compute $\mathbf{x}$ knowing $\mathbf{A}^{-1}$, requires n^2 multiplications and $n(n-1)$ additions, where $n = m2^k$. Thus from (3) and (5) the number of arithmetic operations required is

$$\leqslant 6m^3 7^k/5 + m^2 2^{2k} \text{ multiplications} \tag{8}$$

and $\leqslant 6(5+m)m^2 7^k - 6m^2 2^{2k} - m2^k$ additions $\tag{9}$

Further, using the same approach as in 4.10.1—(7) and assuming that $72(2^k) \leqslant n$, from (8) and (9) it follows that the total number of arithmetic operations required to solve a set of linear equation using the Strassen algorithm is given as

$$\leqslant 20n^{\log_2 7}. \tag{10}$$

4.11 Winograd's Variant of the Strassen Algorithm

Winograd (1973) reduced the number of additions required by Strassen's algorithm to 15. This improved the constant factor in the order of algebraic complexity but not the order-of-magnitude. The algorithm is given as follows.

Let matrices $\mathbf{A}$, $\mathbf{B}$, and $\mathbf{C}$ be as defined in Section 4.10.1. Then to compute the elements of the product matrix $\mathbf{C}$ the following procedure is used.

$(Q1) = \mathbf{A}_{21} - \mathbf{A}_{11},$ $\qquad (Q5) = \mathbf{B}_{22} - \mathbf{B}_{21},$

$(Q2) = \mathbf{A}_{11} + \mathbf{A}_{12},$ $\qquad (Q6) = \mathbf{B}_{12} - \mathbf{B}_{11},$

$(Q3) = \mathbf{A}_{12} - (Q1),$ $\qquad (Q7) = \mathbf{B}_{11} + (Q5),$

$(Q4) = \mathbf{A}_{22} - (Q3),$ $\qquad (Q8) = \mathbf{B}_{21} - (Q7),$

$(P1) = \mathbf{A}_{21}\mathbf{B}_{11},$ $\quad (P5) = (Q4)\mathbf{B}_{22},$ $\quad (Q9) = (P1) + (P7),$

$(P2) = \mathbf{A}_{22}\mathbf{B}_{21},$ $\quad (P6) = \mathbf{A}_{12}(Q8),$ $\quad (Q10) = (Q9) + (P7),$

$(P3) = (Q1)(Q5),$ $\quad (P7) = (Q3)(Q7),$ $\quad (Q11) = (P4) + (P5),$

$(P4) = (Q2)(Q6),$

$\mathbf{C}_{11} = (Q10) + (P6),$ $\qquad\qquad \mathbf{C}_{21} = (P1) + (P2),$

$\mathbf{C}_{12} = (Q10) + (P4),$ $\qquad\qquad \mathbf{C}_{22} = (Q9) + (Q11).$

4.12 The Karatsuba–Makarov Method

At this point we would like to mention another method of multiplication of two matrices which is originally due to Karatsuba (1962) and which is again discussed by Makarov (1975a, 1975b). The method decreases the number of additions required and uses the same number of multiplications as compared with the standard method.

To derive the algorithm, two matrices, **A** and **B**, of order $m2^{k+1}$ and their product, **C**, are presented in the same manner as used by Strassen. The elements $\mathbf{C}_{ij}$ are computed using the following formula:

$$\begin{aligned}
\mathbf{C}_{11} &= \mathbf{A}_{11}(\mathbf{B}_{11} + \mathbf{B}_{21}) + (\mathbf{A}_{12} - \mathbf{A}_{11})\mathbf{B}_{21}, \\
\mathbf{C}_{12} &= \mathbf{A}_{12}(\mathbf{B}_{22} + \mathbf{B}_{12}) - (\mathbf{A}_{12} - \mathbf{A}_{11})\mathbf{B}_{12}, \\
\mathbf{C}_{21} &= \mathbf{A}_{21}(\mathbf{B}_{11} + \mathbf{B}_{21}) + (\mathbf{A}_{22} - \mathbf{A}_{21})\mathbf{B}_{21}, \\
\mathbf{C}_{22} &= \mathbf{A}_{22}(\mathbf{B}_{22} + \mathbf{B}_{12}) - (\mathbf{A}_{22} - \mathbf{A}_{21})\mathbf{B}_{12},
\end{aligned} \tag{1}$$

another possibility is to use

$$\begin{aligned}
\mathbf{C}_{11} &= \mathbf{A}_{12}(\mathbf{B}_{11} + \mathbf{B}_{21}) - (\mathbf{A}_{12} - \mathbf{A}_{11})\mathbf{B}_{11}, \\
\mathbf{C}_{12} &= \mathbf{A}_{11}(\mathbf{B}_{11} + \mathbf{B}_{21}) + (\mathbf{A}_{12} - \mathbf{A}_{11})\mathbf{B}_{11}, \\
\mathbf{C}_{21} &= \mathbf{A}_{22}(\mathbf{B}_{12} + \mathbf{B}_{22}) - (\mathbf{A}_{22} - \mathbf{A}_{21})\mathbf{B}_{12}, \\
\mathbf{C}_{22} &= \mathbf{A}_{21}(\mathbf{B}_{12} + \mathbf{B}_{22}) + (\mathbf{A}_{22} - \mathbf{A}_{21})\mathbf{B}_{12}.
\end{aligned} \tag{2}$$

If, as before, we define by $a(m,k)$ the algorithms which multiply matrices of order $m2^k$, by induction on k, where $a(m,0)$ is the usual algorithm for matrix multiplication, with m^3 multiplication and $m^2(m-1)$ additions, then we can see that in each of the variants, (1) and (2), 8 multiplications of matrices of order $m2^{k-1}$ are required. It follows that the number of multiplications required by the Karatsuba–Makarov algorithm is given by

$$\begin{aligned}
Ma(m,k) &= 8Ma(m,k-1) = 8^k Ma(m,0) \\
&= 8^k m^3 = n^3.
\end{aligned}$$

The number of additions, $Sa(m,k)$, is given by

$$8(m2^{k-1})^2 + 8Sa(m,k-1), \tag{3}$$

since the algorithm requires 8 additions of matrices of order $m2^{k-1}$ and also the 8 matrix multiplications of order $m2^{k-1}$, each require $Sa(m,k-1)$ additions. Relation (3) gives

$$Sa(m,k) = 8^k m^3 - 2(4^k)m^2 = n^3 - 2n^2. \tag{4}$$

This number can be compared with the number of additions, $n^3 - n^2$, required to multiply two matrices of order $n = m2^k$ by the standard method.

4.13 Summary of the Direct Methods Studied

Concerning numerical accuracy, we have seen that in the Gaussian elimination with complete pivoting, a bound for the maximum growth of the pivotal elements is quite small and there are strong reasons for believing (Wilkinson,

1965) that this bound cannot be approached. If partial pivoting is used, the final pivot can be 2^{n-1} times as large as the maximum element in the original matrix. However, although such a bound is attainable it is usually irrelevant for practical purposes, as it is quite unusual for any growth to take place at all and if inner products can be accumulated then the equivalent perturbations in the original matrix are exceptionally small. Hence, Gaussian elimination is considered to be practically stable.

The numerical stability of the Householder method is guaranteed unconditionally in that most satisfactory *a priori* bounds for the equivalent perturbations of the original matrix are available. No dangerous growth in size of elements during the reduction is possible since apart from round-off errors the Euclidean norm of each column of the original matrix is preserved by exact orthogonal transformations.

We have also seen that an error analysis due to Brent of algorithms for matrix multiplications and triangular decomposition using the Winograd identity, shows that the Winograd method can be very bad numerically. However, by employing an appropriate technique of scaling, the method can be made to work with accuracy comparable to the usual method when used without accumulation of inner products. However, the scaling takes the computer time and this lowers the overall efficiency of the method.

Very little is understood yet about error control in the Strassen method. Again, Brent (1970b) has established a satisfactory error bound for floating point matrix multiplication by Strassen's method. However, it seems that no such bounds may be contemplated for matrix inversion or the solution of sets of linear equations by the same method, because no pivoting is possible there.

Concerning the volume of work, Table 4.1 summarizes the number of arithmetic operations required by different methods. The methods of Jordan and of Givens are also included to give a more general picture on the matter.

Practical tests by Brent indicate that Winograd's algorithm, even with the necessary scaling is faster than Strassen's for $n > 250$, though the precise change-over point depends on the machine and compiler used (Brent, 1970a). The Winograd algorithm is also easier to program than Strassen's. And again, one has to bear in mind that in the newer methods the fewer multiplications become of some significance for only quite large n. Also the auxiliary computer storage needs for these methods has to be considered.

Finally, as a practical note, it may be once again pointed out that from the point of view of overall performance, which includes efficiency, accuracy, reliability, generality and ease of use, the conventional methods, like Gaussian elimination, still remain the better methods for most general matrices.

Lower Bound on Matrix Multiplication

By expressing the formulae for matrix multiplication of 2×2 matrices with seven multiplications, Strassen was able to construct an $O(n^{2.81})$ matrix multiplication algorithm. If one could multiply 2×2 matrices in 6 multiplica-

Table 4.1 The number of arithmetic operations required by different direct methods.

Method	Number of multiplications	Number of additions	Number of reciprocals	Number of square roots
Gaussian elimination	$\dfrac{n^3}{3}+n^2-\dfrac{n}{3}$	$\dfrac{n^3}{3}+\dfrac{n^2}{2}-\dfrac{5n}{6}$	n	—
Cholesky decomposition	$\dfrac{n^3}{6}+\dfrac{3n^2}{2}+\dfrac{n}{3}$	$\dfrac{n^3}{6}+n^2-\dfrac{7n}{6}$	—	n
Householder reduction	$\dfrac{2n^2}{3}+\dfrac{5n^2}{2}+\dfrac{17n}{6}-5$	$\dfrac{2n^3}{3}+\dfrac{3n^2}{2}+\dfrac{5n}{6}-3$	$2n-1$	$\left\{ \begin{matrix} n-2 \\ 2n-4 \end{matrix} \right\}$
Jordan elimination	$\dfrac{n^3}{2}+n^2-\dfrac{n}{2}$	$\dfrac{n(n^2-1)}{2}$	n	—
Givens reduction	$\dfrac{4n^3}{3}+\dfrac{5n^2}{2}-\dfrac{17n}{6}$	$\dfrac{2n^3}{3}+n^2-\dfrac{5n}{3}$	n	$\dfrac{n(n-1)}{2}$
Winograd's method	$\sim\dfrac{n^3}{6}$	$\sim\dfrac{n^3}{2}$	—	—
Strassen's method	$\leqslant \tfrac{6}{5}m^3 7^k + m^2 2^{2k}$ $(n=m2^k)$	$\leqslant 6(5+m)m^2 7^k$ $-6m^2 2^{2k}-m2^k$		

tions, we would have an $O(n^{\log_2 6})$ or $O(n^{2.59})$ matrix multiplication algorithm. Actually it has been shown by Hopcroft and Kerr (1971) that seven multiplications is the minimum number required for 2×2 matrix multiplication but, in general, multiplying some small $k\times k$ matrices in m multiplications would result in an $O(n^{\log_k m})$ algorithm. Thus, if $\log_k m < \log_2 7$ we have an improvement. For example, for $\log_3 m < \log_2 7$ to hold, we must have $m\leqslant 21$, that is if an algorithm that multiplies two 3×3 matrices in no more than 21 multiplications, can be found then we shall have a matrix multiplication algorithm which is 'better' than Strassen's.

Thus, what is the minimum number of arithmetic operations that is needed to multiply two $n\times n$ matrices? At present it can only be said that since the problem has $2n^2$ inputs and n^2 outputs, a lower bound on the number of arithmetics must be of $O(n^2)$, and we can write

$$O(n^2) \leqslant \text{number of arithmetic operations} \leqslant O(n^{\log_2 7}).$$

4.14 Epilogue

As Borodin (1973) remarked, to many people, the matrix multiplication problem is the problem of algebraic complexity. The impetus that Strassen's $O(n^{\log_2 7})$ algorithm gave to the area is hard to overemphasize. However the intrinsic difficulty of the problem still eludes researchers. One approach which is used at present in an attempt to throw some light on the problem is to view matrix multiplication (or polynomial multiplication) as the computation of a set of bilinear forms. These are defined as

$$\Psi_k = \sum_{i=1}^{s} \sum_{j=1}^{r} \alpha_{ijk} x_i y_j, \qquad k = 1, 2, \ldots, t,$$

where α's are constants.

Matrix product, complex product and polynomial product all define sets of bilinear forms. One is then interested in the algebraic complexity of computing the values of bilinear forms. The question asked is: Is it possible to infer new 'economical' ways of computing these values using the fact that ψ_k are bilinear? The approach is discussed in Fiduccia (1972), Strassen (1972), Hopcroft and Musinski (1973), Brockett and Dobkin (1973), Borodin (1973). Some useful results have been established using this approach but many more unresolved questions still remain.

The most recent result due to Pan reduces the computational complexity of matrix multiplication to $O(n^{2.795})$ or $O((n^2)^{1.3975})$ arithmetic operations. The approach makes use of both bilinear and trilinear forms. With the new result the subject has been injected with new vitality and ideas.

Exercises

4.1 The set $\mathbf{Ax} = \mathbf{b}$ is solved by (i) the standard Gaussian elimination and (ii) computing the inverse $\mathbf{A}^{-1}$ first, using Gaussian elimination, and then forming the product $\mathbf{A}^{-1}\mathbf{b}$. Show by an operation count that the method (ii) is more costly even if we have to solve $\mathbf{Ax} = \mathbf{b}$ for many different $\mathbf{b}$ vectors. Analyse how the situation will change if in (ii) the technique of complexity $O(n^{2.81})$ were used for computing $\mathbf{A}^{-1}$.

4.2 Show that the solution of $\mathbf{Ax} = \mathbf{b}$ where $\mathbf{A}$ is a tradiagonal matrix, by Gaussian elimination without pivoting can be done using $3(n-1)$ operations for the elimination procedure and $3(n-1)+1$ operations in the back-substitution process.

4.3 With reference to Exercise 4.2, show that partial pivoting results in a triangular matrix with 3 nonzero diagonals instead of 2. Hence, back-substitution requires $5(n-2)+4$ operations.

4.4 Estimate the number of arithmetic operations required to compute a matrix determinant using Gaussian elimination.

4.5 Give step-by-step algorithm for the matrix inversion using Methods 1 and 2 of Section 4.3, respectively.

4.6 Using Choleski's method, find the inverse of the matrix

$$\mathbf{A} = \begin{bmatrix} 2 & 4 & 1 & 2 \\ 4 & 10 & 0 & -4 \\ 1 & 0 & 4.5 & 15 \\ 2 & -4 & 15 & 54 \end{bmatrix}.$$

4.7 What happens if an LU decomposition is applied to a singular matrix?

4.8 Prove that the inverse of an upper (lower) triangular matrix is upper (lower) triangular.

4.9 Find (a) **LU** decomposition, (b) the inverse and (c) the determinant of the matrix

$$\begin{bmatrix} 2.12 & 0.42 & 1.34 & 0.88 & 0.75 & 1.21 & 0.39 & -2.11 \\ 1.34 & 0.95 & 1.87 & 0.43 & 0.54 & 1.25 & 0.75 & 0.91 \\ 0.88 & 1.87 & 0.98 & 0.46 & 1.71 & 0.41 & 1.87 & 0.19 \\ 0.75 & 0.43 & 0.46 & 2.44 & 1.14 & 0.36 & 1.13 & -2.54 \\ 1.21 & 0.54 & 1.71 & -2.14 & 1.49 & 1.71 & 0.39 & 0.87 \\ 0.39 & 1.25 & 0.41 & 0.36 & 1.71 & -1.51 & 0.42 & 1.14 \\ -2.11 & 0.75 & 1.87 & 1.13 & 0.39 & 0.42 & 1.91 & 0.45 \\ 1.52 & 0.91 & 0.19 & -2.54 & 0.87 & 1.14 & 0.45 & 2.35 \end{bmatrix}$$

using standard methods.

4.10 Solve Exercise 4.9 using the techniques which have $O(n^{2.81})$ complexity.

4.11 Compute the product

$$\begin{bmatrix} 3 & 9 \\ 2 & 7 \end{bmatrix} \begin{bmatrix} 1 & 6 \\ 4 & 5 \end{bmatrix}$$

using (a) Strassen's algorithm with 7 multiplications and 18 additions, (b) Winograd's variant with 7 multiplications and 15 additions and (c) Karatsuba-Makarov algorithm with 8 multiplications and 12 additions.

4.12 It is required to invert an $n \times n$ matrix **A**. **A** may be viewed as an $m \times m$ matrix whose entries are $k \times k$ matrices. The Gaussian elimination is then performed on this $m \times m$ matrix and whenever a multiplication of two $k \times k$ matrices is required, use is made of the Winograd identity. Deduce the minimum number of multiplications required by this process.

4.13 We have shown in Section 4.10.2 that solution of linear equations $\mathbf{Ax} = \mathbf{b}$ where **A** is an $n \times n$ matrix is a process of complexity $O(n^{2.81})$. Find the best possible constant factor for this problem, assuming that Strassen's algorithm is used for the multiplications and n is a power of 2.

4.14 Show that the recursive application of the Strassen algorithm to multiply two $n \times n$ matrices requires an auxiliary memory of $n^2 + 8(n^2/4 + n^2/16 + \ldots) = (11/3)n^2$ locations.

4.15 Strassen expressed a matrix decomposed as shown, as the product of three matrices

$$\mathbf{A} = \begin{bmatrix} \mathbf{A}_{11} & \mathbf{A}_{12} \\ \mathbf{A}_{21} & \mathbf{A}_{22} \end{bmatrix} = \begin{bmatrix} \mathbf{I} & \mathbf{0} \\ \mathbf{A}_{21}A_{11}^{-1} & \mathbf{I} \end{bmatrix} \begin{bmatrix} \mathbf{A}_{11} & \mathbf{0} \\ \mathbf{0} & \mathbf{V} \end{bmatrix} \begin{bmatrix} \mathbf{I} & \mathbf{A}_{11}^{-1}\mathbf{A}_{12} \\ \mathbf{0} & \mathbf{I} \end{bmatrix}$$

where $\mathbf{V} = \mathbf{A}_{22} - \mathbf{A}_{21}\mathbf{A}_{11}^{-1}\mathbf{A}_{12}$.

(i) Using the given formula derive an expression for computing the inverse $\mathbf{A}^{-1}$.

(ii) Assuming that **A** is a general matrix of size n, show that computing the inverse by the recursive use of the process suggested in (i) is an algorithm of order $M(n)$, where $M(n)$ denotes the number of operations required for multiplication of two $n \times n$ matrices.

(iii) Explain why the algorithm always works (theoretically) in the case when the original matrix is upper or lower triangular but may fail in the case of a general matrix.

4.16 Generalise Strassen's matrix multiplication algorithm to matrices of dimension $m \times m$ where $m = p2^k$ for integers p and k. Bound the number of arithmetic operations needed.

4.17 Obtain a straight-line program that computes the determinant of a 3×3 matrix.

4.18 Write a program for the Householder algorithm.

4.19 Consider the polynomial $P(x) = \sum\limits_{i=0}^{n} a(i)x^i$ where n is a perfect square. The Borodin–Munro algorithm for evaluation of the polynomial at two points, x_1 and x_2, is given as follows:

Write
$$\mathbf{A} = \begin{bmatrix} a(1) & a(2) \ldots a(\sqrt{n}) \\ a(\sqrt{n}+1) & \ldots & \ldots \\ \ldots & & \\ \ldots & & a(n) \end{bmatrix} \text{ and } \mathbf{X} = \begin{bmatrix} x_1 & x_2 \\ x_1^2 & x_2^2 \\ x_1^{\sqrt{n}} & x_2^{\sqrt{n}} \end{bmatrix},$$

then compute $\quad \mathbf{Y} = \{y_{ik}\} = \mathbf{AX}$
and

$$P(x) = a(0) + y_{1k} + \sum_{i=2}^{\sqrt{n}} y_{ik} x_k^{(k-1)\sqrt{n}}, \qquad k = 1, 2.$$

(a) Estimate the number of multiplications required by the algorithm.
(b) By extending the algorithm for $\sqrt{n}$ points, $x_1, \ldots, x_{\sqrt{n}}$, show that a polynomial of degree n may be evaluated at $\sqrt{n}$ points in $O(n^{\log 7/2})$ arithmetic operations.

Chapter 5

The Fast Fourier Transform

5.1 Introduction

In solving a mathematical problem, it is often convenient to transform the function, work with the transform, and then 'untransform' the result, to deduce non-obvious properties of the original function.

Among others, the Fourier transform is of significant importance in practical work. Simultaneous visualization of a function and its Fourier transform is often the key to successful problem solving.

Probably the best known application of this mathematical technique is the analysis of linear time-invariant systems. But it has also long been a principal analytical tool in such diverse fields as optics, theory of probability, quantum physics, antennas and signal analysis.

Such a wide and successful application was, however, an attribute of the continuous Fourier transform, i.e. the transform of the function(s) defined on variable(s) taking values in a continuum.

On the other hand, in numerical mathematics and data analysis, where one usually deals with ordered sets of numbers, functions of a discrete variable, which are also finite in extent, the discrete form of the Fourier transform is a more convenient tool. Unfortunately, until the mid-sixties the computation necessary to produce the discrete Fourier transforms were known to be notoriously time-consuming even with the tremendous computing speeds available on modern computers.

In 1965 Cooley and Tukey published a new method for computing the Fourier transforms which offered remarkable savings in computing time compared with all previously known methods. It became known as the Fast Fourier Transform or the FFT.

Because of the exceptional importance of the discrete Fourier transform (DFT) in practical applications, discovery of the FFT must be rated as one of the most significant achievements of computational mathematics in recent time.

With the Cooley and Tukey publication it became apparent that the technique of the FFT has been used *ad hoc* here and there, and, in fact, the algorithm has a fascinating history. Its general approach may be traced back to Runge and König (1924), Danielson and Lanczos (1942), and Good (1958, 1960). It has to be emphasized, however, that only with the brilliant contribution by Cooley and Tukey was the full potential of the approach realized.

Our concern in this chapter will be the computation of the discrete Fourier transform using the FFT algorithm. Accordingly we shall only briefly introduce the continuous Fourier transform and then move on to discuss the discrete Fourier transform. We shall then proceed to outline the FFT algorithm and consider its relative efficiency, storage requirements and error control.

5.2 The Continuous Fourier Transform

The Fourier Transform of a Function of One Variable and Its Inverse

The following formulae yield mutual and dual relations between a pair of functions.

If for a given function $f(v)$ we set

$$F(u) = \frac{1}{\sqrt{2\pi}} \int_{-\infty}^{\infty} f(v)e^{-iuv}dv, \qquad i = \sqrt{-1} \tag{1}$$

then the following holds:

$$f(v) = \frac{1}{\sqrt{2\pi}} \int_{-\infty}^{\infty} F(u)e^{+viu}du, \qquad i = \sqrt{-1} \tag{2}$$

where the integrals are interpreted as

$$\lim_{l \to \infty} \int_{-l}^{l} g(y)e^{+ixy}dy \tag{3}$$

with appropriate plus or minus sign used in the exponential function.

The function $F(u)$ is called *the Fourier transform* of $f(v)$ and is normally denoted by $\mathscr{F}[f]$, i.e. $\mathscr{F}[f] = F(u)$. If the function $f(v)$ is integrable in the interval $(-\infty, +\infty)$, then the function $F(u)$ exists for every value of v. Typically $f(v)$ is referred to as a function of the variable time and $F(u)$ is referred to as a function of the variable frequency.

The functions $F(u)$ and $f(v)$ are called the pair of the Fourier transform or the Fourier transform pair.

One also distinguishes between the Fourier transform, $F(u)$, and its inverse, $f(v)$, so that we have $\mathscr{F}[f(v)] = F(u)$ and $\mathscr{F}^{-1}[F(u)] = f(v)$.

Elementary Properties of the Fourier Transform

(a) The Fourier transform is a linear operator.

An operator P being linear on a specified set of functions means that

$$P(\alpha f + \beta g) = \alpha P f + \beta P g$$

for any two functions f and g of the set with α and β arbitrary constants. Now, if $f(v)$ and $g(v)$ have the Fourier transforms $F(u)$ and $G(u)$, respectively, then the sum $f(v) + g(v)$ has the Fourier transform $F(u) + G(u)$, and in general the following holds:

$$\mathscr{F}[\alpha f + \beta g] = \alpha \mathscr{F}[f] + \beta \mathscr{F}[g].$$

(b) The Fourier transform is symmetric.
   If $f(v)$ and $F(u)$ are a Fourier transform pair then $F(v)$ and $f(-u)$ are a Fourier transform pair.
(c) Scaling of the Fourier transform.
   If the Fourier transform of $f(v)$ is $F(u)$ then the Fourier transform of $f(kv)$, where k is a real constant, is given as:

$$\frac{1}{|k|} F\left(\frac{u}{k}\right).$$

If the inverse Fourier transform of $F(u)$ is $f(v)$, then the inverse Fourier transform of $F(ku)$, where k is a real constant, is given by:

$$\frac{1}{|k|} f\left(\frac{v}{k}\right).$$

(d) The property of parity.
   If $f(v)$ is an even function, i.e. $f(v) = f(-v)$, then its Fourier transform is an even function and is real.
   If $f(v)$ is an odd function, i.e. $f(v) = -f(-v)$, then its Fourier transform is an odd and imaginary function.

The Convolution of Two Functions

Let functions $F(u)$ and $G(u)$ be the Fourier transforms of the functions $f(v)$ and $g(v)$ respectively.

Formally we have

$$\frac{1}{\sqrt{2\pi}} \int_{-\infty}^{\infty} F(u)G(u)e^{-ivu}\,du = \frac{1}{2\pi} \int_{-\infty}^{\infty} F(u)e^{-ivu}\left\{\int_{-\infty}^{\infty} g(\tau)e^{iu\tau}\,d\tau\right\}du \qquad (1)$$

$$= \frac{1}{2\pi} \int_{-\infty}^{\infty} g(\tau)f(v-\tau)\,d\tau,$$

i.e. the function

$$F(u)G(u) \text{ and } h(v) = \frac{1}{\sqrt{2\pi}} \int_{-\infty}^{\infty} g(\tau)f(v-\tau)\,d\tau \qquad (2)$$

constitute a Fourier transform *pair*.

The function $h(v)$ is called *the convolution* of the functions $f(v)$ and $g(v)$. We thus have the important result that the convolution of two functions *is*

158

equal to the inverse Fourier transform of the product of the Fourier transforms of the functions.

5.3 The Discrete Fourier Transform

The transform we have first discussed may be called the infinite continuous Fourier transform since it deals with continuous functions which are integrated over an infinite range. As was mentioned earlier, such transforms are well known and widely used. However, in practice one frequently works with data that is given as a set of discrete quantities, and then the finite discrete Fourier transform may be a useful device to simplify the mathematics of the processes studied.

The finite discrete Fourier transform may be visualized as a special case of the infinite continuous Fourier transform when the latter is amenable to machine computation. (For a somewhat different interpretation of the DFT, as a special case of the Fourier series of a periodic function, see Cooley, Lewis and Welch (1977).)

Definition of the DFT

Let $f(v)$, $v = 0, 1, \ldots, N-1$ be a sequence of N complex numbers. The finite discrete Fourier transform or DFT of $f(v)$ is defined as

$$F(u) = \sum_{v=0}^{N-1} f(v) w_N^{-uv}, \qquad u = 0, 1, \ldots, N-1 \tag{1}$$

where here and in sequel, $w_N = \exp(2\pi i/N)$ with $i = \sqrt{-1}$. Similarly, its inverse is defined as

$$f(v) = \frac{1}{N} \sum_{u=0}^{N-1} F(u) w_N^{uv}, \qquad v = 0, 1, \ldots, N-1. \tag{2}$$

The latter formula may also be considered as

$$f(v) = \frac{1}{N} \left[\sum_{u=0}^{N-1} F^*(u) w_N^{-uv} \right]^*, \qquad v = 0, 1, \ldots, N-1, \tag{3}$$

which is read as 'the complex conjugate of the Fourier transform of the complex conjugate of $F(v)$ divided by N gives $f(v)$'. (1) and (2) are a transform pair, that is, substituting $F(u)$ from (1) into (2) gives back $f(v)$, i.e.

$$f(v) = \frac{1}{N} \sum_{u=0}^{N-1} \left[\sum_{k=0}^{N-1} f(k) w_N^{-uk} \right] w_N^{uv}$$

$$= \frac{1}{N} \sum_{k=0}^{N-1} f(k) \sum_{u=0}^{N-1} w_N^{u(v-k)} = f(v) \tag{4}$$

since the exponential function w_N^{uv} satisfies the following orthogonality relationships

$$\sum_{u=0}^{N-1} w_N^{vu} w_N^{-ku} = \sum_{u=0}^{N-1} e^{i(2\pi/N)(v-k)u}$$

$$= \begin{cases} N \text{ if } k \equiv v \bmod N \\ 0 \text{ otherwise.} \end{cases} \tag{5}$$

Note that in the finite discrete Fourier transform both functions, $f(v)$ and $F(u)$, are defined over finite sets of discrete points only, i.e. $f(v)$ is given as $(f(0), f(1),\ldots,f(N-1))$ and $F(u)$ is given as $(F(0), F(1), F(2),\ldots,F(N-1))$. Consequently, we deal with finite sums also. Hence, the name.

Periodicity of the DFT

The exponential function w_N^{uv}, as a function of u and v, is periodic of period N, i.e.

$$w_N^{uv} = w_N^{(u+N)v} = w_N^{u(v+N)}.$$

It follows immediately that the sequences $(f(v), v = 0,\ldots,N-1)$ and $(F(u), u = 0,\ldots,N-1)$ as defined by their transforms (1) and (2) are periodic of period N.

Hence, in relation to (1) and (2), the most significant change required to modify the theorems about the infinite continuous Fourier transform, so that they apply to the finite discrete case is that indexing must be considered modulo N, where N is the number of data points. One useful interpretation of this, put forward by Gentleman and Sande (1966) is to think of the sequence $(f(v),$ $v = 0, 1,\ldots,N-1)$ as a *function* defined on equispaced integer points of a circle with circumference N.

However, it is more convenient to consider $F(u)$ and $f(v)$ to be defined for all integers, i.e. to have

$$F(u), u = 0, \pm 1, \pm 2,\ldots$$

and

$$f(v), v = 0, \pm 1, \pm 2,\ldots$$

with

$$F(u) = F(kN + u), \qquad k = 0, \pm 1, \pm 1,\ldots$$

and

$$f(v) = f(kN + v), \qquad k = 0, \pm 1, \pm 2,\ldots.$$

The periodic extension (5) of the finite sequences to infinite sequences makes easier visualization of the properties of the discrete Fourier transform and, in some cases, their use in computational work. The finite sequences can, of course, be recovered by considering the values of the finite sequences at any N consecutive points.

Elementary Properties of the DFT

The DFT possesses the same properties as the continuous Fourier transform, the properties being appropriately modifed to suit the discrete and periodic character of the DFT. For example, the fundamental property of linearity of the transform is preserved, that is if

$$f(v), \qquad v = 0, 1, \ldots, N-1 \quad \text{and} \quad F(u), \qquad u = 0, 1, \ldots, N-1$$

and

$$g(v), \qquad v = 0, 1, \ldots, N-1 \quad \text{and} \quad G(u), \qquad u = 0, 1, \ldots, N-1 \qquad (6)$$

are two DFT pairs then for any complex constants c and d,

$$cf v) + dg(v) \quad \text{and} \quad cF(u) + dG(u) \qquad (7)$$

is a DFT pair.

The properties of symmetry, scaling and parity of the transform are similarly preserved for the DFT.

Convolution Theorem and Term-by-Term Products of Sequences

Let $f(v)$ and $g(v)$, and $F(u)$ and $G(u)$ be two sequences and their DFT's, respectively, defined as in (6). The term-by-term product of the sequences $F(u)$ and $G(u)$ is the sequence whose kth term is $F(k)G(k)$. The inverse transform of this product, in similarity to the convolution property of the continuous Fourier transform, turns out to be a convolution of the sequences $f(v)$ and $g(v)$. More explicitly, we have that the sequences

$$h(v) = \sum_{\tau=0}^{N-1} f(\tau)g(v-\tau) = \sum_{\tau=0}^{N-1} f(v-\tau)g(\tau), \qquad v = 0, 1, \ldots, N-1 \qquad (8)$$

and $H(u) = F(u)G(u), \qquad u = 0, 1, \ldots, N-1$

constitute a discrete Fourier transform pair.
Similarly,

$$k(v) = f(v)g(v), \qquad v = 0, 1, \ldots, N-1$$

and $K(u) = \dfrac{1}{N} \sum_{\tau=0}^{N-1} F(\tau)G(u-\tau) = \dfrac{1}{N} \sum_{\tau=0}^{N-1} F(u-\tau)G(\tau), \qquad u = 0, 1, \ldots, N-1 \qquad (9)$

constitute a discrete Fourier transform pair.

It is important to note that the convolutions defined by (8) and (9) are cyclic; that is, when one sequence moves over the end of the other, it does not encounter zeros, but rather the periodic extension of the sequence. This is consistent with the periodic extension property of the DFT which was noted earlier.

The convolution theorem corresponding to 5.2–(1) can, thus, be stated as

$$\sum_{\tau=0}^{N-1} f(\tau)g(v-\tau) = \dfrac{1}{N} \sum_{u=0}^{N-1} F(u)G(u) \, w_N^{uv} \qquad (10)$$

where, as before, $w_N = \exp(2\pi i/N)$ with $i = \sqrt{-1}$.

That is, we have a theorem which states that the convolution of two discrete functions is equal to the inverse Fourier transform of the product of the discrete Fourier transforms of the functions. The convolution theorem plays a fundamental role in applications of the discrete Fourier transform.

Proof of the relations given by (8), (9), and (10) is easily obtained using the orthogonality relationships (5).

5.4 The Fourier Transform and Operations on Polynomials

We shall now treat the subject of a rather remarkable relationship between the DFT and operations on polynomials. These results will subsequently be used in some applications of the DFT (cf. Chapter 6).

Let $(a_q, q = 0, \ldots, N - 1)$ be a sequence of real or complex numbers. The discrete Fourier transform of the sequence is defined as

$$b_p = \sum_{q=0}^{N-1} a_q w_N^{pq}, \qquad p = 0, 1, \ldots, N - 1. \tag{1}$$

where, as before, $w_N = \exp(2\pi i/N)$.

Now, denote the sequences $(a_q, q = 0, \ldots, N - 1)$ and $(b_p, p = 0, \ldots, N - 1)$ by two vectors $\mathbf{a}$ and $\mathbf{b}$, respectively, of length N each. Furthermore, let $\mathbf{A}$ be an $N \times N$ matrix such that

$$\mathbf{A} = \{\alpha_{pq}\} = \{w_N^{pq}\}. \tag{2}$$

Then the vector

$$\mathbf{b} = \mathbf{A}\mathbf{a} \tag{3}$$

whose pth component b_p as given in formula (1) is called the DFT of vector $\mathbf{a}$. The matrix $\mathbf{A}$ is non-singular and thus the inverse $\mathbf{A}^{-1}$ exists. Using the orthogonality relationships 5.3–(5) it can be easily shown that inverse $\mathbf{A}^{-1}$ has the simple form given by

$$\mathbf{A}^{-1} = \{\beta_{pq}\} = \{w_N^{-pq}\}. \tag{4}$$

On the basis of definitions (1) and (2) a close relationship may be established between the discrete Fourier transform and polynomial evaluation and interpolation.

Let

$$P(x) = \sum_{k=0}^{N-1} a_k x^k \tag{5}$$

be an $(n - 1)$st degree polynomial. This polynomial can be uniquely represented in one of two ways: either by a set of its coefficients $(a_k, k = 0, \ldots, N - 1)$, or by a set of its values at N distinct points $(Px_j), j = 0, \ldots, N - 1)$.

(Given the values of $P(x)$ at $x_0, x_1, \ldots x_{N-1}$, the process of finding the coefficient representation of a polynomial is called interpolation.) When dealing with polynomials it is often desirable to be able to quickly convert from one form of representation to another.

We shall now show that computing the DFT of a vector $\mathbf{a} = (a_0, a_1, \ldots, a_{N-1})$ is equivalent to converting the coefficient representation of the polynomial (5) to its value representation at the N points $x_0 = w_N^0, x_1 = w_N^1, \ldots, x_{N-1} = w_N^{N-1}$, where $w_N = \exp(2\pi i/N)$. Similarly, the inverse Fourier transform is equivalent to interpolating a polynomial given its values at the points $x_k = w_N^k, k = 0, \ldots, N - 1$. One could define a discrete transform which evaluates a polynomial at a set of points other than the (w_N^k). For example, the integers $1, 2, \ldots, N$ could be used, cf. Aho, Hopcroft, and Ullman (1974).

Assume, first, that a given polynomial $P(x)$ is represented by a set of its coefficients, $(a_k, k = 0, \ldots, N - 1)$, and we wish to convert this representation to the polynomial value representation, $(P(x_j), j = 0, \ldots, N - 1)$. To do this we can choose N arbitrary distinct values of the argument x and evaluate the polynomial at these points. The result will be a set of N points $(P(x_j))$ which is a value representation of the polynomial. Now, suppose that the points x_j are chosen to be equal to

$$x_j = w_N^j = \exp(2\pi ij/N), \qquad j = 0, \ldots, N - 1, \qquad \text{with } i = \sqrt{-1} \tag{6}$$

We thus have

$$\begin{aligned} P(x_0) &= a_{N-1}(w_N^0)^{N-1} + a_{N-2}(w_N^0)^{N-2} + \ldots + a_0 \\ P(x_1) &= a_{N-1}(w_N^1)^{N-1} + a_{N-2}(w_N^1)^{N-2} + \ldots + a_0 \\ &\vdots \\ P(x_{N-1}) &= a_{N-1}(w_N^{N-1})^{N-1} + a_{N-2}(w_N^{N-1})^{N-2} + \ldots + a_0 \end{aligned} \tag{7}$$

The matrix form of set (7) is given as

$$\begin{bmatrix} P(x_0) \\ P(x_1) \\ \vdots \\ P(x_{N-1}) \end{bmatrix} = \begin{bmatrix} 1 & w_N^0 & (w_N^0)^2 & \ldots & (w_N^0)^{N-1} \\ 1 & w_N^1 & (w_N^1)^2 & \ldots & (w_N^1)^{N-1} \\ & & & \vdots & \\ 1 & w_N^{N-1} & (w_N^{N-1})^2 & \ldots & (w_N^{N-1})^{N-1} \end{bmatrix} \begin{bmatrix} a_0 \\ a_1 \\ \vdots \\ a_{N-1} \end{bmatrix} \tag{8}$$

Comparing (8) with (3) we see that the vector

$$\mathbf{P(x)} = (P(x_0), P(x_1), \ldots, P(x_{N-1}))$$

is a DFT of the vector $\mathbf{a} = (a_0, a_1, \ldots, a_{N-1})$.

The statement concerning the inverse DFT and a polynomial interpolation can be proved in a similar way.

The Convolution of Two Vectors and Multiplication of Two Polynomials

One of the important applications of the discrete Fourier transform is computing the convolution of two vectors.

Let $\mathbf{a} = (a_0, a_1, \ldots, a_{N-1})^T$ and $\mathbf{b} = (b_0, b_1, \ldots, b_{N-1})^T$ be two column vectors. We define the convolution of $\mathbf{a}$ and $\mathbf{b}$ as the vector $\mathbf{c} = (c_0, c_1, \ldots, c_{2N-2})^T$ where

$$c_i = \sum_{j=0}^{N-1} a_j b_{i-j} \tag{9}$$

and take $a_k = b_k = 0$ if $k < 0$ or $k \geqslant N$.
Thus, for example, for $N = 3$ we have

$$
\begin{aligned}
c_0 &= a_0 b_0 \\
c_1 &= a_0 b_1 + a_1 b_0 \\
c_2 &= a_0 b_2 + a_1 b_1 + a_2 b_0 \\
c_3 &= a_1 b_2 + a_2 b_1 \\
c_4 &= a_2 b_2
\end{aligned}
\tag{10}
$$

Now, consider the representation of a polynomial by its coefficients. The product of two $(N - 1)$st degree polynomials

$$
P(x) = \sum_{j=0}^{N-1} a_j x^j \quad \text{and} \quad Q(x) = \sum_{j=0}^{N-1} b_j x^j
\tag{11}
$$

is the $(2N - 2)$nd degree polynomial

$$
P(x)Q(x) = \sum_{k=0}^{2N-2} \left(\sum_{j=0}^{k} a_j b_{k-j} \right) x^k.
\tag{12}
$$

We note that the coefficients of the product polynomial are exactly the components of the convolution of the coefficient vectors **a** and **b** of the original polynomials. Hence, if the two $(N - 1)$st degree polynomials are represented by their coefficients then the coefficient representation of their product can be obtained by computing convolutions of the coefficient vectors of the two polynomials-factors.

If $P(x)$ and $Q(x)$ are represented by their values at some N points, then to compute the values representation of their product we can multiply pairs of values at the corresponding points.

The properties of the discrete Fourier transform discussed in this Section are used in the derivation of the main result of Chapter 6.

5.5 The Fast Fourier Transform

We now turn to discuss a highly efficient method for calculating the discrete Fourier transform of a given discrete function $f(v)$. First, recall the definition of the finite DFT, i.e.
given $f(v), v = 0, 1, \ldots, N - 1$, its DFT is defined as

$$
F(u) = \sum_{v=0}^{N-1} f(v) w_N^{-vu}, \qquad u = 0, 1, \ldots, N - 1
\tag{1}
$$

where

$$
w_N = \exp(2\pi i/N) \quad \text{with} \quad i = \sqrt{-1}.
$$

In practical problems, where the DFT is used to find the solution, the number of points, N, may easily be anything between several thousand and a million.

For decades discrete Fourier transforms were computed by a method based on a direct implementation of the definition, thus requiring N^2 operations of multiplication and addition. This fact explains the reputation acquired by the DFT for being a lengthy time-consuming calculation.

The Cooley and Tukey algorithm of the Fast Fourier Transform reduces this number of operations to the number proportional to $N \log N$.

We shall now study this remarkable algorithm in greater detail.

5.5.1 The FFT Algorithm: Elementary Derivation

First, we shall give an elementary derivation of the Fast Fourier Transform (FFT) algorithm. This elementary derivation reveals the essential idea behind the algorithm. In later Sections a more detailed derivation will be given. Consider formula 5.5–(1). For the detailed mathematical analysis which follows, it is convenient to slightly modify notation in the formula: we denote the exponential function w_N^{-vu} by $e\left(\dfrac{vu}{N}\right)$, so that

$$e\left(\frac{vu}{N}\right) = w_N^{-vu} = \exp(-2\pi i vu/N). \tag{1}$$

We note two important properties of $e(x)$, namely that

$$e(r + s) = e(r)e(s)$$

and

$$e(rN) = 1, \quad \text{if } r \text{ is an integer.} \tag{2}$$

Suppose now that N is a product of two factors, r_1 and r_2, i.e. $N = r_1 r_2$. Then the parameters v and u can be expressed in the form:

$$
\begin{aligned}
v = p_0 + p_1 r_2, \quad &\text{with} \quad p_0 = 0, 1, \ldots, r_2 - 1 \\
&\text{and} \quad p_1 = 0, 1, \ldots, r_1 - 1, \\
u = q_0 + q_1 r_1 \quad &\text{with} \quad q_0 = 0, 1, \ldots, r_1 - 1 \\
&\text{and} \quad q_1 = 0, 1, \ldots, r_2 - 1
\end{aligned}
\tag{3}
$$

We thus have for 5.5–(1)

$$
\begin{aligned}
F(q_0 + q_1 r_1) &= \sum_{p_0=0}^{r_2-1} \sum_{p_1=0}^{r_1-1} f(p_0 + p_1 r_2) e\left(\frac{(p_0 + p_1 r_2)(q_0 + q_1 r_1)}{N}\right) \\
&= \sum_{p_0=0}^{r_2-1} \sum_{p_1=0}^{r_1-1} f(p_0 + p_1 r_2) e\left(\frac{q_0 p_0}{r_1 r_2}\right) e\left(\frac{q_0 p_1}{r_1}\right) e\left(\frac{q_1 p_0}{r_2}\right),
\end{aligned}
$$

since $p_1 q_1$ is integer and so $e(p_1 q_1) = 1$,

$$
= \sum_{p_0=0}^{r_2-1} e\left(\frac{q_1 p_0}{r_2}\right) \left\{ e\left(\frac{q_0 p_0}{r_1 r_2}\right) \sum_{p_1=0}^{r_1-1} f(p_0 + p_1 r_2) e\left(\frac{q_0 p_1}{r_1}\right) \right\}. \tag{4}
$$

If we now define the r_2 different sequences

$$\{c_{p_0}(p_1) = f(p_0 + p_1 r_2), \quad p_1 = 0, 1, \ldots, r_1 - 1\} \tag{5}$$

and the r_1 different sequences

$$\left\{ c_{q_0}(p_0) = e\left(\frac{q_0 p_0}{r_1 r_2}\right) \sum_{p_1=0}^{r_1-1} f(p_0 + p_1 r_2) e\left(\frac{q_0 p_1}{r_1}\right), \quad p_0 = 0, 1, \ldots, r_2 - 1 \right\}, \tag{6}$$

we then have

$$F(q_0 + q_1 r_1) = \sum_{p_0=0}^{r_2-1} e\left(\frac{p_0 q_1}{r_2}\right) c_{q_0}(p_0), \qquad \begin{array}{l} q_0 = 0, \ldots, r_1 - 1, \\ q_1 = 0, \ldots, r_2 - 1, \end{array} \tag{7}$$

where

$$c_{q_0}(p_0) = e\left(\frac{q_0 p_0}{r_1 r_2}\right)\left\{ \sum_{p_1=0}^{r_1-1} e\left(\frac{q_0 p_1}{r_1}\right) c_{p_0}(p_1) \right\}, \qquad \begin{array}{l} q_0 = 0, \ldots, r_1 - 1, \\ p_0 = 0, \ldots, r_2 - 1. \end{array} \tag{8}$$

The expressions

$$\sum_{p_0=0}^{r_2-1} e\left(\frac{p_0 q_1}{r_2}\right) c_{q_0}(p_0) \quad \text{and} \quad \sum_{p_1=0}^{r_1-1} e\left(\frac{q_0 p_1}{r_1}\right) c_{p_0}(p_1)$$

are themselves Fourier transforms, but are applied to shorter sequences than the original $N = r_1 r_2$.

Consider an example of $N = 4 \times 3 = r_1 \times r_2 = 12$.

Figs. 5.1 and 5.2 illustrate decomposition, for this case, of a sequence of twelve points, $f(v), v = 0, 1, \ldots, 11$, in short subsequences of four and three points, respectively, for the application of the FFT algorithm. First, in Fig. 5.1, the complete sequence $(f(v))$ shown in diagram (a), is decomposed into three subsequences of $r_1 = 4$ points each; the latter subsequences are shown in diagrams (b), (c) and (d). Consequently, three DFT $\left(\text{times factor } e\left(\frac{q_0 p_0}{12}\right)\right)$ sequences over four points each, are calculated. The result is the sequence

$$\left(c_{q_0}(p_0) = e\left(\frac{q_0 p_0}{12}\right) \sum_{p_1=0}^{3} e\left(\frac{q_0 p_1}{4}\right) c_{p_0}(p_1), \ q_0 = 0, 1, 2, 3 \right) \tag{9}$$

for $p_0 = 0, 1, 2$.

Next, in Fig. 5.2 the sequence of twelve points, $c_{q_0}(p_0)$, as given by (9), is again broken into four subsequences of $r_2 = 3$ points each; the subsequences are shown in diagrams (b), (c), (d) and (e). Finally, four DFT sequences over three points each, are calculated, giving

$$\left(F(q_0 + q_1 r_1) = \sum_{p_0=0}^{2} e\left(\frac{p_0 q_1}{3}\right) c_{q_0}(p_0), \quad q_1 = 0, 1, 2 \right) \tag{10}$$

for $q_0 = 0, 1, 2, 3$.

In general, the Fourier transform of the complete $N = r_1 r_2$ point sequence may thus be accomplished by doing the r_2 different r_1-point Fourier transforms, multiplying through by the appropriate factors, and then doing the r_1 different r_2-point Fourier transforms.

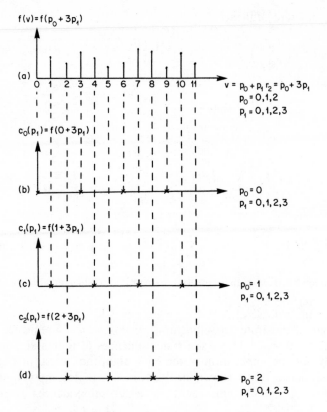

Fig 5.1. The decomposition for $N = 12$, $r_1 = 4$ and $r_2 = 3$ of $f(v)$ in subsequences for the application of the FFT algorithm

For each sequence $\{c_{p_0}(p_1), p_1 = 0, \ldots, r_1 - 1, p_0 \text{ is fixed}\}$,
we compute sequence $\{c_{q_0}(p_0), q_0 = 0, \ldots, r_1 - 1, p_0 \text{ is fixed}\}$.

There are r_2 such sequences altogether.

Then for each sequence $\{c_{p_0}(p_0), p_0 = 0, \ldots, r_2 - 1, q_0 \text{ is fixed}\}$,
we compute sequence $\{F(q_0 + q_1 r_2), q_1 = 0, \ldots, r_2 - 1, q_0 \text{ is fixed}\}$,

and there are r_1 such sequences altogether.

We thus have a recursive procedure which defines the larger Fourier transform in terms of smaller ones. If, following Cooley and Tukey, we use the term 'operation' to mean a complex multiplication followed by a complex addition, then we can easily see that the total number of operations required by the FFT algorithm may be obtained as:

to produce each $c_{q_0}(p_0)$ $r_1 + 1$ operations are required
and there are $N = r_1 r_2$ such elements,

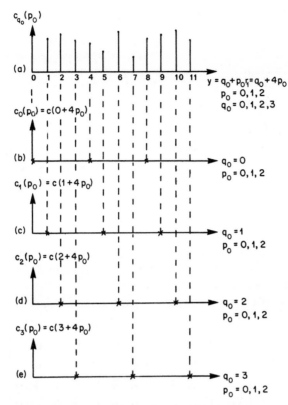

Fig. 5.2. The decomposition for $N = 12, r_1 = 4$ and $r_2 = 3$ of intermediate sequence $c_{q_0}(p_0)$ in subsequences for the purpose of completion of the FFT calculation

hence at this stage	$N(r_1 + 1)$	operations are required;
to produce each $F(q_0 + q_1 r_2)$	r_2	operations are required
and there are $N = r_1 r_2$ such	elements,	
hence at this stage	Nr_2	operations are required.
This gives in total	$N(r_1 + r_2 + 1)$	operations.

The number obtained should be contrasted with N^2 operations required to do the same job using a method based on a direct implementation of the definition of the discrete Fourier transform.

For example, for $N = 10^3$ we may have

$r_1 = 10, r_2 = 100$ giving 111,000 multiplication operations as against 10^6 operations using direct method, a saving of a factor of nearly 10.

5.5.2 Matrix Form of the FFT

The FFT can be conveniently expressed in a matrix form, which reveals in explicit terms the linearity of this transformation process. The matrix form is also useful in error analysis of the computations.

Consider again the example for $N = 4 \times 3 = r_1 r_2 = 12$. We have the following input array:

$$\mathbf{f} = (f_0, f_1, f_2, f_3, f_4, f_5, f_6, f_7, f_8, f_9, f_{10}, f_{11}). \tag{1}$$

The $r_2 = 3$ different subsequences of the sequence $(c_{p_0}(p_1))$ given by 5.5.1–(5) are as follows:

$$\begin{aligned}
c_0(p_1) &= (f_0, f_3, f_6, f_9) \\
c_1(p_1) &= (f_1, f_4, f_7, f_{10}) \\
c_2(p_1) &= (f_2, f_5, f_8, f_{11}).
\end{aligned} \tag{2}$$

(Obtaining a subsequence by starting at the p_0th element and taking every r_2th element thereafter in the manner that $c_{p_0}(p_1)$ is obtained from $f(p_0 + p_1 r_2)$, is called decimating by r_2.)

The matrix form of (2) may be expressed as:

$$
\mathbf{c}^{(1)} =
\begin{bmatrix}
\mathbf{c}_0(p_1) \\
\\
\mathbf{c}_1(p_1) \\
\\
\mathbf{c}_2(p_1)
\end{bmatrix}
\begin{bmatrix}
1 & 0 & 0 & 0 & 0 & 0 & 0 & 0 & 0 & 0 & 0 & 0 \\
0 & 0 & 0 & 1 & 0 & 0 & 0 & 0 & 0 & 0 & 0 & 0 \\
0 & 0 & 0 & 0 & 0 & 0 & 1 & 0 & 0 & 0 & 0 & 0 \\
0 & 0 & 0 & 0 & 0 & 0 & 0 & 0 & 0 & 1 & 0 & 0 \\
0 & 1 & 0 & 0 & 0 & 0 & 0 & 0 & 0 & 0 & 0 & 0 \\
0 & 0 & 0 & 0 & 1 & 0 & 0 & 0 & 0 & 0 & 0 & 0 \\
0 & 0 & 0 & 0 & 0 & 0 & 0 & 1 & 0 & 0 & 0 & 0 \\
0 & 0 & 0 & 0 & 0 & 0 & 0 & 0 & 0 & 0 & 1 & 0 \\
0 & 0 & 1 & 0 & 0 & 0 & 0 & 0 & 0 & 0 & 0 & 0 \\
0 & 0 & 0 & 0 & 0 & 1 & 0 & 0 & 0 & 0 & 0 & 0 \\
0 & 0 & 0 & 0 & 0 & 0 & 0 & 0 & 1 & 0 & 0 & 0 \\
0 & 0 & 0 & 0 & 0 & 0 & 0 & 0 & 0 & 0 & 0 & 1
\end{bmatrix}
\begin{bmatrix}
f_0 \\
f_1 \\
f_2 \\
f_3 \\
f_4 \\
f_5 \\
f_6 \\
f_7 \\
f_8 \\
f_9 \\
f_{10} \\
f_{11}
\end{bmatrix}
\tag{3}
$$

The matrix in (3) is known as a permutation matrix (by a permutation matrix one calls a particular type of a square matrix with unity and zero entries which has exactly one unity entry in each row and in each column); we denote it by $\mathbf{P}$ and thus can write

$$\mathbf{c}^{(1)} = \mathbf{P}\mathbf{f}. \tag{4}$$

Next, we obtain the $r_1 = 4$ different subsequences $\{c_{q_0}(p_0), p_0 = 0, 1, \ldots, r_2 - 1\}$ using equation 5.5.1–(8). (Observe that these subsequences are not quite the Fourier transforms of decimated sequences (2) but are these transforms multiplied by factor $e(q_0 p_0/r_1 r_2)$, (called a 'twiddle factor' by Gentleman and Sande.)

A matrix form of equation 5.5.1–(8) may be deduced as follows:
First, for a *fixed* integer p_0, the matrix form of the equation is

$$
\underbrace{\begin{bmatrix} c_0(p_0) \\ c_1(p_0) \\ c_2(p_0) \\ c_3(p_0) \end{bmatrix}}_{\mathbf{c}(p_0)} = \underbrace{\begin{bmatrix} e\left(\dfrac{0 \times p_0}{12}\right) & & & \\ & e\left(\dfrac{1 \times p_0}{12}\right) & & \\ & & e\left(\dfrac{2 \times p_0}{12}\right) & \\ & & & e\left(\dfrac{3 \times p_0}{12}\right) \end{bmatrix}}_{\mathbf{D}}
$$

$$
\times \underbrace{\begin{bmatrix} e\left(\dfrac{0 \times 0}{4}\right) & e\left(\dfrac{0 \times 1}{4}\right) & e\left(\dfrac{0 \times 2}{4}\right) & e\left(\dfrac{0 \times 3}{4}\right) \\ e\left(\dfrac{1 \times 0}{4}\right) & e\left(\dfrac{1 \times 1}{4}\right) & e\left(\dfrac{1 \times 2}{4}\right) & e\left(\dfrac{1 \times 3}{4}\right) \\ e\left(\dfrac{2 \times 0}{4}\right) & e\left(\dfrac{2 \times 1}{4}\right) & e\left(\dfrac{2 \times 2}{4}\right) & e\left(\dfrac{2 \times 3}{5}\right) \\ e\left(\dfrac{3 \times 0}{4}\right) & e\left(\dfrac{3 \times 1}{4}\right) & e\left(\dfrac{3 \times 2}{4}\right) & e\left(\dfrac{3 \times 3}{4}\right) \end{bmatrix}}_{\mathbf{B}} \underbrace{\begin{bmatrix} c_{p_0}(0) \\ c_{p_0}(1) \\ c_{p_0}(2) \\ c_{p_0}(3) \end{bmatrix}}_{\mathbf{c}_{p_0}}
$$

giving

$$\mathbf{c}(p_0) = \mathbf{DBc}_{p_0}. \tag{4}$$

The matrix form of the complete set of 12 values of $((c_{q_0}(p_0), q_0 = 0, 1, 2, 3)$, $p_0 = 0, 1, 2)$ may, thus, be given as

$$\mathbf{c}^{(2)} = \mathbf{D}_1 \mathbf{B}_1 \mathbf{c}^{(1)} = \mathbf{D}_1 \mathbf{B}_1 \mathbf{P}_1 \mathbf{f}, \tag{5}$$

where $\mathbf{c}^{(2)}$ is column vector

$$(c_0(0), c_1(0), c_2(0), c_3(0), c_0(1), c_1(1), c_2(1), c_3(1), c_0(2), c_1(2), c_2(2), c_3(2)),$$

$\mathbf{D}_1$ is the diagonal matrix of 12 twiddle factors $e(q_0 p_0 / r_1 r_2)$,
$\mathbf{B}_1$ is the block-diagonal matrix whose blocks are identical square submatrices $\mathbf{B}$, each being the Fourier transform of dimension $r_1 = 4$,
$\mathbf{P}_1$ is the permutation matrix, and
$\mathbf{f}$ is column vector given by (1).

Finally, the set of Fourier transforms $F(u)$ given by 5.51–(4) in terms of the input data set and in matrix form, may be represented as:

$$\mathbf{F} = \mathbf{B}_2 \mathbf{P}_2 \mathbf{c}^{(2)} = (\mathbf{B}_2 \mathbf{P}_2)(\mathbf{D}_1 \mathbf{B}_1 \mathbf{P}_1)\mathbf{f}, \tag{6}$$

170

where $\mathbf{D}_1, \mathbf{B}_1$ and $\mathbf{P}_1$ are as before,

$\mathbf{B}_2$ is a block-diagonal matrix whose blocks are identical square submatrices,

$$
\begin{bmatrix}
e\left(\dfrac{0 \times 0}{3}\right) & e\left(\dfrac{1 \times 0}{3}\right) & e\left(\dfrac{2 \times 0}{3}\right) \\
e\left(\dfrac{0 \times 1}{3}\right) & e\left(\dfrac{1 \times 1}{3}\right) & e\left(\dfrac{2 \times 1}{3}\right) \\
e\left(\dfrac{0 \times 2}{3}\right) & e\left(\dfrac{1 \times 2}{3}\right) & e\left(\dfrac{2 \times 2}{3}\right)
\end{bmatrix},
$$

(it is easy to see that each such block is the matrix of a complex Fourier transform of dimension $r_2 = 3$, and $\mathbf{P}_2$ is an appropriate permutation matrix).

The FFT algorithm described here differs somewhat from the original algorithm due to Cooley and Tukey. The difference which is more conveniently displayed in the case when N may be decomposed into three or more factors relates to improving the efficiency performance of the algorithm when different data storage environments are used during the calculation.

5.5.3 The FFT Algorithm for a Non-Uniform Factorization

Consider now the case when N can be factorized into three generally unequal factors, i.e. $N = r_1 r_2 r_3$.

First, we have as before

$$
F(u) = \sum_{v=0}^{N-1} f(v) w_N^{-uv} = \sum_{v=0}^{N-1} f(v) e\!\left(\frac{uv}{N}\right), \qquad u = 0, 1, \ldots, N-1 \tag{1}
$$

where $e\!\left(\dfrac{uv}{N}\right) = w_N^{-uv} = \exp(-2\pi i uv/N)$ with $i = \sqrt{-1}$.

As an obvious extension of the two-factor case we now express the parameters u and v in the form:

$$
\begin{aligned}
v = p_0 + p_1 r_1 + p_2 r_1 r_2, \qquad & p_0 = 0, \ldots, r_1 - 1 \\
& p_1 = 0, \ldots, r_2 - 1 \\
& p_2 = 0, \ldots, r_3 - 1 \\
u = q_0 + q_1 r_3 + q_2 r_2 r_3, \qquad & q_0 = 0, 1, \ldots, r_3 - 1 \\
& q_1 = 0, 1, \ldots, r_2 - 1 \\
& q_2 = 0, 1, \ldots, r_1 - 1
\end{aligned} \tag{2}
$$

We shall now distinguish between the two FFT algorithms, generally known as the decimation-in-time FFT, i.e. the Cooley–Tukey (1965) variant, and the decimation-in-frequency FFT i.e. the Gentleman–Sande (1966) variant.

The basic features of both alogorithms are the same, but the formulae differ somewhat, and this latter fact leads to different ways of indexing the intermediate iteration arrays. Since the FFT often handles quite large arrays of data, it is of importance of keep the indexing of all intermediate as well as the final results as systematic and as simple as possible.

Depending on the data storage environment, one way of indexing intermediate results may be preferable to the other.

The FFT Algorithm which Uses Auxiliary Storage

We shall study the indexing system of the Cooley–Tukey algorithm assuming that the data are given as a sequence stored in a serial order and that an auxiliary array of the same size, N, is available to store the output.
From (1) and (2) we have

$$F(q_0 + q_1 r_3 + q_2 r_2 r_3) = \sum_{p_0=0}^{r_1-1} \sum_{p_1=0}^{r_2-1} \sum_{p_2=0}^{r_3-1} f(p_0 + p_1 r_1 + p_2 r_1 r_2)$$
$$\times e\left(\frac{(q_0 + q_1 r_3 + q_2 r_2 r_3)(p_0 + p_1 r_1 + p_2 r_1 r_2)}{r_1 r_2 r_3}\right) \quad (3)$$

We shall now factorize the exponential function in terms of the parameters p_i, $i = 0, 1, 2$, that is in relation to the time variable v. The factorized expression is

$$F(u) = \sum_{p_0=0}^{r_1-1} \sum_{p_1=0}^{r_2-1} \sum_{p_2=0}^{r_3-1} f(p_0 + p_1 r_1 + p_2 r_1 r_2) e\left(\frac{p_0(q_0 + q_1 r_3 + q_3 r_2 r_3)}{r_1 r_2 r_3}\right)$$
$$\times e\left(\frac{p_1(q_0 + q_1 r_3 + q_2 r_2 r_3)}{r_2 r_3}\right) e\left(\frac{p_2(q_0 + q_1 r_3 + q_2 r_2 r_3)}{r_3}\right)$$
$$= \sum_{p_0=0}^{r_1-1} e\left(\frac{p_0(q_0 + q_1 r_3 + q_2 r_2 r_3)}{r_1 r_2 r_3}\right) \sum_{p_1=0}^{r_2-1} e\left(\frac{p_1(q_0 + q_1 r_3)}{r_2 r_3}\right)$$
$$\times \sum_{p_2=0}^{r_2-1} e\left(\frac{p_2 q_0}{r_3}\right) f(p_0 + p_1 r_1 + p_2 r_1 r_2).$$

Here we used results such as, for example,

$$e\left(\frac{p_1(q_0 + q_1 r_3 + q_2 r_2 r_3)}{r_2 r_3}\right) = e\left(\frac{p_1(q_0 + q_1 r_3)}{r_2 r_3}\right) e(q_2) = e\left(\frac{p_2(q_0 + q_1 r_3)}{r_2 r_3}\right)$$

since $e(q_2) = e^{i2\pi q_2} = 1$.
Thus,

$$F(q_0 + q_1 r_3 + q_2 r_2 r_3) = \sum_{p_0=0}^{r_1-1} e\left(\frac{p_0 q_2}{r_1}\right) e\left(\frac{p_0(q_0 + q_1 r_3)}{r_1 r_2 r_3}\right) \sum_{p_1=0}^{r_2-1} e\left(\frac{p_1 q_1}{r_2}\right)$$
$$\times e\left(\frac{p_1 q_0}{r_2 r_3}\right) \sum_{p_2=0}^{r_2-1} e\left(\frac{p_2 q_0}{r_3}\right) f(p_0 + p_1 r_1 + p_2 r_1 r_2). \quad (4)$$

The computational procedure which calculates the sequence of the Fourier transforms, given by (4) is as follows:

1. Compute the sequence

$$
c_{q_0}(p_0,p_1) = e\left(\frac{p_1 q_0}{r_2 r_3}\right) \sum_{p_2=0}^{r_3-1} e\left(\frac{p_2 q_0}{r_3}\right) f(p_0 + p_1 r_1 + p_2 r_1 r_2),
$$

$$
p_0 = 0, 1, \ldots, r_1 - 1; \quad p_1 = 0, 1, \ldots, r_2 - 1; \quad q_0 = 0, 1, \ldots, r_3 - 1. \tag{5}
$$

This is an r_3–point Fourier transform of points spaced $r_1 r_2$ apart. The *twiddle* factor in front of the sum in (5) depends on q_0 and p_1; and p_0 is called a 'free' index. The values obtained are stored at the serial locations denoted by $q_0 + p_0 r_3 + p_1 r_1 r_3$.

2. Compute the sequence

$$
c_{q_0} + q_1 r_3(p_0) = e\left(\frac{p_0(q_0 + q_1 r_3)}{r_1 r_2 r_3}\right) \sum_{p_1=0}^{r_2-1} e\left(\frac{p_1 q_1}{r_2}\right) c_{q_0}(p_0 p_1),
$$

$$
p_0 = 0, 1, \ldots, r_1 - 1; \quad q_1 = 0, 1, \ldots, r_2 - 1; \quad q_0 = 0, 1, \ldots, r_3 - 1. \tag{6}
$$

This is an r_2-point Fourier transform of points spaced $r_1 r_3$ apart. The *twiddle* factor depends on $q_0 + q_1 r_3$ and p_0. There is no 'free' index, i.e. the *twiddle* factor depends on all three indices by which the sequence (6) is indexed. The values obtained are stored at the serial locations denoted by $q_0 + q_1 r_3 + p_0 r_2 r_3$.

3. Compute the sequence

$$
F(q_0 + q_1 r_3 + q_2 r_2 r_3) = \sum_{p_0=0}^{r_1-1} e\left(\frac{p_0 q_2}{r_1}\right) c_{q_0} + q_1 r_3(p_0)
$$

$$
q_2 = 0, 1, \ldots, r_1 - 1; \quad q_1 = 0, 1, \ldots, r_2 - 1; \quad q_0 = 0, 1, \ldots, r_3 - 1. \tag{7}
$$

This is an r_1-point Fourier transform of points spaced $r_2 r_3$ apart. There is no twiddle factor at this step. The final results are stored in serial order at points denoted by $q_0 + q_1 r_3 + q_2 r_2 r_3$.

The total number of operations required by this algorithm is given by

$$
(r_3 + 1)N + (r_2 + 1)N + r_1 N = N(r_1 + r_2 + r_3 + 2).
$$

The algorithm described performs efficiently in the cases when an auxiliary array is readily available. It may happen, however, that the input array is so large that provision of the necessary auxiliary storage may significantly slow down the process of computation. This would be the case, for example, when backing store facilities which are relatively slowly accessed have to be used. We shall now show that, in fact, the FFT may be carried out *in situ* (i.e. without using any auxiliary storage) though at the price of giving up the serial order of storage of the final Fourier transforms.

5.5.4 The FFT Algorithm *in situ*

In the cases when the FFT algorithm has to be carried out without using an auxiliary storage, its variant developed by Gentleman and Sande is found to be more flexible and convenient to use.

To obtain this algorithm, factorization of the exponential function in equation 5.5.3–(3) is carried out in relation to the parameters q_i, $i = 0, 1, 2$, that is, in relation to the variable u. The corresponding formula is found to be:

$$F(q_0 + q_1 r_3 + q_2 r_2 r_3) = \sum_{p_0=0}^{r_1-1} e\left(\frac{p_0 q_2}{r_1}\right) e\left(\frac{p_0 q_1}{r_1 r_2}\right) \sum_{p_1=0}^{r_2-1} e\left(\frac{p_1 q_1}{r_2}\right)$$

$$\times e\left(\frac{q_0(p_0 + p_1 r_1)}{r_1 r_2 r_3}\right) \sum_{p_2=0}^{r_3-1} e\left(\frac{p_2 q_0}{r_3}\right)$$

$$\times f(p_0 + p_1 r_1 + p_2 r_1 r_2) \tag{8}$$

As before we will suppose that the FFT algorithm is applied to data given as a sequence stored in serial order.

First, for convenience in notation, which will become apparent later, we denote the input array, $(f(v), \; v = p_0 + p_1 r_1 + p_2 r_1 r_2, \; 0 \leqslant v \leqslant N - 1)$, by $(c(p_0, p_1, p_2))$ where as defined earlier $p_0 = 0, \ldots, r_1 - 1$, $p_1 = 0, \ldots, r_2 - 1$, $p_2 = 0, \ldots, r_3 - 1$, and $N = r_1 r_2 r_3$.

The procedure for computing the sequence of the Fourier transforms given by (8) is as follows:

1. Compute the sequence

$$c_{q_0}(p_0 + p_1 r_1) = e\left(\frac{q_0(p_0 + p_1 r_1)}{r_1 r_2 r_3}\right) \sum_{p_0=0}^{r_3-1} e\left(\frac{p_2 q_0}{r_3}\right) c(p_0, p_1, p_2),$$

$$p_0 = 0, 1, \ldots, r_1 - 1; \quad p_1 = 0, 1, \ldots, r_2 - 1; \quad q_0 = 0, 1, \ldots, r_3 - 1. \tag{9}$$

This is an r_3-point Fourier transform of points spaced $r_1 r_2$ apart. Consider the subset of values $c_{q_0}(p_0 + p_1 r_1)$ given by this equation, when the index q_0 assumes its complete range of values and the integer $p_0 + p_1 r_1$ is fixed. To produce this subset, we have used the r_3 input data values $c(p_0, p_1 p_2) = f(p_0 + p_1 r_1 + p_2 r_1 r_2)$, where $p_2 = 0, \ldots, r_3 - 1$. These values will not be needed in further computations and, hence, we can vacate the r_3 corresponding locations. The vacated locations have so far been denoted by $p_0 + p_1 r_1 + p_2 r_1 r_2$. They can now be used to store the subset of values just computed. The new notation for these locations may conveniently be chosen as $p_0 + p_1 r_1 + q_0 r_1 r_2$, where $q_0 = 0, 1, \ldots, r_3 - 1$. Such a notation used for the storage locations is consistent and is not contradictory or overlapping, since the summation on p_2 represents an r_3-point Fourier transform of values spaced $r_1 r_2$ locations apart and the twiddle factor $e(q_0(p_0 + p_1 r_1)/r_1 r_2 r_3)$ depends upon the parameters given as q_0 and $p_0 + p_1 r_1$. The complete set of N values computed may be denoted as

$$(c(p_0, p_1, q_0)) = (c_{q_0}(p_0 + p_1 r_1), \; p_0 = 0, \ldots r_1 - 1, \; p_1 = 0, \ldots, r_2 - 1,$$

$$q_0 = 0, \ldots, r_3 - 1).$$

2. Compute the sequence

$$c_{q_0 + p_0 r_3}(q_1) = e\left(\frac{p_0 q_1}{r_1 r_2}\right) \sum_{p_1=0}^{r_2-1} e\left(\frac{p_1 q_1}{r_2}\right) c(p_0, p_1, q_0), \tag{10}$$

$$p_0 = 0, 1, \ldots, r_1 - 1; \quad q_1 = 0, 1, \ldots, r_2 - 1; \quad q_0 = 0, 1, \ldots, r_3 - 1.$$

Here, we have the summation on p_1 which represents an r_2-point Fourier transform of values $c(p_0, p_1, q_0) = c_{q_0}(p_0 + p_1 r_1)$ placed $r_1 r_3$ locations apart. The twiddle factor depends on the parameters p_0 and q_1 only. Considering the subset of values $c_{q_0 + p_0 r_3}(q_1)$, $q_1 = 0, \ldots, r_2 - 1$ given by this equation, in a manner similar to the above, we note that to produce this subset, we have used the r_2 values $c(p_0, p_1, q_0) = c_{p_0}(p_0 + p_1 r_1)$. These values will not be needed in further Fourier computations and so we can vacate the r_2 corresponding locations which are denoted by $p_0 + p_1 r_1 + q_0 r_1 r_2$ where $p_1 = 0, \ldots, r_2 - 1$. As before, the vacated locations are used to store the subset of values just computed. The new notation for these locations may conveniently be chosen as $p_0 + q_1 r_1 + q_0 r_1 r_2$, where $q_1 = 0, \ldots, r_2 - 1$.

The complete set of N values computed may be denoted as

$$c(p_0, q_1, q_0) = (c_{q_0 + p_0 r_3}(q_1), \ p_0 = 0, \ldots, r_1 - 1, \ q_1 = 0, \ldots, r_2 - 1,$$
$$q_0 = 0, \ldots, r_3 - 1).$$

3. Compute the sequence

$$c(q_2, q_1, q_0) = \sum_{p_0 = 0}^{r_1 - 1} e\left(\frac{p_0 q_2}{r_1}\right) c(p_0, q_1, q_0) \tag{11}$$

$$q_2 = 0, 1, \ldots, r_1 - 1; \quad q_1 = 0, 1, \ldots, r_2 - 1; \quad q_0 = 0, 1, \ldots, r_3 - 1.$$

Here, the summation over p_0 represents N contiguous r_1-point Fourier transforms. These results are successively stored at the locations denoted by $q_2 + q_1 r_1 + q_0 r_1 r_2$, where $q_2 = 0, \ldots, r_1 - 1$. The complete set of N final transforms may now be written as

$$(F(u), u = q_0 + q_1 r_1 + q_2 r_2 r_3) = (c(q_2, q_1, q_0),$$
$$q_2 = 0, \ldots, r_1 - 1, \quad q_1 = 0, \ldots, r_2 - 1, \quad q_0 = 0, \ldots, r_3 - 1). \tag{12}$$

The Cooley–Tukey algorithm can be used in the *in-situ* situation but its indexing of the intermediate results needs a careful handling as otherwise it may present some difficulties in storing the results.

What is more significant, however, is the observation that in the case of *in-situ* calculations of either variant, the final Fourier transform, $F(u)$, with $u = q_0 + q_1 r_3 + q_2 r_2 r_3$, is stored at the array location denoted by $q_2 + q_1 r_1 + q_0 r_1 r_2$. Such a storage scheme is sometimes called a storage with 'digit reversed subscripts'. This feature of the algorithm has to be born in mind, especially since quite frequently the Fourier transforms are first calculated and then immediately used for some further purposes. In this case it may be convenient first to rearrange the Fourier transforms in their proper places in the array so as to obtain the array in the usual serial order form suitable for further use. A computational device for 'unscrambling the Fourier transforms' as it is referred to by Gentleman and Sande, will be given later.

5.6 Optimal Factorization for the FFT Algorithm

We have established that various degrees of saving in computing time may be achieved, using the FFT algorithm for evaluation of the finite discrete Fourier transforms, provided an appropriate factorization of the input data is applied. Denoting, as usual, the size of the input data array by N, we shall now investigate the best way to factorize N so as to achieve the greatest possible saving, using the FFT algorithm.

Let N be factorized into k factors, thus

$$N = r_1 r_2 r_3 \ldots r_k. \tag{1}$$

Using the FFT factorization technique in the way similar to the two- and three-factor cases discussed earlier, it is easy to demonstrate that the required number of operations is equal to

$$N\left(k - 1 + \sum_{j=1}^{k} r_j\right) \quad \text{complex multiplications and}$$

$$N\left(\sum_{j=1}^{k} (r_j - 1)\right) \quad \text{complex additions.}$$

For brevity, we shall use the term 'operation' to mean, as before, a complex multiplication followed by a complex addition. The number of such operations may be given as

$$P = N\left(\sum_{j=1}^{k} r_j\right), \tag{2}$$

where the terms $N(k - 1)$ in the number of complex multiplications and the term Nk in the number of complex additions are ignored as they do not affect the optimality criteria.

Equations (1) and (2) yield

$$\log_2 N = \log_2 r_1 + \log_2 r_2 + \ldots + \log_2 r_k$$

and

$$\frac{P}{N} = r_1 + r_2 + \ldots + r_k,$$

giving

$$\frac{P}{N \log_2 N} = \left(\sum_{j=1}^{k} r_j\right) \bigg/ \left(\sum_{j=1}^{k} \log_2 r_j\right)$$

which may conveniently be presented as

$$\frac{P}{N \log_2 N} = \frac{\sum_{j=1}^{k} \left(\dfrac{r_j}{\log_2 r_j}\right) \log_2 r_j}{\sum_{j=1}^{k} \log_2 r_j}. \tag{3}$$

We now suppose some freedom in the selection of N and its factors, r_j, and determine choices to minimize (3) with respect to r_j, $j = 1, \ldots, k$.

If all of the factors N are equal, $r_1 = r_2 = \ldots = r_k = r$, then (3) can be written as

$$\frac{P}{N \log_2 N} = \frac{r}{\log_2 r},$$

giving

$$P = \left(\frac{r}{\log_2 r}\right) N \log_2 N = rN \log_r N. \tag{4}$$

In this case, we say that we have a radix r, or base r, algorithm. If the factors are not all equal, as in (3), we say that we have a mixed radix algorithm. From (4) we can see that for a radix r algorithm P is proportional to $N \log_2 N$ with a proportionality factor $r/\log_2 r$, whose values are listed in Table 5.1.

Notice that if N is a power of 3 then the number of operations, P, assumes its lowest value equal to

$$P = 1.88 \, N \log_2 N. \tag{5}$$

In this sense, the right-hand side of (5) may be interpreted as the lower bound on the number of operations required to compute the discrete Fourier transform on a set of N points, using the FFT algorithm.

If $N = 2^m$, we may take $r = 2$, obtaining

$$P = 2N \log_2 N, \tag{6}$$

and the same expression holds for P if $N = 4^n$. For given N, this number is quite close to the lower bound on the number of operations, given by (5).

The use of radices 2 and 4 (also 8 and 16) has the advantage that some of the powers of $w_N = \exp(2\pi i/N)$ are simple numbers like ± 1, $\pm i$, $(1 \pm i)/\sqrt{2}$, and multiplications can be avoided.

Bergland (1968) studied in detail the number of additions and multiplications required for radices 2, 4, 8, and 16 if the program written economically, omitting multiplications by simple powers of w_N and combining terms with common factors. The results show some economy in the use of radices 2, 4, 8, and 16 but the savings for 8 and 16 are small compared to the increased complexity of the algorithm. In fact, the radix 2 and 4 algorithms are the ones most widely programmed and used, whenever a choice presents itself of appropriate factorization of N. The FFT algorithm of mixed radices have also been programmed and applied. In particular, the algorithm of mixed radix $4 + 2$ (i.e. all factors equal to 4 except perhaps the last, which may be 2) has been widely used. From Table 5.1 it is seen that the use of factors other than 2 and 4 does not increase

Table 5.1

r	2	3	4	5	6	7	8	9	10	11	16
$\dfrac{r}{\log_2 r}$	2.00	1.88	2.00	2.15	2.31	2.49	2.67	2.82	3.01	3.18	4.00

the time of calculation substantially. For a detailed discussion of the programming of the mixed radix algorithm, see Singleton (1969).

5.7 The FFT Algorithm of Radix 2

Because of its simplicity in programming, the base 2 algorithm has been used extensively in subroutines and in hardware implementations.

Let us consider how it works. We assume that $N = 2^m$ and as an illustration, take the case when $N = 2^4$. The algorithm of radix 2 is then derived by expressing the parameters v and u in binary number form, namely

$$v = 2^{m-1}p_0 + \ldots + 2p_{m-2} + p_{m-1}, \tag{1}$$

in our example

$$v = 2^3 p_0 + 2^2 p_1 + 2p_2 + p_3,$$

and

$$u = 2^{m-1}q_0 + \ldots + 2q_{m-2} + q_{m-1}, \tag{2}$$

in our example

$$u = 2^3 q_0 + 2^2 q_1 + 2q_2 + q_3,$$

where the p's and the q's are equal to 0 or 1 and are termed 'bits'. With this convention, the Fourier transform may be written as a function of its binary parameters:

$$
\begin{aligned}
F(u) &= c(q_0, q_1, q_2, q_3) \\
&= \sum_{p_3} \sum_{p_2} \sum_{p_1} \sum_{p_0} f(p_0, p_1, p_2, p_3) e\left(\frac{up_0 2^3 + up_1 2^2 + up_2 2 + up_3}{N}\right)
\end{aligned} \tag{3}
$$

Transferring certain factors outside the appropriate summations we obtain

$$
\begin{aligned}
F(u) &= c(q_0, q_1, q_2, q_3) \\
&= \sum_{p_3} e\left(\frac{up_3}{2^4}\right) \sum_{p_2} e\left(\frac{up_2}{2^3}\right) \sum_{p_1} e\left(\frac{up_1}{2^2}\right) \sum_{p_0} e\left(\frac{up_0}{2}\right) c_0(p_0, p_1, p_2, p_3)
\end{aligned} \tag{4}
$$

where for uniformity we have denoted $f(p_0, p_1, p_2, p_3)$ by $c_0(p_0, p_1, p_2, p_3)$. We now observe that

$$e\left(\frac{up_0}{2}\right) = e\left(\frac{q_3 p_0}{2}\right),$$

since

$$
\begin{aligned}
e\left(\frac{up_0}{2}\right) &= e^{-i(2\pi/2)(q_0 2^3 + q_1 2^2 + q_2 2 + q_0)p_0} \\
&= e^{-i(2\pi/2)(q_0 2^2 + q_1 2 + q_2)2p_0} e^{-i(2\pi/2)q_3 p_0} = e\left(\frac{q_3 p_0}{2}\right),
\end{aligned}
$$

as

$$e^{-i2\pi(q_0 2^2 + q_1 2 + q_2)p_0} = 1.$$

It follows that the innermost sum of (4), over p_0, depends only on q_3, p_1, p_2, p_3 and can be written as

$$c_1(q_3, p_1, p_2, p_3) = \sum_{p_0} e\left(\frac{up_0}{2}\right) c_0(p_0, p_1, p_2, p_3). \tag{5}$$

There are 16 (or N in the general case) such quantities altogether. The Fourier transform formula (4) now has the form

$$c(q_0, q_1, q_2, q_3) = \sum_{p_3} e\left(\frac{up_3}{2^4}\right) \sum_{p_2} e\left(\frac{up_2}{2^3}\right) \sum_{p_1} e\left(\frac{up_1}{2^2}\right) c_1(q_3, p_1, p_2, p_3) \tag{6}$$

Noting, again, that

$$e\left(\frac{up_1}{2^2}\right) = e\left(\frac{(q_2 2 + q_3)p_1}{2^2}\right),$$

we infer that the innermost sum of (6), over p_1, depends only on q_3, q_2, p_2, p_3 and can be written as

$$c_2(q_3, q_2, p_2, p_3) = \sum_{p_1} c_1(q_3, p_1, p_2, p_3) e\left(\frac{up_1}{2^2}\right). \tag{7}$$

There are, again, N such quantities altogether.

Carrying out two further steps in a manner similar to that just described, we obtain the discrete Fourier transform as

$$F(u) = c(q_0, q_1, q_2, q_3) = c_4(q_3, q_2, q_1, q_0). \tag{8}$$

Note that, as before, the parameter u of $F(u)$ must have its binary bits put in reverse order to yield its index in the array c_4.

Note also that the computational process given by (3)–(8) requires $m = \log_2 N$ steps, and a set of N elements produced at each step.

5.7.1 General Formula and a Flowgraph of the FFT Algorithm of Radix 2

From the relations 5.7–(3)–(8) the kth step of the base 2 algorithm may be expressed as

$$c_{k+1}(q_{m-1}, \ldots, q_{m-1-k}, p_{k+1}, \ldots, p_{m-1})$$
$$= c_k(q_{m-1}, \ldots, q_{m-2-k}, 0, p_{k+1}, \ldots, p_{m-1})$$
$$+ e\left(\frac{q_{m-1} + 2q_{m-2} + \ldots + 2^{k-1}q_{m-k}}{2^{k+1}}\right) c_k(q_{m-1}, \ldots, q_{m-2-k}, 1,$$
$$p_{k+1}, \ldots, p_{m-1}) \tag{1}$$

where $\{c_k(r), r = 0, \ldots, N-1\}$ is the set of N complex numbers,

$$k = 0, \ldots, \log_2 N,$$

$e(x) = e^{-i2\pi x}$, as before, and
all the p's and the q's are either 0 or 1.

Rewriting (1) in a more compact form we get

$$c_{k+1}(r) = \begin{cases} c_k(r) + e(z/2^{k+1})c_k(r + 2^{m-1-k}) & \text{if } q_{m-1-k} = 0, \\ c_k(r - 2^{m-1-k}) + e(z/2^{k+1})c_k(r) & \text{if } q_{m-1-k} = 1. \end{cases} \tag{2}$$

where

$$z = q_{m-1} + 2q_{m-2} + \ldots + 2^k q_{m-1-k} \tag{3}$$

and

$$r = p_{m-1} + 2p_{m-2} + \ldots + 2^{m-2-k}p_{k+1} + 2^{m-1-k}q_{m-1-k} + \ldots + 2^{m-1}q_{m-1}. \tag{4}$$

Note that the arguments $r + 2^{m-1-k}$ and $r - 2^{m-1-k}$ can be obtained from the binary expansion of the r of equation (4), by replacing q_{m-1-k} by its 1-complement. (The 1-complement of a binary number is the result of changing each zero to one and one to zero). The set $(c_m(r), r = 0,\ldots,N-1, m = \log_2 N)$ represents the final set of the discrete Fourier transforms, $(F(u), u = 0,\ldots,N-1)$, though in a rearranged order, so that the Fourier transform $F(u)$, where $u = q_0 + 2q_1 + \ldots + 2^{m-1}q_{m-1}$, is given by $c_m(r)$, where $r = q_{m-1} + 2q_{m-2} + \ldots + 2^{m-1}q_0$, and all the q's are either 0 or 1.

Using the general formula (2) we can now present a flowgraph of the computation of the Fourier transforms. Figure 5.3 illustrates a flowgraph of the actual computation for the case $N = 2^4$. As shown, the data vector or array $(c_0(r)$, $r = 0,\ldots,N-1)$ is represented by a vertical column of nodes on the left of the flowgraph. The second vertical array of nodes is the vector $c_1(r)$ computed in equation (2) for $k = 0$, etc. For general N there will be $m = \log_2 N$ computational arrays. The flowgraph is interpreted as follows. Each node is entered by two solid lines representing directed branches. A branch transmits a quantity from a node in one array, multiplies the quantity by the weight w_N^x, and inputs the result into the node in the next array. The weights, if other than unity, are shown for each branch. Results entering a node from the two branches are combined additively. According to the rules for interpreting the flowgraph we have, for example

$$c_1(11) = c_0(3) + (-1)c_0(11),$$

which is determinded by equations (2)–(4) for $k = 0$ and $r = (11)_2 = 1011$, the latter of which gives, in turn, $q_3 = 1, z = 1$ and

$$e\left(\frac{z}{2^{k+1}}\right) = e^{-2\pi i/2} = e^{-\pi i} = -1.$$

Each node of the flowgraph is expressed similarly.

We can see that the flowgraph is a concise method for representing the computations required in the FFT algorithm.

5.7.2 Analysis of the FFT Flowgraph

We shall now study the flowgraph of Fig. 5.3 in some detail in order to emphasize some useful relations between the nodes. This will, in turn, allow us to simplify the computation formula 5.7.1 − (2) and, what is especially important, to show that the number of actual complex multiplications required to compute and N − length array $(c_k(r) = 0, \ldots, N − 1)$ is equal to $N/2$ and not to N as it may seem from the formula 5.7.1 − (2).

Dual nodes

Consider the node pair $c_1(0)$ and $c_1(8)$. For their calculation only the nodes $c_0(0)$ and $c_0(8)$ are used; these latter nodes are not used in any other calculation in column $k = 1$.

Next consider the node pair $c_1(1)$ and $c_1(9)$. For their calculations only the nodes $c_0(1)$ and $c_0(9)$ are used. Again, these latter nodes are not used in any other calculation in column $k = 1$. Etc.

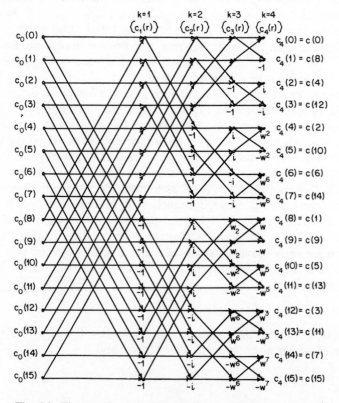

Fig. 5.3. Flowgraph of the Cooley–Tukey FFT algorithm for $N = 2^4$ ($w_N^x = \exp (− 2\pi i x/N)$, e.g. $w_N^8 = − 1$, $w_N^4 = i$, $w_N^{12} = − w_N^4 = − i$, $w_N^0 = 1$). The weights for each branch, if other than unity, are shown immediately below the particular branch

Similar observations are readily done in columns $k = 2$, $k = 3$ and $k = 4$.

We shall call the pairs $c_1(0)$ and $c_1(8)$, $c_1(1)$ and $c_1(9)$, etc., the dual nodes.

We further note that in the kth column the dual nodes are spaced at the distance $r = N/2^k = 2^{m-k}$ nodes, e.g. in column $k = 2$ the dual nodes $c_2(0)$ and $c_2(4)$, $c_2(1)$ and $c_2(5)$, etc., are spaced at the distance $r = 2^{m-k} = 4$ nodes.

These observations enable us to deduce the formula for calculation of the current column nodes in terms of the nodes of the immediately preceding column, as follows:

$$c_k(r) = c_{k-1}(r) + w_N^{z'} c_{k-1}(r + 2^{m-k})$$
$$c_k(r + 2^{m-k}) = c_{k-1}(r) + w_N^{z''} c_{k-1}(r + 2^{m-k}) \tag{1}$$

where the powers z' and z'' have to be determined.

To do this we note that the branches stemming from, say, node $c_1(12)$ are multiplied by factors $w_N^4 = e(4/N) = i$ $(i = \sqrt{-1})$, and

$$w_N^{12} = e\left(\frac{12}{N}\right) = -i \qquad (i = \sqrt{-1}),$$

prior to input at nodes $c_2(8)$ and $c_2(12)$, respectively.

It is important to note that $w_N^4 = -w_N^{12}$ and, thus, in the corresponding formulae given by (1) for computing $c_2(8)$ and $c_2(12)$, only one complex multiplication is required since the same data, $c_1(12)$, is to be multiplied by these factors.

In general, if the weighting factor at one node is

$$w_N^x = e\left(\frac{x}{N}\right),$$

then the weighting factor at the dual node is

$$w_N^{x + (N/2)} = e\left(\frac{x + N/2}{N}\right) = e\left(\frac{x}{N} + \frac{1}{2}\right).$$

And since $e\left(\dfrac{x}{N}\right) = -e\left(\dfrac{x}{N} + \dfrac{1}{2}\right)$, only one multiplication is required in the computation of a dual node pair.

Formulae for computing a dual pair

We, thus, obtain the following formulae for calculating an array $(c_k(r)$, $r = 0, \ldots, N - 1)$:

$$c_k(r) = c_{k-1}(r) + e\left(\frac{z}{2^k}\right) c_{k-1}(r + 2^{m-k}),$$
$$c_k(r + 2^{m-k}) = c_{k-1}(r) - e\left(\frac{z}{2^k}\right) c_{k-1}(r + 2^{m-k}), \tag{2}$$

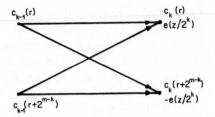

Fig. 5.4. Flowgraph for the basic 'butterfly' operation for the FFT algorithm
of radix 2

where

$$z = q_{m-1} + 2q_{m-2} + \ldots + 2^{k-1}q_{m-k} \tag{3}$$

and

$$r = p_{m-1} + 2p_{m-2} + \ldots + 2^{m-1-k}p_k + 2^{m-k}q_{m-k} + \ldots + 2^{m-1}q_{m-1}, \tag{4}$$

and all the q's and the p's are either 0 or 1.

These formulae are equivalent to 5.7.1–(2)–(4) but are more convenient for practical use. We note that the value of z is determined by

(a) writing the index r in binary form in m bits,
(b) scaling or sliding this binary number $m - 1$ bits to the right and filling in the newly opened bit position on the left with zeros, and
(c) reversing the order of the bits.

A flowgraph section of equation (2) is shown in Fig. 5.4. The diagram is referred to as the basic 'butterfly' operation for the radix 2 algorithm.

Skipping over the second nodes of the dual pairs

In computing an array $(c_k(r), \; r = 0, \ldots, N - 1)$ one normally begins with node $r = 0$ and sequentially works down the array, computing the equation pair (2)–(4). Since the dual of any node in the kth column is always down $N/2^k$ in the column, it follows that one must skip some nodes after every $N/2^k$ node.

For example consider array $k = 2$ of Fig. 5.3. We begin with node $r = 0$. Its dual node is located at $r = N/2^2 = 4$. Next is node $r = 1$, and its dual node is located at $r = 5$, and so forth until we reach node $r = 4$. At this point we have reached a set of nodes previously encountered, as these nodes are the duals for nodes $r = 0, 1, 2$, and 3. So we must skip over the nodes $r = 4, 5, 6$, and 7. We then carry out the computations for nodes $r = 8, 9, 10$, and 11 and again skip over their duals, $r = 12, 13, 14$, and 15. In general, if we work from the top down in array k, then we compute equations (2)–(4) for the first $N/2^k = 2^{m-k}$ nodes, skip the next 2^{m-k}, etc., until we reach a node index greater than $N - 1$.

The FFT radix 2 algorithm is, thus, basically concerned with computing the set of 'butterfly' operations given by (2)–(4), for $k = 1, 2, \ldots, m$, $m = \log_2 N$.

In other words, for every fixed k we compute the array $A(I) = c_k(I - 1)$, $I = 1$, ..., N, using equations (2)–(4). Initially $A(I)$ is set to $c_0(I - 1)$, $I = 1, \ldots, N$. Assuming a fixed k, $1 \leqslant k \leqslant m$, equations (2) may be rewritten in the form which is more convenient for the FFT algorithm specification:

$$A(I) \leftarrow A(I) + E * A(K)$$
$$A(K) \leftarrow A(I) - E * A(K)$$

where

$$I = 1, 2, \ldots, m,$$
$$K = I + 2^k$$

and

$E = \exp(-2\pi i z_k / 2^k)$ with z_k equal to the 'bit-reversed version' of the number $(I - 1)$.

We shall next outline the bit-reversing procedure and a sequence of steps required for unscrambling the Fourier transform sequence, and then shall proceed to give a complete procedure for the radix 2 algorithm.

5.7.3 Unscrambling the Fast Fourier Transform

We shall now consider a procedure for implementing the bit-reversing operation.

Let K be a binary number equal to $b_0 b_1 b_2 b_3$, that is its value is

$$K = 2^3 \times b_0 + 2^2 \times b_1 + 2 \times b_2 + 2^0 \times b_3. \tag{1}$$

We wish to compute the binary number I, such that

$$I = (b_3 b_2 b_1 b_0)_2 = 2^3 \times b_3 + 2^2 \times b_2 + 2 \times b_1 + 2^0 \times b_0. \tag{2}$$

We note that the basic part of the bit-reversing operation consists in determining the binary bits, b_0, b_1, b_2, and b_3, given a number K as in (1). As soon as these bits are determined, the bit-reversing problem is solved, since deduction of the number I as in (2) is then readily done.

Now, to determine the bit b_3, consider the following sequence of operations:

(i)   divide the decimal number K by 2, ⎫ these two operations can be replaced
(ii) truncate the result,   ⎬ by one operation known as the
                                          ⎭ 'integer division' by 2.
(iii) multiply the latest result by 2,
(iv) subtract the value obtained in (iii)
      from the initial value of K.

If the difference in (iv) is zero, then the bit b_3 must be 0, since in this case division by 2, truncation and subsequent multiplication by 2 does not alter K. However, if the difference in (iv) is non-zero, then the bit b_3 must be 1, since in this case truncation changes the value of K.

The other bits, b_2, b_1 and b_0 can be identified in a similar manner.

Example 5.1.

Let $K = 14 = 2^3 \times 1 + 2^2 \times 1 + 2 \times 1 + 2^0 \times 0 = (1\ 1\ 1\ 0)_2 = (b_0 b_1 b_2 b_3)_2$

We wish to determine

$$I = (b_3 b_2 b_1 b_0)_2 = (0\ 1\ 1\ 1) = 2^3 \times 0 + 2^2 \times 1 + 2 \times 1 + 2^0 \times 1 = 7.$$

The bit-reversing is done as follows (the sign $\div$ stands for integer division):

$14 \div 2 = 7;$	$7 \times 2 = 14;$	$14 - 14 = 0 \Rightarrow b_3 = 0.$
$7 \div 2 = 3;$	$3 \times 2 = 6;$	$7 - 6 \neq 0 \Rightarrow b_2 = 1.$
$3 \div 2 = 1;$	$1 \times 2 = 2;$	$3 - 2 \neq 0 \Rightarrow b_1 = 1.$
$1 \div 2 = 0;$	$0 \times 2 = 0;$	$1 - 0 \neq 0 \Rightarrow b_0 = 1.$

Collecting the bits sequentially from left to right as soon as they are identified, we get

$$I = (b_3 b_2 b_1 b_0)_2 = (0\ 1\ 1\ 1)_2 = 7.$$

We finally note that in the process of unscrambling the output array, a situation occurs similar to the one which leads to skipping over in the calculation of dual pairs. Namely, if we proceed down the array, interchanging $c_m(K)$ with appropriate $c_m(I)$, we will eventually encounter a node which has previously been interchanged. To eliminate the possibility of considering such a node again, we simply check to see if I (the integer obtained by bit-reversing K) is less than K. If so, this implies that the node has been interchanged by a previous operation and should not be touched.

A complete procedure for unscrambling the Fourier transform sequence is thus given as follows.

We assume as input the sequence $c_0(I), I = 0, 1, \ldots, N - 1$ and the integer $m = \log_2 N$. The sequence is stored at the array $A(I)$ such that $A(I) = c_m(I - 1)$, $I = 1, 2, \ldots, N$. In the procedure I is used as the normal ('bit-forward') integer and J as the bit-reversed integer.

1. Set $I \leftarrow 1$, $J = 1$. Compute $N \leftarrow 2^m$.
2. Test $I \geqslant J$?
   If true, go to Step 4.
3. Interchange $A(I)$ and $A(J)$.
4. Set $K \leftarrow 2^{m-1}$ (i.e. K is set equal to a word with a 1 in the units position of the J-register and zeros everywhere else).
5. Test $K \geqslant J$?
   If true, go to Step 7.
6. Set $J \leftarrow J - K$, $K \leftarrow K/2$. Go back to Step 5.

7. Set $J \leftarrow J + K, \quad I \leftarrow I + 1.$
8. Test $I \geqslant N$?
   If true, exit; otherwise go back to Step 2.

We are now ready to outline a complete procedure for the FFT radix 2 algorithm.

We assume the input in the form of an array $A(I), I = 1, \ldots, N$, with $N = 2^m$, m an integer. A parameter J is used to enable the organization of the necessary skipping operations. The bit-reversing operation for computing the parameter z is given as a single step. With reference to the flowgraph of Fig. 5.3, for example, the bit-reversing operation is based on the following formulae:

$$\text{for } k = 1, \quad r = p_3 + 2p_2 + 2^2 p_1 + 2^3 q_3,$$
$$z = q_3;$$
$$\text{for } k = 2, \quad r = p_3 + 2p_2 + 2^2 q_2 + 2^3 q_3,$$
$$z = q_3 + 2q_2;$$
$$\text{for } k = 3, \quad r = p_3 + 2p_1 + 2^2 q_2 + 2^3 q_3,$$
$$z = q_3 + 2q_2 + q_1;$$
$$\text{for } k = 4, \quad r = q_0 + 2q_1 + 2^2 q_2 + 2^3 q_3,$$
$$z = q_3 + 2q_2 + 2^2 q_1 + 2^3 q_0.$$

The FFT algorithm of radix 2:

1. Set $k \leftarrow 1.$
2. Set $E \leftarrow 1.$
3. Set $J \leftarrow 1.$
4. Set $I \leftarrow J.$
5. Compute $K \leftarrow I + 2^{m-k}$
6. Test $K > N$?
   If true, go to Step 15.
7. Compute $A(I) \leftarrow A(I) + E * A(K)$
   $$A(K) \leftarrow A(I) - E * A(K)$$
8. Set $I \leftarrow I + 1.$
9. Test $I > 2^{m-k} * J$?
   If false, go back to Step 5.
10. Set $J \leftarrow J + 2.$
11. Test $J > 2^k$?
    If true, go to Step 15.
12. $I \leftarrow I + 2^{m-k}.$
13. Carry out the bit-reversing operation on I-1 to obtain z, i.e. $z =$ IBR $(I - 1)$, which stands for 'integer bit reversed'.
14. Compute $E = \exp(-2\pi i z / 2^k)$,
    then go back to Step 5.
15. $k \leftarrow k + 1.$
16. Test $k > m$? If false go back to Step 2.
17. Carry out the 'unscrambling' of the Fourier transform sequence.

Finally we note that instead of 'post-processing' the data, that is unscrambling the computed Fourier transform sequence, one can 'pre-process' the data, that is rearrange the input sequence $f(v)$, $v = 0, \ldots, N - 1$, in the bit-reversed-index ordering before executing the FFT algorithm. In this case the computed Fourier transform sequence will be output in normal serial order.

5.8 Basic Discrete Fourier Transform (DFT) Computational Algorithms for Different Types of Data

Below we describe three procedures which permit one to use the complex FFT algorithm efficiently for special types of data. The procedures are given in such a form that one can use them by executing a complex DFT subroutine with appropriate rearranging of either the data or the output. The DFT subroutine is assumed to accept as input, an integer N and a sequence of complex numbers $f(v), v = 0, 1, \ldots, N - 1$, and to yield, as output, the DFT sequence

$$F(u) = \sum_{v=0}^{N-1} f(v) w_N^{-uv}, \qquad u = 0, 1, \ldots, N - 1. \tag{1}$$

Also useful is the inverse relationship to (1), i.e.

$$f(v) = \frac{1}{N} \sum_{u=0}^{N-1} F(u) w_N^{uv}, \qquad v = 0, 1, \ldots, N - 1, \tag{2}$$

where $f(v)$ may be referred to as the inverse DFT (IDFT) of $F(u)$.

The Doubling Algorithm

Suppose that an N-point sequence, $f(v)$, $v = 0, 1, \ldots, N - 1$, where N is even, is too long to compute in one application of the DFT algorithm while a sequence of $N/2$ points can be accommodated. In this case, define the two $N/2$-point sequences and their transforms:

$$f_1(v) = f(2v), \qquad\qquad v = 0, 1, \ldots, \frac{N}{2} - 1 \tag{3}$$

$$F_1(u) = \sum_{v=0}^{(N/2)-1} f_1(v) w_{N/2}^{-uv}, \qquad u = 0, 1, \ldots, \frac{N}{2} - 1,$$

and

$$f_2(v) = f(2v + 1), \qquad\qquad v = 0, 1, \ldots, \frac{N}{2} - 1 \tag{4}$$

$$F_2(u) = \sum_{v=0}^{(N/2)-1} f_2(v) w_{N/2}^{-uv}, \qquad u = 0, 1, \ldots, \frac{N}{2} - 1.$$

Now consider the DFT, $F(u)$, of the complete N-point sequence, $(f(v))$:

$$F(u) = \sum_{v=0}^{N-1} f(v) w_N^{-uv}, \qquad u = 0, 1, \ldots, N - 1 \tag{5}$$

wherefrom, separating the odd- and even-indexed terms, we get

$$F(u) = \sum_{v=0}^{(N/2)-1} f(2v)w_N^{-2uv} + \sum_{v=0}^{(N/2)-1} f(2v+1)w_N^{-(2v+1)u}$$

$$= \sum_{v=0}^{(N/2)-1} f(2v)w_{N/2}^{-uv} + \left[\sum_{v=0}^{(N/2)-1} f(2v+1)w_{N/2}^{-uv} \right] w_N^{-u}. \qquad (6)$$

The two sums are the $N/2$-point DFT's of $f_1(v)$ and $f_2(v)$, so

$$F(u) = F_1(u) + F_2(u)w_N^{-u}, \qquad u = 0, 1, \ldots, \frac{N}{2} - 1. \qquad (7)$$

Substituting $u + N/2$ for u and using the fact that $F_1(u)$ and $F_2(u)$ are periodic with period $N/2$, we get

$$F\left(u + \frac{N}{2}\right) = F_1(u) - F_2(u)w_N^{-u}, \qquad u = 0, 1, \ldots, \frac{N}{2} - 1 \qquad (8)$$

where we used the fact that $w_N^{-(u+N/2)} = -w_N^{-u}$.

Therefore, to obtain the DFT of $f(v)$ in this case, the procedure is as follows:

1. Form the sequences

$$f_1(v) = f(2v) \quad \text{and} \quad f_2(v) = f(2v+1), \qquad v = 0, 1, \ldots, \frac{N}{2} - 1.$$

2. Compute

$$F_1(u) = \sum_{v=0}^{(N/2)-1} f_1(v)w_{N/2}^{-uv}, \qquad u = 0, 1, \ldots, \frac{N}{2} - 1$$

and

$$F_2(u) = \sum_{v=0}^{(N/2)-1} f_2(v)w_{N/2}^{-uv}, \qquad u = 0, 1, \ldots, \frac{N}{2} - 1.$$

3. Compute

$$F(u) = F_1(u) + F_2(u)w_N^{-u}$$

and

$$F\left(u + \frac{N}{2}\right) = F_1(u) - F_2(u)w_N^{-u}, \qquad u = 0, 1, \ldots, \frac{N}{2} - 1.$$

The Inversion Algorithm

This algorithm assumes that a DFT sequence

$$F(u), \qquad u = 0, 1, \ldots, N - 1$$

is given as input and that one wishes to recover the sequence

$f(v)$, $v = 0, \ldots, N - 1$, that is to compute the inverse DFT

$$f(v) = \frac{1}{N} \sum_{u=0}^{N-1} F(u)w_N^{uv}. \tag{9}$$

Taking complex conjugates of both sides in (10) we get

$$\overline{f(v)} = \frac{1}{N} \sum_{u=0}^{N-1} \overline{F(u)}w_N^{-uv}, \qquad v = 0, \ldots, N - 1 \tag{10}$$

which may be interpreted as the DFT of the sequence

$$\left(\frac{\overline{F(u)}}{N}, \qquad u = 0, \ldots, N - 1 \right).$$

(Here $\bar{z}$ denotes the complex conjugate of z.)

Therefore, the computational procedure to obtain (9) may be given as follows:

1. Form

$$\frac{\overline{F(u)}}{N}, \qquad u = 0, 1, \ldots, N - 1$$

2. Compute

$$\overline{f(v)} = \sum_{u=0}^{N-1} \frac{\overline{F(u)}}{N} w_N^{-uv}, \qquad v = 0, 1, \ldots, N - 1.$$

3. Form complex conjugates of $(\overline{f(v)})$, i.e.

$$f(v) = (\overline{f(v)}), \qquad v = 0, 1, \ldots, N - 1.$$

Two-at-a-time Algorithm for Real Sequences

Suppose that two data sequences $f_1(v)$ and $f_2(v)$, $v = 0, 1, \ldots, N - 1$, are real, and one wishes to compute their respective DFT's. This can be done with some saving on the number of arithmetic operations by first forming a single complex sequence

$$f(v) = f_1(v) + if_2(v). \qquad v = 0, 1, \ldots, N - 1, \tag{11}$$

from which we let

$$F(u) = \sum_{v=0}^{N-1} f(v)w_N^{-uv},$$

$$F_1(u) = \sum_{v=0}^{N-1} f_1(v)w_N^{-uv} \tag{12}$$

and

$$F_2(u) = \sum_{v=0}^{N-1} f_2(v)w_N^{-uv}, \qquad u = 0, 1, \ldots, N - 1,$$

and then observing that

$$F(u) = F_1(u) + iF_2(u), \qquad u = 0, 1, \ldots, N - 1, \tag{13}$$

where $F_1(u)$ and $F_2(u)$ are conjugate even.

The complex function $f(z)$ is called even if $f(z) = f(-z)$; the complex function $f(z)$ is called conjugate even if $f(z) = \overline{f(-z)}$, which for the function periodic with period p also means that $f(z) = \overline{f(p-z)}$. In our case we, thus, have that $F_r(u) = \overline{F_r(N-u)}$ for $r = 1, 2$).

Next, taking the complex conjugate of the right-hand side of (13) and replacing u by $(N-u)$, we get

$$\overline{F_1(u)} - i\overline{F_2(u)} = \overline{F_1(N-u)} - i\overline{F_2(N-u)}$$
$$= \overline{F_1(u)} - i\overline{F_2(u)} = \overline{F(u)} = \overline{F(N-u)},$$

that is

$$\overline{F(N-u)} = F_1(u) - iF_2(u), \qquad u = 0, 1, \ldots, N-1. \tag{14}$$

Solving (15) and (16) for $F_1(u)$ and $F_2(u)$, we find

$$F_1(u) = \tfrac{1}{2}\left[\overline{F(N-u)} + F(u)\right]$$
$$F_2(u) = \tfrac{1}{2}\left[\overline{F(N-u)} - F(u)\right], \qquad u = 0, 1, \ldots, \frac{N}{2}. \tag{15}$$

The computational procedure, therefore, may be given as:

1. Form

$$f(v) = f_1(v) + if_2(v), \qquad v = 0, 1, \ldots, N-1.$$

2. Compute the sequence

$$F(u) = \sum_{v=0}^{N-1} f(v)w_N^{-uv}, \qquad u = 0, 1, \ldots, N-1.$$

3. Form the sequence

$$\overline{F(N-u)}, \qquad u = 0, 1, \ldots, \frac{N}{2}.$$

4. Compute

$$F_1(u) = \tfrac{1}{2}\left[\overline{F(N-u)} + F(u)\right],$$
$$F_2(u) = \tfrac{1}{2}\left[\overline{F(N-u)} - F(u)\right], \qquad u = 0, 1, \ldots, \frac{N}{2}.$$

Values of $F_1(u)$ and $F_2(u)$ for $u > (N/2)$ need not be computed since the two functions are conjugate even.

5.9 Round-off Errors in the Fast Fourier Transform

When programs for the FFT algorithm began to replace those using conventional methods, it was very soon noticed that not only was the speed improved by a factor proportional to $N/\log_2 N$, but, in general, the accuracy was much greater. The numerical stability of the FFT has been studied theoretically by several authors, all substantiating the empirical results. By comparing upper bounds, Gentleman and Sande (1966) have shown that accumulated floating-point round-off error is significantly less when one uses the FFT than when one computes the discrete Fourier transform directly, by using the DFT defining formula. Welch (1969) has derived approximate upper and lower bounds on the accumulated error in a fixed-point algorithm of radix 2. Weinstein (1969) has used a statistical model for floating-point round-off errors to predict the variance of the accumulated errors. Kaneko and Liu (1970) have also used a statistical approach and derived bounds for the mean squared error in a floating-point algorithm of radix 2. They treated both cases, the rounding and the chopping-off (truncation) of the excess digits in the values computed. Using the matrix form of the factorisation, Ramos (1971) has derived approximate upper bounds for the ratios of the root-average-square (RAS) and the maximum round-off errors in the output to the RAS value of the output for both single and multidimensional transforms.

Following mainly Gentleman and Sande and to a certain extent, Ramos, we shall now derive upper bounds on the accumulated errors generated in computing the discrete Fourier transforms (a) using the standard technique and (b) using the FFT algorithm.

Let us first recall that the matrix form of the discrete Fourier transform is given by

$$F(u) = Af(v) \tag{1}$$

where the matrix $A = \{a_{uv}\} = e^{-i2\pi uv/N}$, and $F(u)$ and $f(v)$ are column vectors of length N.

The matrix factorization of the FFT algorithm is then given as

$$F(u) = P_{m+1}(B_m P_m)(D_{m-1} B_{m-1} P_{m-1})\ldots(D_1 B_1 P_1)f(v) \tag{2}$$

where $P_s, s = 1,\ldots, m+1$, are permutation matrices,

$D_s, s = 1,\ldots, m-1$, are diagonal matrices of complex exponential elements, called the twiddle factors by Gentleman and Sande, rotation factors by Singleton (1969), and

$B_s, s = 1,\ldots, m$, are block-diagonal matrices whose blocks are identical square submatrices, each being a complex Fourier transform of dimension r_s, where $N = r_1 \ldots r_m$.

The formula (2) holds for the Gentleman–Sande algorithm as well as for the Cooley–Tukey algorithm but with different diagonal matrices, D_s.

Let $F(u)$ be the *exact* Fourier transform and $F(u)_{FT}$ and $F(u)_{FFT}$ be the Fourier transforms *computed* using (1) and using (2), respectively, then the errors incurred in computing the transforms by these algorithms may be given as

$$e(u)_{DFT} = F(u)_{DFT} - F(u), \tag{3}$$

and

$$e(u)_{\text{FFT}} = F(u)_{\text{FFT}} - F(u), \qquad u = 0, 1, \dots, N-1, \tag{4}$$

respectively.

We shall now show that the Euclidean norms of the vectors given by (3) and (4), i.e. the sums of the squared errors, $|e(\mathbf{u})|^2$, for $\mathbf{u} = (0, \dots, N-1)$, for algorithms of (1) and (2), are bounded from above as follows:

$$\frac{\| \mathbf{e(u)}_{\text{DFT}} \|_E}{\| \mathbf{F(u)} \|_E} \leqslant 1.06(2N)^{3/2} 2^{-t} \tag{5}$$

and

$$\frac{\| \mathbf{e(u)}_{\text{FFT}} \|_E}{\| \mathbf{F(u)} \|_E} \leqslant 1.06\sqrt{N}\left(\sum_s (2r_s)^{3/2} \right) 2^{-t}, \tag{6}$$

where $\| . \|_E$ denotes the Euclidean norm of a vector,
$N = r_1 \dots r_m$, as before, and
t is the number of binary digits in the mantissa of the floating-point number.

Note that, in particular, from (6) it follows that
if $N = r^m$, then

$$\frac{\| \mathbf{e(u)}_{\text{FFT}} \|_E}{\| \mathbf{F(u)} \|_E} \leqslant 1.06\sqrt{N}m(2r)^{3/2} 2^{-t}; \tag{7}$$

furthermore, if $N = 2^m$, then

$$\frac{\| \mathbf{e(u)}_{\text{FFT}} \|_E}{\| \mathbf{F(u)} \|_E} \leqslant 8.5 (\sqrt{N} \log_2 N) 2^{-t}, \tag{8}$$

which can be compared with the corresponding case of formula (5), i.e.
if $N = 2^m$ then

$$\frac{\| \mathbf{e(u)}_{\text{DFT}} \|_E}{\| \mathbf{F(u)} \|_E} \leqslant 3\sqrt{N} N 2^{-t}. \tag{9}$$

Proof of (5).

Equation (1) expressed in real arithmetic, may be written as

$$\begin{bmatrix} \text{Re}(\mathbf{F(u)}) \\ \text{Im}(\mathbf{F(u)}) \end{bmatrix} = \begin{bmatrix} \mathbf{C} & -\mathbf{S} \\ \mathbf{S} & \mathbf{C} \end{bmatrix} \begin{bmatrix} \text{Re}(\mathbf{f(v)}) \\ \text{Im}(\mathbf{f)(v)}) \end{bmatrix}, \tag{10}$$

where $\mathbf{C}$ and $\mathbf{S}$ are real matrices with elements

$$C(u,v) = \cos\left(\frac{2\pi}{N}(u-1)(v-1) \right) \text{ and } S(u,v) = \sin\left(\frac{2\pi}{N}(u-1)(v-1) \right),$$

$$u, v = 1, 2, \dots, N, \text{ and}$$

$\mathrm{Re}(\mathbf{F(u)})$, $\mathrm{Im}\ (\mathbf{F(u)})$, $\mathrm{Re}(\mathbf{f(v)})$ and $\mathrm{Im}(\mathbf{f(v)})$ are the real and imaginary parts of the vectors $\mathbf{F(u)}$ and $\mathbf{f(v)}$.

Now, in terms of the norms we get

$$\| \mathbf{F(u)}_{\mathrm{DFT}} - \mathbf{F(u)} \|_E = \| (\mathbf{Af(v)})_{\mathrm{DFT}} - \mathbf{Af(v)} \|_E. \tag{11}$$

At this stage we need a lemma proved by Wilkinson which states that

If $\mathbf{A}$ is a real $p \times q$ matrix and $\mathbf{B}$ is a real $q \times r$ matrix then, for floating-point multiplication, the following holds

$$\| (\mathbf{AB})^{\mathrm{comp}} - (\mathbf{AB})^{\mathrm{exact}} \|_E \leqslant 1.06 q\, 2^{-t} \| \mathbf{A} \|_E \| \mathbf{B} \|_E. \tag{12}$$

where each term of the matrix product is the sum of q product terms.

Applying the lemma to our problem and noting that $p = q = 3 = 2N$, and that the Euclidean norm of the $2N \times 2N$ real matrix

$\begin{bmatrix} \mathbf{C} & -\mathbf{S} \\ \mathbf{S} & \mathbf{C} \end{bmatrix}$ is given by

$$\left[\sum_{v=1}^{N} \sum_{u=1}^{N} \left(2\cos^2(u-1)(v-1)\frac{2\pi}{N} + 2\sin^2(u-1)(v-1)\frac{2\pi}{N} \right) \right]^{1/2} = \sqrt{2N},$$

we get

$$\| \mathbf{e(u)}_{\mathrm{DFT}} \|_E = \| \mathbf{F(u)}_{\mathrm{DFT}} - \mathbf{F(u)} \|_E \leqslant 1.06 \sqrt{N}\,(2N)^{3/2}\, 2^{-t} \| \mathbf{f(v)} \|_E$$
$$\leqslant 1.06\,(2N)^{3/2}\, 2^{-t} \| \mathbf{F(u)} \|_E,$$

since

$$\| \mathbf{F(u)} \|_E = \left[\sum_{u=0}^{N-1} \left(\sum_{v=0}^{N-1} \mathrm{e}^{\mathrm{i}2\pi uv/N} f(v) \right)^2 \right]^{1/2} = \sqrt{N} \| \mathbf{f(v)} \|_E \tag{13}$$

This completes the proof of the error bound formula (5) for the ordinary DFT.

Proof of (6).

From equation (2) we have

$$\| \mathbf{F(u)}_{\mathrm{FFT}} - \mathbf{F(u)} \|_E = \| \mathrm{fl}(\mathbf{P}_{m+1}\,\mathrm{fl}(\mathbf{B}_m\mathrm{fl}(\mathbf{P}_m\mathrm{fl}(\mathbf{M}_{m-1}\mathrm{fl}(\mathbf{M}_{m-2}\cdots\mathrm{fl}(\mathbf{M}_1\mathbf{f(v)})\ldots)$$
$$\underbrace{\qquad\qquad\qquad\qquad\qquad}_{m+1\ \text{brackets}}$$
$$- \mathbf{P}_{m+1}\mathbf{B}_m\mathbf{P}_m\mathbf{M}_{m-1}\mathbf{M}_{m-2}\cdots\mathbf{M}_1\mathbf{f(v)} \|_E, \tag{14}$$

where $\mathbf{M}_s$ denotes the matrix product $\mathbf{D}_s\mathbf{B}_s\mathbf{P}_s$, $s = 1,\ldots,m-1$ and $\mathrm{fl}(.)$ denotes as usual, the result computed in floating point arithmetic as opposed to the result computed exactly.

We shall now present relation (14) in the form:

$$\| \mathbf{e(u)}_{FFT} \|_E = \| fl(\mathbf{P}_{m+1} fl(\mathbf{B}_m fl(\mathbf{P}_m fl(\mathbf{M}_{m-1} \cdots fl(\mathbf{M}_1 \mathbf{f(v)}) \underbrace{\ldots)}$$
$$\underbrace{\qquad\qquad\qquad\qquad}_{m+2 \text{ brackets}}$$

$$- \mathbf{P}_{m+1} fl(\mathbf{B}_m fl(\mathbf{P}_m fl(\mathbf{M}_{m-1} \cdots fl(\mathbf{M}_1 \mathbf{f(v)}) \underbrace{\ldots)}$$
$$\underbrace{\qquad\qquad\qquad\qquad}_{m+1 \text{ brackets}}$$

$$+ \mathbf{P}_{m+1} fl(\mathbf{B}_m fl(\mathbf{P}_m fl(\mathbf{M}_{m-1} \cdots fl(\mathbf{M}_1 \mathbf{f(v)}) \underbrace{\ldots)}$$
$$\underbrace{\qquad\qquad\qquad\qquad}_{m+1 \text{ brackets}}$$

$$- \mathbf{P}_{m+1} \mathbf{B}_m fl(\mathbf{P}_m fl(\mathbf{M}_{m-1} \cdots fl(\mathbf{M}_1 \mathbf{f(v)}) \underbrace{\ldots)}$$
$$\underbrace{\qquad\qquad\qquad\qquad}_{m \text{ brackets}}$$

$$+ \ldots + \mathbf{P}_{m+1} \mathbf{B}_m \mathbf{P}_m \mathbf{M}_{m-1} \cdots \mathbf{M}_{s+1} fl(\mathbf{M}_s \cdots fl(\mathbf{M}_1 \mathbf{f(v)}) \underbrace{\ldots)}$$
$$\underbrace{\qquad\qquad\qquad\qquad}_{s \text{ brackets}}$$

$$- \mathbf{P}_{m+1} \mathbf{B}_m \mathbf{P}_m \mathbf{M}_{m-1} \cdots \mathbf{M}_{s+1} \mathbf{M}_s fl(\mathbf{M}_{s-1} \cdots fl(\mathbf{M}_1 \mathbf{f(v)}) \underbrace{\ldots)}$$
$$\underbrace{\qquad\qquad\qquad\qquad}_{s-1 \text{ brackets}}$$

$$+ \ldots + \mathbf{P}_{m+1} \mathbf{B}_m \mathbf{P}_m \mathbf{M}_{m-1} \cdots \mathbf{M}_s \cdots \mathbf{M}_2 fl(\mathbf{M}_1 \mathbf{f(v)})$$
$$- \mathbf{P}_{m+1} \mathbf{B}_m \mathbf{P}_m \mathbf{M}_{m-1} \cdots \mathbf{M}_s \cdots \mathbf{M}_2 \mathbf{M}_1 \mathbf{f(v)} \|_E. \tag{15}$$

In order to proceed we shall need the following information:
(i) For any matrix $\mathbf{A}$ and any vector $\mathbf{x} \neq \mathbf{0}$ we have:

$$\| \mathbf{Ax} \|_E \leq \| \mathbf{A} \|_s \| \mathbf{x} \|_E \tag{16}$$

where the subscripts E and s stand for the Euclidean vector and the spectral matrix norms, respectively. (The spectral norm of a matrix $\mathbf{A}$ is defined as $\| \mathbf{A} \|_s = [\text{modulus of the maximum eigenvalue of } (\mathbf{A}^*\mathbf{A})]^{1/2}$).
(ii) The spectral norms of $\mathbf{D}_s$, $\mathbf{B}_s$ and $\mathbf{P}_s$ are $1, \sqrt{r_s}$ and 1, respectively, since $\mathbf{D}_s^* \mathbf{D}_s = I$, $\mathbf{B}_s^* \mathbf{B}_s = r_s I$ and $\mathbf{P}_s^* \mathbf{P}_s = I$ where I is the $N \times N$ identity matrix and r_s denotes the complex block in the block-diagonal (with identical blocks) matrix $\mathbf{B}_s$.
(iii) The matrices $\mathbf{P}_s$, $s = 1, \ldots, m+1$ are the permutation matrices and, as it has been noted earlier, simply reorder the vector values, introducing no round-off errors.
By virtue of (i), (ii), and (iii) we can rewrite expression (15) as

$$\| \mathbf{e(u)}_{FFT} \|_E \leq \sqrt{r_1 r_2 \cdots r_m} \sum_{s=1}^{m} \| fl(\mathbf{M}_s fl(\mathbf{M}_{s-1} \cdots fl(\mathbf{M}_1 \mathbf{f(v)}) \underbrace{\ldots)}$$
$$\underbrace{\qquad\qquad\qquad\qquad}_{s \text{ brackets}}$$

$$- \mathbf{M}_s fl(\mathbf{M}_{s-1} \cdots fl(\mathbf{M}_1 \mathbf{f(v)}) \underbrace{\ldots)} \|_E$$
$$\underbrace{\qquad\qquad\qquad\qquad}_{s-1 \text{ brackets}} \tag{17}$$

where matrix $\mathbf{M}_m$ denotes the product $\mathbf{B}_m \mathbf{P}_m$ and $r_1 r_2 \ldots r_m = N$, as before.

We now estimate the bound for the round-off error incurred during the calculations involving matrix $\mathbf{M}_s$, $s = 1, \ldots, m$, i.e.

$$\| \mathrm{fl}(\mathbf{M}_s \mathbf{z}_{s-1}) - \mathbf{M}_s \mathbf{z}_{s-1} \|_E$$

where $\mathbf{z}_{s-1}$ is an N-length vector.

We recall that matrix $\mathbf{M}_s = \mathbf{D}_s \mathbf{B}_s \mathbf{P}_s$ may be partitioned into N/r_s disjoint complex blocks and, thus, the error from each block can be bounded separately. Since in real arithmetic each block is $2r_s$ square and since the Euclidean norm of each block is $\sqrt{2r_s}$, we get the error bound given by

$$1.06(2r_s)\sqrt{2r_s}\, 2^{-t} \| \mathbf{z}_{s-1,j} \|_E$$

where the vector $\mathbf{z}_{s-1,j}$ denotes the appropriate part of the vector $\mathbf{z}_{s-1}$.

For the complete matrix $\mathbf{M}_s$, it follows:

$$\begin{aligned} \| \mathrm{fl}(\mathbf{M}_s \mathbf{z}_{s-1}) - \mathbf{M}_s \mathbf{z}_{s-1} \|_E \\ \leqslant 1.06(2r_s)\sqrt{2r_s}\, 2^{-t} (\| \mathbf{z}_{s-1,1} \|_E + \| \mathbf{z}_{s-1,2} \|_E + \ldots + \| \mathbf{z}_{s-1,N/r_s} \|_E) \\ = 1.06(2r_s)^{3/2} \| \mathbf{z}_{s-1} \|_E. \end{aligned} \tag{18}$$

Relation (18) may now be used to obtain the bounds for the right-hand side terms in (17). So, for the matrix $\mathbf{M}_1$ we get

$$\| \mathrm{fl}(\mathbf{M}_1 \mathbf{f}(\mathbf{v})) - \mathbf{M}_1 \mathbf{f}(\mathbf{v}) \|_E \leqslant 1.06(2r_1)^{3/2} \| \mathbf{f}(\mathbf{v}) \|_E. \tag{19}$$

Next, denoting $\mathbf{M}_1 \mathbf{f}(\mathbf{v})$ by $\mathbf{z}_1$, for the matrix $\mathbf{M}_2$ we obtain:

$$\begin{aligned} \| \mathrm{fl}(\mathbf{M}_2 \mathrm{fl}(\mathbf{z}_1)) - \mathbf{M}_2 \mathrm{fl}(\mathbf{z}_1) \|_E &\leqslant 1.06(2r_2)^{3/2} \| \mathrm{fl}(\mathbf{z}_1) \|_E \\ &\leqslant 1.06(2r_2)^{3/2} \sqrt{r_1} \| \mathbf{f}(\mathbf{v}) \|_E, \end{aligned} \tag{20}$$

where we first assume that

$$\| \mathrm{fl}(\mathbf{z}_1) \|_E = \| \mathbf{z}_1 \|_E$$

and then use the bound

$$\| \mathbf{z}_1 \|_E \leqslant \| \mathbf{M}_1 \|_s \| \mathbf{f}(\mathbf{v}) \|_E = \sqrt{r_1} \| \mathbf{f}(\mathbf{v}) \|_E$$

Continuing in this manner, we obtain estimates for each term of the sum on the right-hand side of the expression (17). Then, using these estimates we can write:

$$\begin{aligned} \| \mathbf{e}(\mathbf{u})_{\mathrm{FFT}} \|_E &\leqslant 1.06\sqrt{N} \left(\sum_{s=1}^{m} (2r_s)^{3/2} r_1 \ldots r_{s-1} \right) 2^{-t} \| \mathbf{f}(\mathbf{v}) \|_E \\ &\leqslant 1.06\sqrt{N} \left(\sum_{s=1}^{m} (2r_s)^{3/2} \right) 2^{-t} \sqrt{N} \| \mathbf{f}(\mathbf{v}) \|_E \\ &\leqslant 1.06\sqrt{N} \left(\sum_{s=1}^{m} (2r_s)^{3/2} \right) 2^{-t} \| \mathbf{F}(\mathbf{u}) \|_E. \end{aligned} \tag{21}$$

This completes the proof of formula (6).

The upper bounds on the accumulated errors given by equations (5) and (6) indicate that, in general, the errors incurred in the final transforms computed using the Fast Fourier Transform algorithm are smaller than the errors incurred in the computations involving the ordinary DFT algorithm.

A more careful analysis of the accumulated errors associated with the use of the FFT algorithm is due to Ramos (1970). In this analysis the errors in the elements of the matrices involved, i.e. the errors in the floating-point computations of sines and cosines, are considered directly. The upper bound on the errors derived is significantly lower than the bound given by (6). In terms of the Euclidean norms this bound may be expressed as

$$\frac{\|e(\mathbf{u})_{\text{FFT}}\|}{\|F(\mathbf{u})\|_E} < \left[\sum_{s=1}^{m} 2\sqrt{r_s(r_s + \gamma)} + (m-1)(3 + 2\gamma) \right] 2^{-t}, \tag{22}$$

where γ is an absolute error constant, $\gamma \geq 0$, such that

$$\text{fl}(\text{sinfl}(x)) = \sin x + \gamma 2^{-t},$$
$$\text{fl}(\text{cosfl}(x)) = \cos x + \gamma 2^{-t}.$$

For $N = 2^m$ this bound is reduced to

$$\frac{\|e(\mathbf{u})_{\text{FFT}}\|_E}{\|F(\mathbf{u})\|_E} < [\sqrt{2}\log_2 N + (\log_2 N - 1)(3 + 2\gamma)] 2^{-t} \tag{23}$$

In computational experiments reported by Ramos, the constant γ takes on the values in the range $1.5 < \gamma < 10$.

Kaneko and Liu (1970) studied the Gentleman–Sande FFT algorithm of radix 2 under the assumptions that the round-off errors are random variables uniformly distributed in the interval $(-2^{-t}, 2^{-t})$ and are independent of the numbers in the rounding of which they occur. Under these assumptions the authors established that the sum of the mean squared errors is bounded by

$$\frac{1}{2}\sqrt{\log_2 N}\, 2^{-t} < \frac{\left[\sum_{u=0}^{N-1} \mu[|e(u)_{\text{FFT}}|^2] \right]^{1/2}}{\|F(\mathbf{u})\|_E} < \sqrt{\log_2 N}\, 2^{-t}. \tag{24}$$

Formula (24) shows that the total relative mean square error in the Fourier transforms computed using the FFT radix 2 algorithm, may be expected to increase with N increasing, at most, as $m = \log_2 N$.

5.10 Conclusion

The FFT algorithm makes a remarkable improvement in the practicality with which one can apply discrete Fourier transforms in numerical calculation. In this chapter the algorithm has been discussed in detail. Its merits in terms of the numerical accuracy have also been exposed. Our aim was to present the basic principles of efficient computation of the DFT. With the material presented in this chapter, there should be little difficulty in programming an FFT radix 2 algorithm. Readers interested in the concrete areas of application of the DFT are referred to the work by Cooley, Lewis, and Welch (1977). This work is a good source of further recent references.

Exercises

5.1   Let $f(v)$ and $h(v)$ be periodic discrete functions:

$$f(v) = \begin{cases} 1, & v = 0, 4, \\ 2, & v = 1, 2, 3, \\ 0, & v = 5, 6, 7, \end{cases}$$

$$f(v + 8r) = f(v), \qquad r = 0, \pm 1, \pm 2, \ldots,$$
$$h(v) = f(v),$$
$$h(v + 8r) = h(v), \qquad r = 0, \pm 1, \pm 2, \ldots.$$

Let further $g(v) = f(v) - h(v - 4)$.
(i) Compute the discrete Fourier transforms, $F(u)$, $H(u)$, and $G(u)$, of the functions $f(v)$, $h(v)$, and $g(v)$;
(ii) Demonstrate the (discrete) convolution theorem using $f(v)$ and $h(v)$;
(iii) Compute the inverse DFT of $F(u)$, $H(u)$ and $G(u)$.

5.2   Prove the following convolution properties:
(a) convolution is commutative: $(h(v)f(v)) = (f(v)h(v))$;
(b) convolution is associative: $h(v)[g(v)f(v)] = [h(v)g(v)f(v)]$;
(c) convolution is distributive over addition:
$h(v)[g(v) + f(v)] = h(v)g(v) + h(v)f(v)$.

5.3   In formula 5.5–(1) assume that $N = r_1 r_2 r_3 r_4$ and that the parameters v and u are expressed in the form

$$v = p_0 + p_1 r_1 + p_2 r_1 r_2 + p_3 r_1 r_2 r_3 \quad \text{with} \quad p_0 = 0, 1, \ldots, r_1 - 1,$$
$$p_1 = 0, 1, \ldots, r_2 - 1,$$
$$p_2 = 0, 1, \ldots, r_3 - 1,$$
$$p_3 = 0, 1, \ldots, r_4 - 1,$$

$$u = q_0 + q_1 r_4 + q_2 r_4 r_3 + q_3 r_4 r_3 r_2 \quad \text{with} \quad q_0 = 0, 1, \ldots, r_4 - 1,$$
$$q_1 = 0, 1, \ldots, r_3 - 1,$$
$$q_2 = 0, 1, \ldots, r_2 - 1,$$
$$q_3 = 0, 1, \ldots, r_1 - 1.$$

Derive the FFT algorithm for the case where the exponential function is factorised in the first instance, in terms of the parameters $p_j, j = 0, 1, 2, 3$.

5.4   Consider two polynomials $P_n(x) = \sum_{k=0}^{n} a_{n-k} x^k$ and $Q_m(x) = \sum_{k=0}^{m} b_{m-k} x^k$ with real coefficients and of degree n and m, respectively. Show that the product $W_{n+m}(x) = P_n(x)Q_m(x)$ can be computed in $O((n+m)\log(n+m))$ arithmetic operations.

5.5 Let $N = 2^m$. Describe in detail (perhaps using an ALGOL-like language) the FFT algorithm of radix 2 which would compute the DFT of the discrete function $f(v)$ and output the result in sequential order.

5.6 Derive the basic 'butterfly' operation for the Gentleman–Sande FFT algorithm.

5.7 Obtain an expression similar to 5.7.1–(2) for the kth step of the Gentleman–Sande algorithm of radix 2.

5.8 Suppose that a real N-point sequence $f(v), v = 0, 1, \ldots, N-1$ is given with N even. Show how the computation of the DFT sequence

$$F(u) = \sum_{v=0}^{N-1} f(v)w_N^{-uv}, \qquad u = 0, 1, \ldots, N-1$$

can be arranged with the use of the two-at-a-time and doubling algorithms and assert that in this way the computation time required is halved as compared with the use of direct DFT procedure.

5.9 Develop a flowgraph of the Gentleman–Sande FFT algorithm for $N = 2^3$.

5.10 Show that
(a) The DFT of the real even sequence $f(v), v = 0, 1, \ldots, N-1$, where N is an even integer, is equivalent to the cosine transform of the real sequence $f(v), v = 0, 1, \ldots, N/2$;
(b) The DFT in this case is also real and even and can be related to the coefficients of the cosine series

$$f(v) = \sum_{u=0}^{N/2} \alpha(u)\cos\left(\frac{\pi uv}{N}\right)$$

by the formulae

$$\alpha(u) = F(u)/N, \qquad u = 0, N/2,$$
$$\alpha(u) = 2F(u)/N, \qquad u = 1, 2, \ldots, (N/2)-1;$$

(c) using the results obtained in (a) and (b), deduce a computational procedure for obtaining the DFT of the N-point real even sequence with N being an even integer.

5.11 Develop the FFT algorithm of radix '4 + 2' for the case $N = 8$.

5.12 Prove formula 5.9–(2).

5.13 Develop a computation procedure for DFT of a $2N$-point function by means of an N-point transform.

5.14 Find a polynomial whose values at $1, 2, 3, 4$ are, respectively, $1, 2, 2, 1$.

5.15 Let two $(N-1)$st-degree polynomials, $P(x) = \sum_{i=0}^{N-1} a_i x^i$ and $Q(x) = \sum_{i=0}^{N-1} b_i x^i$ be represented by their values at N points $x_j, j = 0, \ldots, N-1$. For their product $R(x) = P(x)Q(x)$ we then have $R(x_j) = \sum_{i=0}^{2N-2} c_i x_j^i = P(x_j)Q(x_j)$,

$j = 0,\ldots,N - 1$. Show that the vector $\mathbf{c} = (c_0, c_1, \ldots, c_{2N-2})$ can be obtained as the inverse Fourier transform of the componentwise product of the Fourier transforms of the vectors $\mathbf{a} = (a_0, a_1, \ldots, a_{N-1})$ and $\mathbf{b} = (b_0, b_1, \ldots, b_{N-1})$. (Hint. Consider $P(x)$ and $Q(x)$ as $(2N - 2)$ st-degree polynomials where the coefficients of the $(N - 1)$st highest powers of x are zero.)

Chapter 6

Fast Multiplication of Numbers: Use of the Convolution Theorem

The problem of fast multiplication of numbers as a part of the general problem of the minimum computation time of functions is of significant practical and theoretical interest in relation to the design of efficient computer algorithms.

We shall consider the fastest available algorithm for multiplication of two numbers. It is based on the use of the Fast Fourier Transform algorithm for computing the convolution of vectors. The development of ideas follows mainly the exposition in Knuth (1969) and Borodin and Munro (1975).

6.1 On the Minimum Computation Time of Functions

Of the four basic arithmetic operations, addition, subtraction, multiplication and division, the latter two are the most time consuming. Given two integers of, say, length n and m, respectively, we can easily observe that using well known conventional methods it takes longer to multiply or divide the integers than to add or subtract them.

We introduce the following basic operations:
(A) addition or subtraction of one-digit integers, giving a one-digit answer and a carry,
(B) multiplication of a one-digit integer by another one-digit integer, giving a two-digit answer,
(C) division of a two-digit integer by a one-digit integer, provided that the quotient is a one-digit integer, and yielding also a one-digit remainder.

The conventional method for adding two integers requires a number of basic operations proportional to the sum of the number of digits in the addends, i.e. $k(n + m)$, while the conventional method of multiplying requires a number of primitive operations proportional to the product of the numbers of digits in the factors, i.e. cnm, where k and c are constants. Thus, in terms of the basic operations, it takes longer to multiply two numbers than to add them. Fortunately, it turns out that the conventional method of multiplying is far from the best possible one. For simplicity, let us assume that both the integers are of the same length n. We shall call the integer n the dimension of the problem of multiplication of two n-digit numbers, and shall consider the time complexity function, $T(n)$, of various multiplication algorithms. The term 'unit of time' will be used in the sense equivalent to 'one basic operation'.

Toom (1963) and Schönhage (1966) have each devised a different multiplication algorithm for reducing the number of basic operations from n^2 to $n^{1+\varepsilon}$, for arbitrary small $\varepsilon > 0$. Schönhage showed that his method can be executed (using a multi-tape Turing machine model) to multiply in a time proportional to

$$n^{1+(\sqrt{2}+\varepsilon)}/(\log_2 n)^{1/2},$$

and Cook (1966) proved the same result for Toom's method. Just how much further the bound can be reduced remains as an open question.

Karatsuba and Ofman (1962) stated that the process of multiplication of two n-bit binary numbers on an automata which may be obtained as a generalization of the one that they themselves have suggested, can be carried out with the time complexity function of order $n \log_2 n$. But they gave no proof of this.

We shall now show that for large enough n, the product of two n-digit numbers can be computed in $O(n \log_2 n \log_2 \log_2 n)$ time. For convenience, let us assume that we are working with the integers expressed in binary notation. Also, to complete the mathematical model which will be assumed throughout in the following, in addition to the operations (A), (B) and (C) above we define a fourth basic operation:—

(D) shifting a single binary digit left or right d positions, with $d > 0$.

6.2 An Initial Reduction

If we have two binary numbers each of $n = 2q$ bits,

$$u = (u_{n-1} u_{n-2} \ldots u_0)_2 \quad \text{and} \quad v = (v_{n-1} v_{n-2} \ldots v_0)_2,$$

then to obtain their product using the standard multiplication method, would require $O(n^2) = O(4q^2)$ basic operations. We say that the time complexity of the standard method of multiplying two n-bit binary numbers is a function of $O(n^2)$ or $O(4q^2)$.

Let us now write

$$u = 2^q U_1 + U_0, \qquad v = 2^q V_1 + V_0, \tag{1}$$

where $U_1 = (u_{n-1} u_{n-2} \ldots u_q)_2$ is the 'most significant half' of u and $U_0 = (u_{q-1} u_{q-2} \ldots u_0)_2$ is the 'less significant half', and, similarly,

$$V_1 = (v_{n-1} v_{n-2} \ldots v_q)_2, \qquad V_0 = (v_{q-1} v_{q-2} \ldots v_0)_2.$$

We have for the product

$$uv = (2^q U_1 + U_0)(2^q V_1 + V_0) = (2^{2q} + 2^q) U_1 V_1 + 2^q (U_1 - U_0)(V_0 - V_1)$$
$$+ (2^q + 1) U_0 V_0. \tag{2}$$

The computation has now been reduced to three multiplications of q-bit numbers $U_1 V_1$, $(U_1 - U_0)(V_0 - V_1)$ and $U_0 V_0$, plus some simple shifting and adding operations. We thus have got an algorithm for multiplication of two binary numbers of $n = 2q$-bits, with the time complexity function of $O(3q^2)$. The method was originated by Karatsuba and Ofman (1962) and Ofman (1962).

We shall now show that formula (2) defines a recursive process for multiplication of two numbers which for large n is significantly faster than the conventional method of order n^2.

If we let $T(n)$ be the number of basic operations required to perform multiplication of two n-digit binary numbers, then since shifting and adding can be done in $O(n)$ operations, from (2) we have

$$T(n) \leqslant 3T\left(\frac{n}{2}\right) + cn \qquad \text{for some constant } c. \tag{3}$$

We shall now prove by induction that (3) implies that

$$T(2^m) \leqslant c(3^m - 2^m), \qquad m \geqslant 1. \tag{4}$$

Proof.

Consider the multiplying scheme (2) applied to two 2-bit binary integers, i.e. $b = (b_1 b_0)_2 = 2b_1 + b_0$ and $c = (c_1 c_0)_2 = 2c_1 + c_0$, where b_1, b_0, c_1 and c_0 are equal to either 1 or 0. For their product we have

$$bc = (2^2 + 2)b_1 c_1 + 2(b_1 - b_0)(c_0 - c_1) + (2 + 1)b_0 c_0.$$

We see that the computation is composed of three basic multiplication operations on $b_1 c_1$, $(b_1 - b_0)(c_0 - c_1)$ and $b_0 c_0$, and some shifting and adding operations. At this stage we are not interested in the actual number of either shifting or adding operations, we simply denote their total number by, say, c. Hence, we can write

$$T(2) \leqslant 3T(1) + c, \tag{5}$$

which shows that (3) holds for $n = 2$. Let us now choose a constant c 'somewhat larger' than the original constant in (5) and such that we can write

$$T(2) \leqslant c.$$

The inequality obtained can also be written as

$$T(2) \leqslant c(3^1 - 2^1).$$

Now assuming that

$$T(2^k) \leqslant c(3^k - 2^k)$$

holds for some k, $k \geqslant 1$, and using (3) with $n = 2^{k+1}$, we obtain

$$T(2^{k+1}) \leqslant 3T(2^k) + c2^k \leqslant 3c(3^k - 2^k) + c2^k = c(3^{k+1} - 2^{k+1}),$$

which completes the proof.

Finally, from (4) we get

$$T(n) \leqslant T(2^{\lceil \log n \rceil}) \leqslant c(3^{\lceil \log n \rceil} - 2^{\lceil \log n \rceil}) \leqslant 3c3^{\log n} = 3cn^{\log 3} = 3cn^{1.585}. \qquad (6)$$

Formula (6) shows that the recursive process for multiplication of two numbers defined by (2) is of order $n^{1.585}$ and hence is a considerable improvement over the standard method for large enough n.

6.3 The Schönhage–Strassen Algorithm for Fast Multiplication of Integers

6.3.1 Generalization of the Initial Reduction

The time for multiplying two numbers can be reduced still further, in the limit as n approaches infinity, if we generalize the method outlined above by splitting u and v into $(r + 1)$ parts of equal size, q, for fixed $r \geqslant 1$. The time $T(n)$ in this case will yield

$$T((r + 1)q) \leqslant (2r + 1)T(q) + cq \quad \text{for any fixed } r.$$

This general method can be obtained as follows.
Let

$$u = (u_{(r+1)q-1} u_{(r+1)q-2} \cdots u_q u_{q-1} \cdots u_0)_2$$

and

$$v = (v_{(r+1)q-1} v_{(r+1)q-2} \cdots v_q v_{q-1} \cdots v_0)_2$$

be broken into $r + 1$ parts,

$$u = U_r 2^{qr} + U_{r-1} 2^{q(r-1)} + \ldots + U_1 2^q + U_0 = \sum_{j=0}^{r} U_j 2^{qj},$$

$$v = V_r 2^{qr} + V_{r-1} 2^{q(r-1)} + \ldots + V_1 2^q + V_0 = \sum_{j=0}^{r} V_j 2^{qj}, \qquad (1)$$

where each U_j and each V_j is a q-bit integer.
Corresponding to the integers u and v we can form the polynomials

$$P_u(x) = \sum_{j=0}^{r} U_j x^j \quad \text{and} \quad P_v(x) = \sum_{j=0}^{r} V_j x^j, \quad \text{both of degree } r.$$

From (1) it is obvious that

$$u = P_u(2^q), \qquad v = P_v(2^q)$$

and therefore

$$w = uv = P_u(2^q)P_v(2^q). \qquad (2)$$

The method we seek for computing uv may now be described with reference to formulae (1) and (2). Namely, we seek a good way to compute the coefficients of the polynomial

$$W(x) = P_u(x)P_v(x) = W_{2r}x^{2r} + \ldots + W_1 x + W_0. \tag{3}$$

We first note that the coefficients W_s, $s = 0, \ldots, 2r$, form a sequence of integers obtained from two sequences of integers, $(U_0, \ldots, U_r)$ and $(V_0, \ldots, V_r)$, as follows:

$$
\begin{aligned}
W_0 &= U_0 V_0, \\
W_1 &= U_0 V_1 + U_1 V_0, \\
W_2 &= U_0 V_2 + U_1 V_1 + U_2 V_0, \\
&\;\;\vdots \\
W_r &= U_0 V_r + U_1 V_{r-1} + \ldots + U_{r-1} V_1 + U_r V_0, \\
W_{r+1} &= U_1 V_r + U_2 V_{r-1} + \ldots + U_r V_1, \\
W_{r+2} &= U_2 V_r + \ldots + U_r V_2, \\
&\;\;\vdots \\
W_{2r} &= U_r V_r.
\end{aligned}
\tag{4}
$$

We shall refer to the W_j's as the product coefficients.

Recalling the convolution theorem of Section 5.3 we see that it can be used to compute the product coefficients W_j and this computation process would entail:

(i) calculation of the respective Fourier transforms of the sequences $(U_0, \ldots, U_r)$ and $(V_0, \ldots, V_r)$,

i.e. $\displaystyle F_k = \sum_{s=0}^{r} U_s e^{-isk(2\pi/(r+1))}, \qquad k = 0, \ldots, r,$

$$
G_k = \sum_{s=0}^{r} V_s e^{-isk(2\pi/(r+1))}, \qquad k = 0, \ldots, r, \quad i = \sqrt{-1}.
\tag{5}
$$

(ii) calculation of the product of the Fourier transforms obtained in (i), by componentwise multiplication,

i.e. $F_0 G_0, F_1 G_1, \ldots, F_r F_r.$ \tag{6}

(iii) calculation of the inverse Fourier transform of the products,

i.e.

$$
\begin{aligned}
W_0 &= F_0 G_0 e^{it0(2\pi/1)}, \\
W_1 &= F_0 G_0 e^{it0(2\pi/1)} + F_1 G_1 e^{it1(2\pi/2)}, \\
&\;\;\vdots \\
W_r &= F_0 G_0 e^{it0(2\pi/1)} + F_1 G_1 e^{it1(2\pi/2)} + \ldots + F_r G_r e^{itr(2\pi/(r+1))}, \\
W_{r+1} &= \qquad\qquad\quad F_1 G_1 e^{it1(2\pi/2)} + \ldots + F_r G_r e^{itr(2\pi/(r+1))}, \\
&\;\;\vdots \\
W_{2r} &= \qquad\qquad\qquad\qquad\qquad\qquad\qquad F_r G_r e^{itr(2\pi/(r+1))},
\end{aligned}
\tag{7}
$$

Assuming that the Fast Fourier Transform algorithm is used to compute Fourier transforms required in the above process, we may expect that, in virtue of the economies in the number of arithmetic operations which are associated with the use of the FFT algorithm as compared to the standard computation algorithms, the method outlined in (i)–(iii) is one of the efficient ways to compute the product coefficients W_j's.

However even more startling efficiency in the calculation of the product coefficients can be achieved if resort is made to the convolution theorem, to the so called modular arithmetic and to the special properties of the numbers involved in computation, namely the numbers being integers. In the next section we give a brief discussion of the concepts of modular arithmetic and its suitability for developing some fast algorithms.

6.3.2 Basic Modular (or Residue) Arithmetic

Modular (or residual) arithmetic is an alternative to conventional arithmetic and in some situations it is preferable for use on large integer numbers. Instead of working directly with the number q, one uses several 'moduli' $p_1, p_2, \ldots, p_r$ which contain no common factors, and works with 'residues' $q \bmod p_1, q \bmod p_2, \ldots, q \bmod p_r$. We shall employ the following notation for the 'residues'

$$q_1 = 1 \bmod p_1, \quad \ldots, \quad q_r = q \bmod p_r.$$

The sequence $(q_1, q_2, \ldots, q_r)$ can easily be computed from an integer number q by means of divisions. However, the more important thing is that provided q is an integer within certain bounds determined by the capacity of a particular computer, no information is lost in the process of obtaining the residues as q can be computed from $(q_1, q_2, \ldots, q_r)$. More precisely, assuming q to be in the range $0 \leqslant q < p_1 p_2 \ldots p_r$ (sometimes it is more convenient to consider a completely symmetric range, $-\frac{1}{2} p_1 p_2 \ldots p_r < q < \frac{1}{2} p_1 p_2 \cdots p_r$), the representation $(q_1, \ldots, q_r)$ is unique and no other integer s, $0 \leqslant s < p_1 p_2 \ldots p_r$ will have the same residues as q. This fact is a consequence of the important 'Chinese Remainder Theorem' which states that

If $p = p_1 p_2 \cdots p_r$ with $p_1, p_2, \ldots, p_r$ positive integers that are relatively prime in pairs, and $a, q_1, q_2, \ldots, q_r$ are integers, then there is exactly one integer q that satisfies the conditions

$$a \leqslant q < a + p, \quad \text{and} \quad q \equiv q_j \bmod p_j \quad \text{for} \quad 1 \leqslant j \leqslant r. \tag{1}$$

Proof.

Assume that there are two integers q and s, $q \neq s \bmod p$ which satisfy the conditions

$$a \leqslant q < a + p, \quad q \equiv q_j \bmod p_j, \quad j = 1, \ldots, r$$
$$a \leqslant s < a + p, \quad s \equiv q_j \bmod p_j, \quad j = 1, \ldots, r.$$

Then $q = q_j + c_j p_j$,   where c_j is some integer constant
and $s \equiv (q - c_j p_j) \bmod p_j = q \bmod p_j$.
Hence, $s - q$ is a multiple of p_j for all j and, since the $p_j, j = 1, \ldots, r$, are relatively
prime, the $s - q$ is a multiple of $p = p_1 p_2 \ldots p_r$. It follows that $q = s \bmod p$. This
contradicts the assumption made initially and shows that the problem (1)
has at most one solution.

It remains to show the existence of at least one solution. We shall not dwell
upon a strict proof of this second half of the theorem which may be found
for example Knuth (1969)) but rather describe a practically usable method to
convert from $(q_1, \ldots, q_1)$ to q. One such method was suggested by Garner.

Garner's method for conversion from $(q_1, q_2, \ldots, q_r)$ to q:—

I. Compute the constants $c_{ij}, 1 \leqslant i < j \leqslant r$, such that

$$c_{ij} p_i \equiv 1 \bmod p_j. \tag{2}$$

The c_{ij} is called the multiplicative inverse of the p_i, modulo p_j and is sometimes
expressed as

$$c_{ij} = p_i^{-1} \bmod p_j.$$

For our purposes in the present text we shall make use of only multiplicative
inverses that are equal to unity. We shall, therefore, restrict ourselves to this
specific case and refer the reader for methods of computing c_{ij} in a general
case to Knuth (1969).

II. Set

$$a_1 = q_1 \bmod p_1,$$
$$a_2 = (q_2 - a_1) c_{12} \bmod p_2,$$
$$a_3 = ((q_3 - a_1) c_{13} - a_2) c_{23} \bmod p_3, \tag{3}$$
$$\vdots$$
$$a_r = (\ldots ((q_r - a_1) c_{1r} - \ldots - a_{r-1}) c_{r-1,r} \bmod p_r,$$

then

$$q = a_1 p_{r-1} \ldots p_1 + a_{r-1} p_{r-2} \ldots p_1 + a_3 p_2 p_1 + a_2 p_1 + a_1 \tag{4}$$

is a number satisfying the conditions

$$0 \leqslant q < p, \qquad q \equiv q_j \bmod p_j: \qquad 1 \leqslant j \leqslant r.$$

Example 6.1

Given a number $q = 51$ and a set of mutually prime numbers

$$p_1 = 7, \quad p_2 = 13, \quad p_3 = 17 \quad \text{and} \quad p_4 = 29.$$

Compute the residues

$$q_1 = 51 \bmod 7 = 2, \qquad q_3 = 51 \bmod 17 = 0,$$
$$q_2 = 51 \bmod 13 = 12, \qquad q_4 = 51 \bmod 29 = 22.$$

Thus, a 'modular representation' of the number 51 in this case is given as (2, 12, 0, 22).

We now wish to recover the number 51 from its modular representation (2, 12, 0, 22).

For this, we first compute the constants

$$c_{ij}, 1 \leqslant i < j \leqslant 4$$

from the equations

$$7c_{12} = 1 \bmod 13,$$
$$7c_{13} = 1 \bmod 17, \quad 13c_{23} = 1 \bmod 17,$$
$$7c_{14} = 1 \bmod 29, \quad 13c_{24} = 1 \bmod 29, \quad 17c_{34} = 1 \bmod 29.$$

We obtain

$$c_{12} = 2,$$
$$c_{13} = 5, \quad c_{23} = 4,$$
$$c_{14} = 25, \quad c_{24} = 9, \quad c_{34} = 12.$$

Next we compute

$$a_1 = 2 \bmod 7 = 2,$$
$$a_2 = ((12 - 2)2) \bmod 13 = 7,$$
$$a_3 = (((0 - 2)5 - 7)4) \bmod 17 = 0,$$
$$a_4 = ((((22 - 2)25 - 7)9 - 0)12) \bmod 29 = 0.$$

Finally, we have

$$q = a_4 p_3 p_2 p_1 + a_3 p_2 p_1 + a_2 p_1 + a_1 = 51.$$

The sequence $(q_1, q_2, \ldots, q_r)$ may, thus, be considered as a new type of internal computer representation of the number q. It is called a 'modular representation' of the number q. Given q as $(q_1, q_2, \ldots, q_r)$ and s as $(s_1, s_2, \ldots, s_r)$ and assuming single-precision arithmetic, we have
for addition

$$(q_1, q_2, \ldots, q_r) + (s_1, s_2, \ldots, s_r) = ((q_1 + s_1) \bmod p_1, \ldots, (q_r + s_r) \bmod p_r) \quad (5)$$

for subtraction

$$(q_1, q_2, \ldots, q_r) - (s_1, s_2, \ldots, s_r) = ((q_1 - s_1) \bmod p_1, \ldots, (q_r - s_r) \bmod p_r) \quad (6)$$

for multiplication

$$(q_1, q_2, \ldots, q_r)(s_1, s_2, \ldots, s_r) = (q_1 s_1 \bmod p_1, \ldots, q_r s_r \bmod p_r), \quad (7)$$

where

$$(q_j + s_j) \bmod p_j = \begin{cases} q_j + s_j, & \text{if } q_j + s_j < p_j, \\ q_j + s_j - p_j, & \text{if } q_j + s_j \geqslant p_j, \end{cases} \quad (8)$$

and

$$(q_j - s_j) \bmod p_j = \begin{cases} q_j - s_j, & \text{if } q_j - s_j \geqslant 0, \\ q_j - s_j + p_j, & \text{if } q_j - s_j < 0, \end{cases} \quad (9)$$

$q_j s_j \bmod p_j$ is formed by first multiplying $q_j s_j$ and then dividing by p_j.

It is usually convenient to let p_1 be the largest odd number that fits in a computer word, to let p_2 be the largest odd number smaller than p_1, that is relatively prime to p_1, and so on until enough p_j's have been formed to give the desired range p.

In the sequel we will be using n-bit numbers, modulo $(2^n + 1)$. Such numbers lie in the range 0 to 2^n. The number 2^n requires $n + 1$ bits for its representation and, for convenience, we shall represent such a number by a special symbol -1, i.e.

$$2^n = -1 \bmod (2^n + 1). \tag{10}$$

Such a special case is easily handled by modular arithmetic. For example, if $q = 2^n$ and $s < 2^n$ then $(q + s) \bmod (2^n + 1)$ and $(q - s) \bmod (2^n + 1)$ are computed as in (8) and (9) with q replaced by -1, and $qs \bmod (2^n + 1)$ is obtained by computing $(2^n + 1 - s) \bmod (2^n + 1)$. We also note a useful formula for binary multiplication:

$$2^r(u_{n-1} \ldots u_0)_2 = \left[(u_{n-1-r} \ldots u_0 0 \ldots 0)_2 - (0 \ldots 0 u_{n-1} \ldots u_{n-r})_2 \right]$$
$$\bmod (2^n + 1) \tag{11}$$

for $0 \leqslant r \leqslant n$.

It shows that multiplication of a binary number by a power of 2 may be obtained using simple shift and subtraction/addition operations.

Our next step on the way to an efficient calculation of the product coefficients W_j's is to introduce the so called integer Fourier transform which is especially convenient for computations involving integers.

6.3.3 The Integer Fourier Transform

We define the integer Fourier transform for a sequence of integers, $\{b_s, s = 0, 1, \ldots, N - 1\}$ as follows:

$$c_k = \left(\sum_{s=0}^{N-1} w^{ks} b_s \right) \bmod K,$$

where w is an integer such that $w^N = 1 \bmod K$ and the computation of c_k is carried out modulo K. The integer Fourier transform may be treated in the same way as the finite Fourier transform except that all results are obtained modulo K.

The convolution theorem for the integer Fourier transform, among other cases, is valid when N and w are powers of 2. Under these conditions on N and w, the theorem renders a fast and conveniently programmable algorithm for computing convolution of sequences of integers mod $(2^{N/2} + 1)$ by (i) an integer Fourier transform, (ii) componentwise multiplication and (iii) an inverse integer Fourier transform.

We may note that the convolution of two N-vectors with integer components can be computed exactly, provided the components of the convolution are in the range 0 to $2^{N/2}$. If the components of the convolution are outside the range 0 to $2^{N/2}$ then they will be correct, modulo $(2^{N/2} + 1)$. These results follow from the properties of modular arithmetic.

6.3.4 Multiplication of Two Numbers Using Modular Arithmetic

Let us look again at the problem formulated by equations (6.3.1–(1)–(4)) in the light of the computation using modular arithmetic.

(1) The two $n = (r + 1)q$-bit integers, u and v, the product of which we seek to compute, are presented as

$$u = U_r 2^{rq} + U_{r-1} 2^{(r-1)q} + \ldots + U_0 = (U_r U_{r-1} \ldots U_0)_{2^q},$$
$$v = V_r 2^{rq} + V_{r-1} 2^{(r-1)q} + \ldots + V_0 = (V_r V_{r-1} \ldots V_0)_{2^q}, \tag{1}$$

where U_j's and V_j's are q-bit integers each.
Since u and v are n-bit integers each, their values, in general, are in the range

$$0 \leqslant u, v < 2^n. \tag{2}$$

(2) The product $w = uv$ in view of (1) is, generally, a $2n$-bit number defined in the range

$$0 \leqslant w < 2^{2n}. \tag{3}$$

The product's representation is given as

$$w = uv = W_{2r} 2^{2rq} + W_{2r-1} 2^{(2r-1)q} + \ldots + W_0 = (W_{2r} W_{2r-1} \ldots W_0)_{2^q}, \tag{4}$$

where the product coefficients W_j's are $2q$-bit integers each and are expressed in terms of the U_k's and V_k's as given by 6.3.1–(4).

(3) Now, suppose that all we want to do is to compute value of w, modulo $(2^n + 1)$. This value will then be in the range

$$0 \leqslant w \bmod (2^n + 1) < 2^n. \tag{5}$$

Note that the complete product, w, can be uniquely determined using, say, the Garner method for conversion.

We can write

$$w \bmod (2^n + 1) = (uv) \bmod (2^n + 1).$$

This may be expressed as

$$w \bmod (2^n + 1) = (F_r 2^{rq} + F_{r-1} 2^{(r-1)q} + \ldots + F_1 2^q + F_0)$$
$$= (F_r F_{r-1} \ldots F_0)_{2^q}. \tag{6}$$

Using the fact that

$$2^{sq} \bmod (2^{(r+1)q} + 1) = \begin{cases} 2^{sq}, & \text{if } s < r + 1, \\ -2^{(s-r-1)q} & \text{if } s \geqslant r + 1. \end{cases}$$

the coefficients F_s's in (6) are readily expressed as

$$F_s = W_s - W_{r+s+1}$$
$$= (U_s V_0 + U_{s-1} V_1 + \ldots + U_0 V_s) - (U_r V_{s+1} + \ldots + U_{s+1} V_r), \tag{7}$$

where W_{2r+1} set equal to 0. (See also Example 6.2).
We shall refer to the F_s's as the reduced product coefficients.

Example 6.2

Let

$$N = 2^2 q = (r+1)q$$

and

$$u = U_3 2^{3q} + U_2 2^{2q} + U_1 2^q + U_0, \quad v = V_3 2^{3q} + V_2 2^{2q} + V_1 2^q + V_0.$$

The product of u and v is given as

$$
\begin{aligned}
w = uv = {} & U_3 V_3 2^{6q} + (V_3 U_2 + V_2 U_3)2^{5q} + (V_3 U_1 + V_2 U_2 + V_1 U_3)2^{4q} \\
& + (V_3 U_0 + V_2 U_1 + V_1 U_2 + V_0 U_3)2^{3q} \\
& + (V_2 U_0 + V_1 U_1 + V_0 U_2)2^{2q} + (V_1 U_0 + V_0 U_1)2^q + V_0 U_0,
\end{aligned}
$$

and

$$
\begin{aligned}
w \bmod (2^{4q}+1) = {} & (uv) \bmod (2^{4q}+1) \\
= {} & (V_3 U_0 + V_2 U_1 + V_1 U_2 + V_0 U_3)2^{3q} \\
& + [(V_2 U_0 + V_1 U_1 + V_0 U_2) - U_3 V_3]2^{2q} \\
& + [(V_1 U_0 + V_0 U_1) - (V_3 U_2 + V_2 U_3)]2^q \\
& + [V_0 U_0 - (V_3 U_1 + V_2 U_2 + V_1 U_3)] \\
= {} & F_3 2^{3q} + F_2 2^{2q} + F_1 2^q + F_0,
\end{aligned}
$$

where we used the fact that

$$
\begin{aligned}
2^{6q} \bmod (2^{4q}+1) &= (2^{4q} 2^{2q}) \bmod (2^{4q}+1) \\
&= 2^{4q} \bmod (2^{4q}+1) 2^{2q} \bmod (2^{4q}+1) = -2^{2q},
\end{aligned}
$$

and, similarly, that

$$2^{5q} \bmod (2^{4q}+1) = -2^q,$$

$$2^{4q} \bmod (2^{4q}+1) = -1.$$

Recalling again that, generally, 2^p requires $p + 1$ bits for its binary representation, we conclude that the product of two q-bit numbers must be less than 2^{2q} and, further, since W_s and W_{r+s+1} are sums of $s + 1$ and $r - s$ such products, respectively, where $0 \leqslant s \leqslant r$, then F_s must be in the range

$$(r - s)2^{2q} < F_s < (s + 1)2^2 q, \quad s = 0, \ldots, r. \tag{8}$$

From (8) it follows that there are at most $(r + 1)2^{2q}$ possible values which F_s may assume.

(4) As soon as the values F_s's are computed, the product $w \bmod (2^n + 1) = (uv)$ $\bmod (2^n + 1)$ given by (6) can be computed in $(r + 1)(2q + \log_2 (r + 1))$ additional basic operations of shifting, since the reduced coefficients lie in the range 0 to

$(r + 1)2^{2q} = 2^{\log_2(r + 1) + 2q}$, and, hence, in binary representation will be the integers of length of $(\log_2 (r + 1) + 2q)$ bits, at most, and thus the $(r + 1)$ such integers can be added together using just $(r + 1) (\log_2 (r + 1) + 2q)$ basic shifting operations.

(5) Noting that the time complexity function $(r + 1)$ $(\log_2 (r + 1) + 2q)$ is of order n and, further, recalling that the aim of our current analysis is to produce an algorithm to multiply two n-bit integers in significantly more efficient time than that of the standard multiplying algorithm which is of order n^2, we can see that the success of the analysis depends on fast computation of the reduced coefficients.

We now turn to discuss how these coefficients can be computed.

6.3.5 Calculation of the Reduced Product Coefficients Exactly

Since the reduced product coefficients are defined in the range

$$0 \leqslant F_s < (r + 1)2^{2q}, \qquad s = 0, \ldots, r, \tag{1}$$

we can carry out their computation, modulo $(r + 1)2^{2q}$. This process of computing will, on the one hand, result in no loss of information on the coefficients, and on the other hand, enable us to carry out the computation rapidly, by computing the F_s's twice, once modulo $(r + 1)$ and once modulo $(2^{2q} + 1)$. Strictly speaking we should have the second computation, modulo 2^{2q} instead of modulo $(2^{2q} + 1)$, but the latter is more convenient to perform and conversion to the exact values is then readily brought about, as we shall now see.
If we denote by

$$F_s^{(1)} = F_s \bmod (r + 1)$$

and $\tag{2}$

$$F_s^{(2)} = F_s \bmod (2^{2q} + 1),$$

then the exact values F_s can be obtained from the $F_s^{(1)}$ and $F_s^{(2)}$ by the formula

$$F_s = (2^{2q} + 1)\left[F_s^{(1)} - F_s^{(2)} \bmod (r + 1)\right] + F_s^{(2)}. \tag{3}$$

Formula (3) follows directly from the Garner method for conversion to the integer from its modular representation, noting that

(i) $(r + 1)$ and $(2^{2q} + 1)$ are relatively prime and $(r + 1) < (2^{2q} + 1)$ by definition, and
(ii) since $(r + 1) \leqslant 2^{2q}$ and $(r + 1)$ is a power of 2, $(r + 1)$ divides 2^{2q} and thus the multiplicative inverse of $(2^{2q} + 1)$, modulo $(r + 1)$ is equal to 1, i.e. $(2^{2q} + 1) = 1 \bmod (r + 1)$.

From (3) it follows that the three steps needed to calculate the F_s are:

1. Determine $F_s^{(1)} = F_s \bmod (r + 1)$, $\qquad s = 0, \ldots, r.$

2. Determine $F_s^{(2)} = F_s \bmod (2^{2q} + 1)$, $\qquad\qquad s = 0, \ldots, r$.

3. Let $F_s^{(3)} = (2^{2q} + 1)[(F_s^{(1)} - F_s^{(2)}) \bmod (r + 1)] + F_s^{(2)}$

and compute

$$F_s = \begin{cases} F_s^{(3)}, & \text{if } F_s^{(3)} < (r + 1)2^{2q}, \\ F_s^{(3)} - (r + 1)(2^{2q} + 1), & \text{if } F_s^{(3)} \geqslant (r + 1)2^{2q}, \end{cases}$$
$$s = 0, \ldots, r.$$

This last step is introduced to satisfy the requirement that F_s should be taken modulo $(r + 1)2^{2q}$ while $F_s^{(3)}$ is a number modulo $(r + 1)(2^{2q} + 1)$.

We shall now proceed to outline the method to compute the $F_s^{(1)}$'s and $F_s^{(2)}$,s. In the analysis to follow it is convenient to assume that the integer n and its factors $(r + 1)$ and q, are all powers of 2, i.e. we set

$$n = 2^m = (r + 1)q = 2^k 2^l,$$

so that

$$r + 1 = 2^k, \quad q = 2^l, \quad k + l = m. \tag{4}$$

We leave the parameters k and l undetermined at this stage and will defer the choice of their most convenient values till Section 6.3.6.

1. $F_s^{(1)} = F_s \bmod (r + 1)$

$\qquad = [(U_s V_0 + U_{s-1} V_1 + \ldots + U_0 V_s) - (U_r V_{s+1} + \ldots + U_{s+1} V_r)]$
$\qquad \bmod 2^k$.

Here we wish to carry out some computations modulo (power of 2) on binary integers. Such computations are simple and can be done quickly. We write

$F_s^{(1)} = [(U_s V_0 \bmod 2^k + U_{s-1} V_1 \bmod 2^k + \ldots + U_0 V_s \bmod 2^k)$
$\qquad - (U_r V_{s+1} \bmod 2^k + \ldots + U_{s+1} V_r \bmod 2^k)] \bmod 2^k$

$\qquad = [((U_s \bmod 2^k)(V_0 \bmod 2^k) + (U_{s-1} \bmod 2^k)(V_1 \bmod 2^k) + \ldots$
$\qquad + (U_0 \bmod 2^k)(V_s \bmod 2^k)) - ((U_r \bmod 2^k)(V_{s+1} \bmod 2^k) + \ldots$
$\qquad + (U_{s+1} \bmod 2^k)(V_r \bmod 2^k))] \bmod 2^k$

$\qquad = [(U_s' V_0' + U_{s-1}' V_1' + \ldots + U_0' V_s') - (U_r' V_{s+1}' + \ldots + U_{s+1}' V_r')]$
$\qquad \bmod 2^k, \tag{5}$

where we use the notation $U_j' = U_j \bmod 2^k$ and $V_j' = V_j \bmod 2^k$.

Since U_j' and V_i' are each k-bit long, their products, $U_j' V_i'$, are $2k$-bit long and the convolutions are sums of at most $r + 1 = 2^k$ such products, then each of the convolutions is at most $3k$-bit long. It follows that the convolutions $(U_s' V_0' + U_{s-1}' V_1' + \ldots + U_0' V_s')$ and $(U_r' V_{s+1}' + \ldots + U_{s+1}' V_r')$ can be 'read off' from the binary representation of the respective products

$$P_r^{(1)} = (U_r' U_{r-1}' \ldots U_1' U_0')_{2^{3k}} \times (V_r' V_{r-1}' \ldots V_1' V_0')_{2^{3k}}, \qquad (6)$$

$$P_r^{(2)} = (U_0' U_1' \ldots U_r')_{2^{3k}} \times (V_0' V_1' \ldots V_r')_{2^{3k}},$$

where, as before, the bracket $(Y_q' Y_{q-1}' \ldots Y_0')_{2^{3k}}$ denotes the integer

$$Y_q' 2^{3kq} + Y_{q-1}' 2^{3k(q-1)} + \ldots Y_1' 2^{3k} + Y_0'. \qquad (7)$$

To illuminate the point let us consider a simple example, for $r + 1 = 4 = 2^2$. We have the products

$$P_3^{(1)} = (U_3' U_2' U_1' U_0')_{2^{3k}} \times (V_3' V_2' V_1' V_0')_{2^{3k}},$$

$$P_3^{(2)} = (U_0' U_1' U_2' U_3')_{2^{3k}} \times (V_0' V_1' V_2' V_3')_{2^{3k}},$$

and their respective binary representation may be given as

$$
\begin{aligned}
P_3^{(1)} &= (U_3' 2^{(3k)3} + U_2' 2^{(3k)2} + U_1' 2^{3k} + U_0') \\
&\quad \times (V_3' 2^{(3k)3} + V_2' 2^{(3k)2} + V_1' 2^{(3k)} + V_0') \\
&= U_3' V_3' 2^{(3k)6} + (U_2' V_3' + U_3' V_2') 2^{(3k)5} + (U_1' V_3' + U_2' V_2' + U_3' V_1') 2^{(3k)4} \\
&\quad + (U_0' V_3' + U_1' V_2' + U_2' V_1' + U_3' V_0') 2^{(3k)3} \\
&\quad + (U_0' V_2' + U_1' V_1' + U_2' V_0') 2^{(3k)2} + (U_0' V_1' + U_1' V_0') 2^{3k} + U_0' V_0' \\
&= (P_6^{(1)} P_5^{(1)} P_4^{(1)} P_3^{(1)} P_2^{(1)} P_1^{(1)} P_0^{(1)})_{2^{3k}}, \text{ say} \\
P_3^{(2)} &= (U_0' 2^{(3k)3} + U_1' 2^{(3k)2} + U_2' 2^{(3k)} + U_3') \\
&\quad \times (V_0' 2^{(3k)3} + V_1' 2^{(3k)2} + V_2' 2^{(3k)} + V_3') \\
&= U_0' V_0' 2^{(3k)6} + (U_1' V_0' + U_0' V_1') 2^{(3k)5} + (U_2' V_0' + U_1' V_1' + U_0' V_2') 2^{(3k)4} \\
&\quad + (U_3' V_0' + U_2' V_1' + U_1' V_2' + U_0' V_3') 2^{(3k)3} \\
&\quad + (U_3' V_1' + U_2' V_2' + U_1' V_3') 2^{(3k)2} + (U_3' V_2' + U_2' V_3') 2^{3k} + U_3' V_3' \\
&= (P_6^{(2)} P_5^{(2)} P_4^{(2)} P_3^{(2)} P_2^{(2)} P_1^{(2)} P_0^{(2)})_{2^{3k}}, \text{ say}
\end{aligned}
$$

Thus,

$$
\begin{aligned}
F_0' &= F_0 \bmod (2^{4q} + 1) = P_0^{(1)} - P_2^{(2)} = P_6^{(2)} - P_4^{(1)}, \\
F_1' &= F_1 \bmod (2^{4q} + 1) = P_1^{(1)} - P_1^{(2)} = P_5^{(2)} - P_5^{(1)}, \\
F_2' &= F_2 \bmod (2^{4q} + 1) = P_2^{(1)} - P_0^{(2)} = P_4^{(2)} - P_6^{(1)}, \\
F_3' &= F_3 \bmod (2^{4q} + 1) = P_3^{(1)} \qquad = P_3^{(2)}.
\end{aligned}
$$

The products given by (6) can be computed, using the multiplication method of Section 6.2, with the number of basic operations of order $(r + 1)^{1.585}$.

2. $F_s^{(2)} = F_s \bmod (2^{2q} + 1), \qquad s = 0, \ldots, r.$

This is the most time consuming of the three steps. Savings on the number of operations required (as compared with the standard computation procedures) are achieved by using the (integer) convolution theorem. We shall first outline the sequence of computations that are performed in Step 2 and then give a formal justification of these computations.

The computational procedure is as follows:

(a) Set $\psi = 2^{2q/(r+1)}$, so that $\psi^{r+1} = 2^{2q}$, and compute the integer Fourier transforms with $w = \psi^2$, modulo $2^{2q} + 1$, of the sequences

$$(U_0, \psi U_1, \psi^2 U_2, \ldots, \psi^r U_r) \quad \text{and} \quad (V_0, \psi V_1, \ldots, \psi^r V_r),$$

i.e.

$$a_t = \left[\sum_{s=0}^{r} w^{st}(\psi^s U_s) \right] \bmod (2^{2q} + 1)$$

and

$$b_t = \left[\sum_{s=0}^{r} w^{st}(\psi^s V_s) \right] \bmod (2^{2q} + 1), \qquad t = 0, \ldots, r. \tag{8}$$

(b) Compute the pairwise products of the Fourier transforms computed in (a), modulo $(2^{2q} + 1)$, i.e.

$$c_t = a_t b_t \bmod (2^{2q} + 1), \qquad t = 0, \ldots, r. \tag{9}$$

(c) Compute the inverse integer Fourier transform, modulo $2^{2q} + 1$, of the sequence of pairwise products from step (b), i.e.

$$d_s = \left[\sum_{t=0}^{r} w^{-st} c_t \right] \bmod (2^{2q} + 1), \qquad s = 0, \ldots, r. \tag{10}$$

The result of this computation will be

$$(d_0, d_1, \ldots, d_r) = (2^k F_0^{(2)}, 2^k \psi F_1^{(2)}, \ldots, 2^k \psi^r F_r^{(2)}) \bmod (2^{2q} + 1). \tag{11}$$

Note that

$$2^k = r + 1. \tag{12}$$

Compute $F_s^{(2)}$ by multiplying d_s by $2^{-k} \psi^{-s} \bmod (2^{2q} + 1)$. In fact, since ψ is a power of 2, the $F_s^{(2)}$ may be obtained by an appropriate shifting operation analogous to 6.3.2–(11).

We shall now justify the procedure given in Step 2.

By definition we have

$$F_s^{(2)} \equiv \sum_{0 \leqslant \alpha, \beta \leqslant r} U_\alpha V_\beta \bmod (2^{2q} + 1). \tag{13}$$

Multiplying (13) by ψ^s we get

$$\psi^s F_s^{(2)} \equiv \psi^s \left[\sum_{0 \leqslant \alpha, \beta \leqslant r} U_\alpha V_\beta \bmod (2^{2q} + 1) \right]$$

$$= \sum_{\substack{0 \leqslant \alpha, \beta \leqslant r \\ \alpha + \beta \equiv s \bmod (r+1)}} (\psi^\alpha U_\alpha)(\psi^\beta V_\beta) \bmod (2^{2q} + 1). \tag{14}$$

For Step 2(c) we can write

$$d_s = \left[\sum_{t=0}^{r} \psi^{-2st} c_t \right] \bmod (2^{2q} + 1)$$

$$= \left[\sum_{t=0}^{r} \psi^{-2st} a_t b_t \bmod (2^{2q} + 1) \right] \bmod (2^{2q} + 1)$$

$$= \left[\sum_{t=0}^{r} \psi^{-2st} \left(\sum_{j=0}^{r} \psi^{2jt} (\psi^j U_j) \right) \bmod (2^{2q} + 1) \right.$$

$$\times \left. \left(\sum_{i=0}^{r} \psi^{2it} (\psi^i V_i) \right) \bmod (2^{2q} + 1) \right] \bmod (2^{2q} + 1)$$

$$= \left[\sum_{0 \leqslant j, i \leqslant r} (\psi^{2j} U_j)(\psi^{2i} V_i) \sum_{t=0}^{r} \psi^{2(-s+j+i)t} \right] \bmod (2^{2q} + 1).$$

Now, assuming that (we shall later prove this statement)

$$\sum_{t=0}^{r} \psi^{2\gamma t} \equiv \begin{cases} r + 1, & \text{if } \gamma \bmod (r+1) = 0, \\ 0, & \text{if } \gamma \bmod (r+1) \neq 0, \end{cases} \qquad (15)$$

we get

$$d_s = \left[(r+1) \sum_{\substack{0 \leqslant j, i \leqslant r \\ 2(j+i-s) \equiv 0 \bmod (r+1)}} (\psi^{2j} U_j)(\psi^{2i} V_i) \right] \bmod (2^{2q} + 1). \qquad (16)$$

From (14), (12), and (16) it follows that

$$F_s^{(2)} = [2^{-k} \psi^{-s} \bmod (2^{2q} + 1)] d_s, \qquad s = 0, \ldots, r,$$

which completes justification of computation procedure given in Step 2, except for a proof of statement (15).

In order to prove that

$$\left(\sum_{t=0}^{r} \psi^{2\gamma t} \right) \bmod (r+1) = \begin{cases} r + 1 = 2^k, & \text{if } \gamma \bmod (r+1) = 0, \\ 0, & \text{if } \gamma \bmod (r+1) \neq 0, \end{cases}$$

we shall consider both cases in detail.

(i) If $\gamma \bmod (r+1) = 0$, then γ is a multiple of $(r+1)$ and we may set $\gamma = c(r+1)$, where c is a constant. We get

$$\psi^{2\gamma t} = \psi^{2c(r+1)t} = (\psi^{r+1})^{2ct} = (2^{2q})^{2ct} = (-1)^{2ct} \bmod (2^{2q} + 1).$$

If follows

$$\left(\sum_{\substack{t=0 \\ \gamma \bmod (r+1) = 0}}^{r} \psi^{2\gamma t} \right) \bmod (r+1) = \left(\sum_{t=0}^{r} (-1)^{2ct} \right) \bmod (r+1) = r + 1.$$

(ii) If $\bmod (r+1) \neq 0$, then γ must be an odd number, since $r+1$ is by definition a power of 2. Let $\gamma \bmod (r + 1) = 2^\kappa \lambda$, where λ is odd and $0 \leqslant \kappa < k$, $r + 1 = 2^k$. Setting $T = 2^{k-1-\kappa}$, we get

$$\psi^{2\gamma T} = \psi^{2\gamma 2^{k-1-\kappa}} = \psi^{\gamma 2^k 2^{-\kappa}} = \psi^{2^k \lambda} = (\psi^{(r+1)})^\lambda = (2^{2q})^\lambda \equiv -1 \bmod (2^{2q} + 1).$$

Noting that

$$T = 2^{k-1-\kappa} = 2^k 2^{-1-\kappa} = (r + 1)2^{-1-\kappa}, \text{ or that } 2T = (r + 1)2^{-\kappa},$$

we can write

$$\left(\sum_{t=0}^{r} \psi^{2\gamma t} \right) \bmod (r + 1) \equiv 2^\kappa \sum_{t=0}^{2T-1} \psi^{2\gamma t} = 2^\kappa \sum_{t=0}^{T-1} (\psi^{2\gamma t} + \psi^{2\gamma(t+T)})$$

$$\equiv 2^\kappa \sum_{t=0}^{T-1} (\psi^{2\gamma t} + \psi^{2\gamma t}(-1)) \equiv 0.$$

This completes the proof.

6.3.6 Estimation of the Work Involved

The computational procedure is nearly complete. It remains only to specify the values k and l, where by definition $n = 2^m = 2^{k+1}$ (cf. Section 6.3.5–(4)). To do this we shall first evaluate the total amount of work involved as a function of k and l, and then specify k and l in such a way as to make this amount as small as possible. We already know that the work required to compute the product $uv \bmod (2^n + 1)$, after the reduced product coefficients have been computed, is of order n. We also know that Step 1 may be completed in $O(n)$ time, at most. We turn now to estimate the work involved in computation of Step 2. In this step three Fourier transforms, modulo $2^{2q} + 1$, are used. Each of them consists of $k = \log_2 (r + 1)$ steps, and each step requires an amount of work of order n, since the operations involved are simple shifts of the $(6q + \log_2 q)$-bit numbers, modulo $(2^{2q} + 1)$. Thus, each Fourier transform requires the work of $O(kn)$. Further, Step 2 also requires $r + 1$ ($= 2^k$) multiplications of $a_t b_t \bmod (2^{2q} + 1)$, where the integers a_t and b_t are in the range 0 to 2^{2q}, and, thus, are of length $2q (= 2^{l+1})$-bit each.

Now, let $T(n)$ denote the time it takes to multiply two n-bit numbers, modulo $2^n + 1$, by the method given. Let also $T'(n) = T(n)/n$. Then, for Step 2 we have

$$T(n) = 2^k T(2^{l+1}) + O(kn). \tag{1}$$

Dividing (1) by $n(= 2^{k+1})$, we get

$$\frac{T(n)}{n} = \frac{2^k}{2^{k-1}} \frac{T(2^{l+1})}{2^{l+1}} + O\left(\frac{kn}{n}\right),$$

giving

$$T'(n) = 2T'(2^{l+1}) + O(k). \tag{2}$$

To yield the lowest possible time in (2), the l has to be chosen as small as possible.

However, the choice of values for l is restricted by the relation between l and k which follows from the definition of ψ in Step 2(a). Namely, since

$$\psi = 2^{2q/(r+1)} = 2^{2^{l+1-k}}$$

should be a power of 2, then $l + 1 - k \geqslant 0$, or $l + 1 \geqslant k$. $\qquad\qquad$ (3)

Condition (3) implies that the best we can do to minimise the $T'(n)$ in (2) is to set

$$l = \begin{cases} \dfrac{m}{2}, & \text{if } m \text{ is even,} \\[2mm] \dfrac{m-1}{2}, & \text{if } m \text{ is odd,} \end{cases} \qquad\qquad (4)$$

and

$$k = \begin{cases} \dfrac{m}{2}, & \text{if } m \text{ is even,} \\[2mm] \dfrac{m+1}{2}, & \text{if } m \text{ is odd.} \end{cases} \qquad\qquad (5)$$

(4) and (5) give the final specification of l and k.

The conditions on l and k mean that there is a constant C such that

$$T'(n) \leqslant 2T'(2\sqrt{n}) + C \log_2 n \qquad \text{for all } n \geqslant 3, \qquad\qquad (6)$$

since $2^l \leqslant 2^{m/2} = \sqrt{n}$ and $k = \log_2(r+1) < \log_2((r+1)q) = \log_2 n$.

We shall, finally, show that (6) implies that

$$T'(n) \leqslant C' \log_2 n \log_2 \log_2 n \qquad \text{for suitable } C'. \qquad\qquad (7)$$

In the proof we shall use the method of induction.

For $n = 4$ in (6) we have

$$T'(4) \leqslant 2T'(4) + 2C = 2(T'(4) + C). \qquad\qquad (8)$$

Denoting $T'(4) + C$ by C', we can present (8) as

$$T'(4) \leqslant 2C' = C' \log_2 4 \log_2 \log_2 4.$$

Now assuming that (7) is valid for all $k \leqslant n - 1$ we shall prove its validity for $k = n$.

Substituting $C' \log_2 2\sqrt{n} \log_2 \log_2 2\sqrt{n}$ for $T'(2\sqrt{n})$ in (6) we get

$$\begin{aligned} T'(n) &\leqslant 2C' \log_2 2\sqrt{n} \log_2 \log_2 2\sqrt{n} + C \log_2 n \\ &= 2C'(1 + \tfrac{1}{2}\log_2 n) \log_2(1 + \tfrac{1}{2}\log_2 n) + C \log_2 n \\ &= 2C' \log_2(1 + \tfrac{1}{2}\log_2 n) + C' \log_2 n \log_2(1 + \tfrac{1}{2}\log_2 n) + C \log_2 n. \end{aligned}$$

For large n, $1 + \tfrac{1}{2}\log_2 n \leqslant \tfrac{2}{3}\log_2 n$, so we can write

$$T'(n) \leqslant 2C'(\log_2(\tfrac{2}{3}) + \log_2\log_2 n) + C'\log_2 n(\log_2(\tfrac{2}{3}) + \log_2\log_2 n) + C\log_2 n$$
$$= 2C'(1 - \log_2 3) + 2C'\log_2\log_2 n + C'\log_2 n(1 - \log_2 3)$$
$$\quad + C'\log_2 n\log_2\log_2 n + C\log_2 n$$
$$= C'\log_2 n\log_2\log_2 n + 2C'\log_2\log_2 n + \log_2 n(C + C' - C'\log_2 3)$$
$$\quad + 2C'(1 - \log_2 3)$$
$$= O(\log_2 n\log_2\log_2 n).$$

Finally, we get

$$T(n) = nT'(n) \leqslant O(n\log_2 n\log_2\log_2 n). \tag{9}$$

This completes estimation of the work involved in Step 2.

We note that Step 3 requires only a few operations of negligible cost, and we denote this cost by $\varepsilon(n)$.

The total amount of work involved in multiplication of two n-bit numbers, using the method outlined in Sections 6.3.4 and 6.3.5, is thus estimated by

$$W_{\text{total}} \leqslant O(n) + O(n) + O(n\log_2 n\log_2\log_2 n) + \varepsilon(n), \tag{10}$$

which proves that

it is possible to multiply two n-bit numbers in
$O(n\log_2 n\log_2\log_2 n)$ time.

This rather remarkable result is due to Schönhage and Strassen (1971).

Exercises

6.1 Assuming that necessary Fourier transforms are carried out using the FFT algorithm, estimate the total number of multiplication operations required by the computational process given in Section 6.3.1, for obtaining the product of two binary integers.

6.2 Give a 'modular representation' for one million with respect to the set of moduli $p_1 = 11$, $p_2 = 19$, $p_3 = 23$, $p_4 = 31$ and $p_5 = 43$.

6.3 Let the moduli be 2, 3, 5, 7. Given the modular representation $(1, 2, 3, 4)$ recover the original number.

6.4 Find $(11101001100111)_2 \bmod (2^5 + 1)$.

6.5 Compute the product of the binary numbers $(1101010)_2$ and $(1000001)_2$ using the formulae 6.2–(1), (2).

6.6 Compute the product in Exercise 6.5 using the computational process of Section 6.3.1 and assuming that $q = 1$ and $r = 6$.

6.7 Write a program to find residues of a number modulo a collection of mutually prime moduli.

Chapter 7

Internal Sorting

In this chapter we intend to analyse some of the better known computer methods to sort a collection of N items into a specified order. Sorting is one of the most frequently encountered tasks in processing. It may often occupy a significant amount of the total computer time needed by a particular data processing system. Hence, efficient sorting methods are eagerly sought.

Two kinds of sorting are generally distinguished, internal sorting when the items are stored in the computer's high-speed memory, also referred to as sorting of arrays; and external sorting when there are more items than can be kept in the memory at once, referred to as sorting of sequential files. Methods of internal sorting are not always readily extendable for external sorting since internal sorting allows a fair amount of flexibility in the structuring and accessing the data while external sorting often imposes strict constraints on the accessing of the data. As a result, different approaches and criteria are used for establishing algorithms suitable for a particular kind of sorting.

In the discussion to follow we shall refer to the entire collection of N items as an array or a file and assume that a key, $K(j)$, is associated with each item and that this key governs the sorting process. The goal of sorting is to determine a permutation $p_1, p_2, \ldots, p_N$ of the items which sets the keys in ascending order

$$K(p_1) \leqslant K(p_2) \leqslant \ldots \leqslant K(p_N).$$

For notational clarity, we shall not distinguish between an item and its key. In other words, we shall speak of an item K with key K. Furthermore, unless specified otherwise, we shall generally assume that the keys of a given file are all distinct.

Our discussion of sorting will not aim at an exhaustive study of this and related problems but rather at presenting a selected choice of methods which render an educational analysis of sorting. The approach intended is that in the spirit of Knuth.

Finally, we restrict ourselves to one particular class of methods based on comparison of the keys. For such methods the number of comparisons required is one of the characteristics on which the criteria of optimal sorting depend.

218

Indeed, if we assume that a bounded number of arithmetic operations occur between execution of comparisons, it follows that the total number of steps required by the method in question is proportional to the number of comparisons executed in a computation. We also note that the number of comparisons affords a machine independent criterion for studying various comparison sorts.

We shall first introduce examples of straightforward but not necessarily efficient sorting techniques which however illustrate a particular idea on which a whole family of sorting methods is based. Then, we will proceed to discuss more efficient sorting methods and general criteria for optimal sorting.

7.1 Introduction

When analysing a sorting algorithm one normally obtains some estimates of the running time which is made up by the number of times that each step of the algorithm is executed (frequency analysis) and the computer memory required by the algorithm (storage analysis). The most important part of the analysis is a study of the expected or average performance of the algorithm on a randomly ordered array. Some basic definitions and facts concerning the laws of probability and statistical analysis that may be useful for this purpose can be found in Appendix B.

7.2 Find-the-Largest Algorithm

Consider an algorithm that finds the largest key of a given array of keys.

We are given an array of unordered keys, $K(1), K(2), \ldots, K(n)$, generally not necessarily all different.

We wish to find the largest key of the array, $P = K(k) = \max_{1 \le j \le n} K(j)$, and such that k is as large as possible.

Algorithm Find-the-Largest

Introduce the pivot P and set it initially to be equal to $K(n)$. Compare P with $K(n-1)$:

(a) if $P < K(n-1)$, then set P equal to $K(n-1)$,
(b) if $P > K(n-1)$, then leave P unchanged.

Next, compare P with $K(n-2)$ and repeat the step (a) or (b) where $n-1$ is replaced by $n-2$. Etc.

The process is continued until the first key of the array is encountered.

7.2.1 Frequency Analysis of the Algorithm

The first observation, which is easily made, is that the algorithm requires a fixed amount of storage equal to the length of the input array, so that we do not need to concern ourselves with a storage analysis of the algorithm. The frequency analysis, on the other hand, amounts to counting the number of steps in the algorithm and the number of times that each step is executed. We notice in this respect that the only quantity which is not easily deduced is the number of times that the value of the pivot is changed. We denote this number by M which stands for key moves. To obtain complete information on this quantity, we need to know its minimum, average and maximum values.

We start the analysis with an obvious fact that if the last key of the array is $K(n) = \max_k K(k)$, then the value of M is zero. This is the minimum value. Furthermore, if we are unlucky enough to use the algorithm on an array which has the keys related as $K(1) > K(2) > \ldots > K(n)$, then the pivot value is changed after each comparison giving thus $M = n - 1$. This is the maximum value.

To find the average value of M, we enter the field governed by the laws of probability and so we have to specify conditions on the input array in order to set up a stochastic model of the computation. We assume that the $K(k)$ are distinct values (which is different from the initial formulation of the problem) and that each of the $n!$ permutations of these values is equally likely. (Note that the assumptions made here are not the only ones possible.)

We next observe that the algorithm find-the-largest concerns the relative order of the keys and not their precise values. For example, take the case of $n = 3$. Given a set of three different keys we find that whatever the actual values of the keys the following six possibilities are equally probable.

Permutation	Situation	Value of M
1 2 3	$K(1) < K(2) < K(3)$	0
1 3 2	$K(1) < K(3) < K(2)$	1
2 3 1	$K(2) < K(3) < K(1)$	1
3 1 2	$K(3) < K(1) < K(2)$	1
3 2 1	$K(3) < K(2) < K(1)$	2
2 1 3	$K(2) < K(1) < K(3)$	0

It follows that the $K(1), \ldots, K(n)$ may be taken to be the numbers $1, \ldots, n$ in some order; and under our assumptions the probability that M has the value k is

$$p_{nk} = \frac{\text{Number of permutations of } n \text{ items for which } M = k}{n!}. \tag{1}$$

Then in the usual manner, the average value is defined to be

$$\mu[M] = M_n = \sum_k k p_{nk}, \tag{2}$$

the variance, V_n, is defined as

$$\text{var}[M] = V_n = \sum_k (k - M_n)^2 p_{nk} = \sum_k k^2 p_{nk} - 2 M_n \sum_k k p_{nk} + M_n^2$$

$$= \sum_k k^2 p_{nk} - M_n^2, \tag{3}$$

and the standard deviation, σ_n, is defined as

$$\sigma_n = (V_n)^{1/2}.$$

Thus, the behaviour of M can be determined by determining the probabilities p_{nk}.

Equation (1) shows that to determine the probabilities p_{nk} we must enumerate the number of permutations of n keys for which $M = k$.

Consider the permutations $q_1, q_2, \ldots, q_n$ on $\{1, 2, \ldots, n\}$. If $q_1 = n$, the value of M is one higher than the value obtained on $q_2, \ldots, q_n$; if $q_1 \neq n$, the value of M is exactly the same as its value on $q_2, \ldots, q_n$.

Now, we have Prob $(q_1 = n) = 1/n$ and Prob $(q_1 \neq n) = (n-1)/n$. For the probability of M being equal to k we consider the simultaneous occurrence of $q_1 = n$ and M being equal to $k - 1$ for the remaining elements $q_2, \ldots, q_n$, i.e. $(1/n)p_{(n-1)(k-1)}$, plus the simultaneous occurrence of $q_1 \neq n$ and M being equal to k for the remaining elements $q_2, \ldots, q_n$, i.e.

$$\frac{n-1}{n} p_{(n-1)k}.$$

Therefore

$$p_{nk} = \frac{1}{n} p_{(n-1)(k-1)} + \frac{n-1}{n} p_{(n-1)k} \qquad (4)$$

To complete the formulation of the problem for determining the p_{nk} we provide initial conditions for equation (4):

$$\begin{aligned} p_{1k} &= \delta_{0k} && \text{where } \delta_{ij} \text{ is Kronecker's } \delta, \\ p_{nk} &= 0, && \text{if } k < 0. \end{aligned} \qquad (5)$$

Value of M_n and σ_n can now be determined for any n by first solving for p_{nk} the recurrence relations (4)–(5), and then substituting values of p_{nk} into the sums of (2) and (3).

For example, for $n = 4$ we get

$$p_{10} = 1;$$

$$p_{20} = \frac{1}{2}, \quad p_{21} = \frac{1}{2};$$

$$p_{30} = \frac{1}{3}, \quad p_{31} = \frac{1}{2}, \quad p_{32} = \frac{1}{6};$$

$$p_{40} = \frac{1}{4}, \quad p_{41} = \frac{11}{24}, \quad p_{42} = \frac{1}{4}, \quad p_{43} = \frac{1}{24};$$

and

$$M_4 = \sum k p_{4k} = \frac{13}{12} = 1.08, \qquad \delta_4 = \sqrt{V_4} = \frac{\sqrt{285}}{432} \approx 0.81.$$

The method, however, becomes very time consuming even for moderate values of n. So, instead, a more efficient technique is used to obtain information about the quantities p_{nk}. This technique makes use of the so called generating functions.

Definition of the Generating Function

Given a sequence of numbers $\{a_j, j = 0, \ldots, \infty\}$ one can set up an infinite sum in terms of a parameter z,

$$G(z) = a_0 + a_1 z + a_2 z^2 + \ldots = \sum_j a_j z^j.$$

The function G is a single quantity which represents the whole sequence $\{a_j\}$. It is called the generating function for the sequence $\{a_j, j = 0, \ldots, \infty\}$.

The generating function possesses several important properties which make it easy to manipulate into forms from which we may deduce conclusions concerning the number sequence itself.

For our problem we let

$$G_n(z) = p_{n0} + p_{n1} z + \ldots = \sum_k p_{nk} z^k. \tag{6}$$

Actually, $G_n(z)$ is a polynomial, since $M \leqslant n - 1$ and so $p_{nk} = 0$ for $k \geqslant n$. It is convenient, however, to specify the $G_n(z)$ as an infinite sum.

The average, M_n, and the variance, V_n, can now be determined from the generating function, $G_n(z)$.

By the definition we have $G_n(1) = 1$, i.e. the sum of all probabilities is equal to 1. Differentiating (6) we get

$$G_n'(z) = \sum_{k=0}^{n-1} k p_{nk} z^{k-1}$$

giving

$$G_n'(1) = \sum_{k=0}^{n-1} k p_{nk} = M_n, \text{ the average number of changes,} \tag{7}$$

and

$$G_n''(z) = \sum_{k=0}^{n-1} k(k-1) p_{nk} z^{k-2} = \sum_{k=0}^{n-1} k^2 p_{nk} z^{k-2} - \sum_{k=0}^{n-1} k p_{nk} z^{k-2},$$

giving

$$G_n''(1) = \sum_{k=0}^{n-1} k^2 p_{nk} - \sum_{k=0}^{n-1} k p_{nk} = \sum_{k=0}^{n-1} k^2 p_{nk} - G_n'(1).$$

The standard deviation of M is, then, expressed as

$$\sigma_n^2 = \sum_k k^2 p_{nk} - M_n^2 = G_n''(1) + G_n'(1) - [G_n'(1)]^2. \tag{8}$$

It remains to determine $G_n'(1)$ and $G_n''(1)$.

From (4) we have

$$P_{n0} = \frac{1}{n}0 + \frac{n-1}{n}P_{(n-1)0},$$

$$P_{n1} = \frac{1}{n}P_{(n-1)0} + \frac{n-1}{n}P_{(n-1)1},$$

$$P_{n2} = \frac{1}{n}P_{(n-1)1} + \frac{n-1}{n}P_{(n-1)2},$$

$$\vdots \qquad ,$$

giving the recurrence relation

$$G_n(z) = \frac{1}{n}zG_{n-1}(z) + \frac{n-1}{n}G_{n-1}(z)$$

$$= \frac{z+n-1}{n}G_{n-1}(z). \qquad (9)$$

Differentiating equation (9) we get

$$G_n'(z) = \frac{1}{n}G_{n-1}(z) + \frac{z+n-1}{n}G_{n-1}'(z),$$

so that

$$G_n'(1) = \frac{1}{n} + G_{n-1}'(1). \qquad (10)$$

Initial condition (5) yields

$$G_1(z) = 1,$$

giving

$$G_1'(z) = 0$$

and

$$G_1'(1) = 0.$$

Thus, from (10) we have

$$M_n = \sum_k kp_{nk} = G_n'(1) = \frac{1}{n} + \frac{1}{n-1} + \ldots + \frac{1}{2} = H_n - 1,$$

where $H_n = 1 + \frac{1}{2} + \frac{1}{3} + \ldots + \frac{1}{n}$ is the notation used for the harmonic numbers.
(For large n, H_n is close to the natural logarithm of n, i.e. $H_n \approx \log_e n$.) The value of σ_n is obtained in a way similar to the above by differentiating (9) twice.

We are ready to sum up the final result.

The number of times that the pivot is changed in the find-the-largest algorithm

224

is

$$M = \begin{cases} \min & 0, \\ \text{ave} & (H_n - 1), \text{dev } (H_n - H_n^{(2)}), \\ \max & (n - 1), \end{cases} \tag{11}$$

where

$$H_n^{(2)} = 1 + \frac{1}{2^2} + \frac{1}{3^2} + \ldots + \frac{1}{n^2}.$$

The analysis is now complete. We shall proceed to show how this algorithm and its analysis can be used to solve a more general problem, i.e. that of sorting an arbitrary array of keys into order.

7.3 Comparison Sorting by Selection

Having established the largest key of the array, $K(j)$, we exchange the places of two keys, the last in the array, $K(n)$, and the largest, $K(j)$. The find-the-largest algorithm is now repeated on the keys $K(1), \ldots, K(n-1)$, etc. Eventually, the complete array will be reordered into non-decreasing order.

This idea of repeated selection is basic to a whole group of sorting techniques which may be identified as sorting by selection.

7.3.1 The Straight Selection Sort

The algorithm just described can be formalized as follows:

1. Find the largest key. Transfer it into its proper position by exchanging it with the key currently occupying that position. This position is not considered again in the subsequent selections.
2. Repeat Step 1 on the remaining keys. This time the second largest key will be selected.
3. Continue this process until all n keys have been selected.

Following Knuth, we shall refer to this algorithm as the straight selection sort.

7.3.2 Analysis of the Method

Assuming that every selection of the current largest key is carried out using the find-the-largest algorithm, the performance of the straight selection sort is fully characterized by the following two quantities:

the number of comparisons of the keys, C, and
the number of exchanges of the keys, M.

Both quantities can be estimated using the results of the analysis of the find-the-largest algorithm.

So, the total number of comparisons required by the straight selection sort is given as

$$C = (n-1) + (n-2) + \ldots + 1 = \tfrac{1}{2}n(n-1). \tag{1}$$

This number is constant, whatever the initial permutation of the relative order of the keys.

The number of exchanges, on the other hand, is dependent on the relative order of the keys in the initial array. Hence, we need to estimate its minimum, average and maximum values.

If the initial array is in order then the straight selection sort will require no exchanges of the keys. Thus, the number of exchanges in this case equals zero. This is, clearly, the minimum value.

If the initial array is in reverse order then at each execution of the find-the-largest algorithm, the maximum number of exchanges is carried out. Noting that on each execution of the find-the-largest algorithm the length of the unordered subarray is reduced by one key at each end, we obtain the maximum value for the number of exchanges as

$$(n-1) + (n-3) + \ldots + 1 = \tfrac{1}{4}n^2. \tag{2}$$

Now, to determine the average number of exchanges we first note that

(a) If the straight selection sort starts with a random permutation of $\{1, 2, \ldots, n\}$, then the first execution of the find-the-largest algorithm yields a random permutation of $\{1, 2, \ldots, n-1\}$ followed by n. This follows from the observation that the permutation $q_1 q_2 \ldots q_{n-1} n$ is produced from each of the inputs

$$\begin{aligned} & n q_2 q_3 \cdots q_{n-1} q_1, \\ & q_1 n q_3 \cdots q_{n-1} q_2, \\ & \quad\vdots \\ & q_1 q_2 \ \cdots q_{n-1} n. \end{aligned} \tag{3}$$

(b) Since, as shown in (a), the occurrence of each permutation of $\{1, 2, \ldots, n-1\}$ in $\{K(1), K(2), \ldots K(n-1)\}$ is equally likely, and since the average number of times that the pivot is changed during the first application of the find-the-largest algorithm is $H_n - 1$, the average number of changes of current pivot in the straight selection algorithm, M_n, satisfies the recurrence relation

$$M_n = H_n - 1 + M_{n-1}. \tag{4}$$

It follows that

$$M_n = (H_n - 1) + (H_{n-1} - 1) + \ldots + (H_2 - 1)$$

$$= \sum_{j=2}^{n} H_j - n + 1 = [(n+1)H_n - n - H_1] - n + 1$$

$$\left(\text{where we have used the fact that } \sum_{j=1}^{n} H_j = (n+1)H_n - n \right),$$

giving

$$M_n = (n + 1)H_n - 2n. \tag{5}$$

The determination of the standard deviation of the number of exchanges in the straignt selection sort is an unsolved problem for general n.

Summarizing our findings we write

Number of comparisons, $C = \frac{1}{2}n(n - 1)$,

$$\text{Number of exchanges, } M = \begin{cases} \min 0, \\ \text{ave } [(n + 1)H_n - 2n], \quad \text{dev ?,} \\ \max \frac{1}{4}n^2. \end{cases} \tag{6}$$

We have found that the straight selection sort, on average, requires $O(n \log n)$ exchanges. This order of complexity, as it happens, is optimal for this class of sorting algorithms.

However, the number of comparisons in the straight selection sort is fixed and of order n^2. This raises the overall time complexity of the algorithm to $O(n^2)$. Thus, the question arises of what modifications to the method can be made so as to achieve an overall time complexity of order $n \log n$. Of immediate concern in this respect is to reduce the number of comparisons required.

We note that on the first search among the keys, at least $n - 1$ comparisons are needed as otherwise we cannot claim that we have found the largest key of the array. But on the subsequent searches ways may be found to utilize information about the relative order of the keys which has been obtained as a by-product during the first search.

7.3.3 Heapsort—a Comparison Sort of Complexity $n \log_2 n$.

The algorithm which is designed to make the most of this information is known as the Heapsort. The method was invented by Williams (1964) and Floyd (1964). This important algorithm consists of two phases. In the first phase the algorithm rearranges an arbitrary array into a heap. In the second phase it repeatedly removes the top of the heap and transfers it to its proper final position.

It is at the heap creation phase that the information obtained at every comparison of the keys is utilized in an optimal way. Denote the input array by $A(1 : n)$. With this array we associate a binary tree, a structure which is extremely convenient and used in considering many sorting (as well as searching) algorithms.

There are several ways to represent a tree structure. We shall use the graphical representation which explicitly illustrates the branching relationships (which, incidentally, led to the very name 'tree'). A graph is generally defined to be a set of points (called nodes) together with a set of lines (called branches) joining certain pairs of distinct nodes. A binary tree structure is shown in Fig. 7.1.

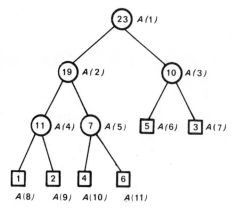

Fig. 7.1. A heap

We draw our trees as if they grew downwards, and the top node, i.e. $A(1)$ in Fig. 7.1, is called the root. A convenient way of numbering the nodes is to start at the root and to number the nodes from left to right on each level in turn so that if $A(k)$ is the root of a subtree, then $A(2k)$ and $A(2k + 1)$ are the two nodes immediately below it.

The nodes $A(2k)$ and $A(2k + 1)$ are called descendants of $A(k)$; if $A(k)$ is at level i then $A(2k)$ and $A(2k + 1)$ are said to be at level $i + 1$. Inversely, node $A(k)$ is said to be the (direct) ancestor of $A(2k)$ and $A(2k + 1)$.

We define the root of a tree to be at level 0. If a node has no descendants, it is called a terminal node, and denoted by $\square$; a node which is not terminal is an internal node, and denoted by $\bigcirc$.

The number of branches which have to be traversed in order to proceed from the root to a node $A(j)$ is called the path length of $A(j)$. The root has path length 0, its direct descendants have path lengths 1, etc. In general, a node at level i has path length i. Further properties of the trees will be introduced when and as necessary. Now, let us return to the definition of a heap. We define a heap as a binary tree with nodes $A(1)$ to $A(n)$ inclusive, so that all terminal nodes are on the same level or on two adjacent levels (See Fig. 7.2). The nodes of the tree are labelled with the keys of the array $A(1:n)$. Heapsort then rearranges the keys in the tree until the key associated with each node is greater than or equal to the keys associated with the node's two direct descendants. Such a labelled tree we call a heap. Once again, study Fig. 7.2. The sequence of keys on the path from each terminal node to the root is linearly ordered vis., $1 < 11 < 19 < 23$. Also the largest key in a subtree is always at the root of that subtree.

The heap property is, thus, defined as

$$A(k) \geqslant A(2k) \qquad \text{for } 1 \leqslant k \leqslant \frac{n}{2}$$

$$n \geqslant 2,$$

$$A(k) \geqslant A(2k + 1) \qquad \text{for } 1 \leqslant k < \frac{n}{2},$$

228

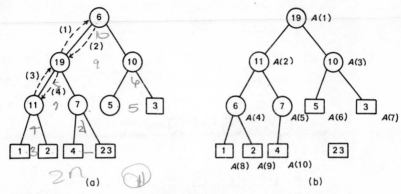

Fig. 7.2. The root removal and tree up-dating (or selection) phase of Heapsort (a) The keys 6 and 23 in the heap of Fig. 7.1 are exchanged (b) The heap is rearranged and key 23 is excluded

and the heap creation algorithm is as follows:

1. For $k = \left\lfloor \dfrac{n}{2} \right\rfloor, \left\lfloor \dfrac{n}{2} \right\rfloor - 1, \ldots, 1$ do Step 2 through 6.
2. Branch on the larger of two direct descendants:

    $A(2k) \geqslant A(2k + 1)$?

    (If n is even, skip the comparison test.)
    Denote the larger of the two keys by $A(p)$, i.e. let p assume the index of the larger key.
3. Smaller than the ancestor?

    $A(k) \geqslant A(p)$?

    If true, continue the do-loop of Step 1.
4. Exchange $A(k)$ and $A(p)$.
5. Test for the end of the input array:

    $(2p) > n$?

    If false set $k \leftarrow 2k$ and go to Step 2 otherwise continue the do-loop of Step 1.

A heap creation process is illustrated in Example 7.1.

Example 7.1

	$A(1)$	$A(2)$	$A(3)$	$A(4)$	$A(5)$	$A(6)$	$A(7)$	$A(8)$	$A(9)$	$A(10)$	$A(11)$
	11	2	3	1	4	5	10	23	19	7	6
$k=5$	11	2	3	1	(7)	5	10	23	19	(4	6)
$k=4$	11	2	3	(23)	7	5	10	(1	19)	4	6
$k=3$	11	2	(10)	23	7	(5	3)	1	19	4	6
$k=2$	11	(23)	10	(2	7)	5	3	1	19	4	6
	11	23	10	(19)	7	5	3	1	(2	4)	6
$k=1$	(23)	(11	10)	19	7	5	3	1	2	4	6
	23	(19)	10	(11	7)	5	3	1	2	4	6
	23	19	10	(11)	7	5	3	(1	2)	4	6

The last row in the table gives an initial heap. This heap is shown in Fig. 7.2.

In the second phase Heapsort removes and stores the largest key from the heap by interchanging $A(1)$ and $A(n)$. Thereafter location n of the array is considered no longer part of the heap. To rearrange the tree in locations $1, 2, \ldots, n-1$ into a heap, the new key $A(1)$ is sunk as far down a path in the tree as necessary. The process of interchanging $A(1)$ and $A(n-1)$ is then repeated and the tree is considered to occupy locations $1, 2, \ldots, n-1$, and so on. Removal of the largest key from the heap of Fig. 7.1 and the appropriate up-dating of the tree are demonstrated in Fig. 7.2. The array of Fig. 7.2 (a) is

6	19	10	11	7	5	3	1	2	4	23

and the resulting array of Fig. 7.2 (b) is

19	11	10	6	7	5	3	1	2	4	23

The selection phase of Heapsort is given as follows.

Let $A(1:s)$ be the current heap array.

1. Remove the (current) largest key from the heap:
   exchange $A(1)$ and $A(s)$ and from then onwards disregard the $A(s)$.
(Up-dating the heap array $A(1:(s-1))$.)
2. Set $k \leftarrow 1$.
3. Find the largest of two direct descendants:

   $A(2k) \geqslant A(2k+1)$?

   Denote the larger of the two keys by $A(p)$, i.e. let p assume the index of the larger key.
4. Smaller than the ancestor?

   $A(k) \geqslant A(p)$?

   If true, terminate.
5. Set $k \leftarrow p$ and go back to Step 3.

7.3.4 The Number of Comparisons Required by the Heapsort

At the creation of the heap for the first time we need $\leqslant C(n-2)$, where $1 < C < 2$, comparisons. At the second (selection) phase of Heapsort, the maximum number of comparisons is easily deduced from the binary (heap) tree. Consider such a tree:

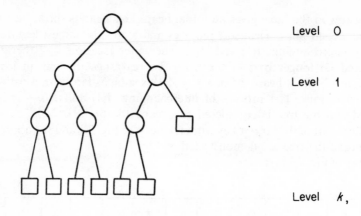

Level 0

Level 1

Level k,

where $k = \lfloor \log_2 n \rfloor$ since the total number of nodes, n, is given as

$$2^k \leqslant n \leqslant 2^{k+1}.$$

At every up-dating of the tree, after the removal of the current largest key, we need $\leqslant 2k$ comparisons, and $n - 1$ such up-datings are required to complete the Heapsort. Thus, the number of comparisons used at the second phase of the method is $2(n - 1) \log_2 n$, and the total number of comparisons required by the Heapsort is

$$\leqslant 2(n - 2) + 2(n - 1)\log_2 n,$$

that is, a quantity of $O(n \log n)$. This is an upper bound on the number of comparisons.

7.3.5 The Number of Exchanges Required by the Heapsort

In the creation of the heap for the first time, at most $n/2$ exchanges are needed; every removal of the current largest key means one exchange, giving in total $n - 1$;
the up-dating of the tree at the selection phase altogether uses $\leqslant n \log n$ exchanges.
Hence, the total number of exchanges required to sort n keys by the Heapsort is

$$\leqslant \frac{n}{2} + (n - 1) + n \log n,$$

that is a quantity of $O(n \log n)$.

Thus we have shown that in terms of the number of comparisons and the number of exchanges, the Heapsort is a comparison sort of $O(n \log n)$. However the problem of finding the average values for these numbers is still unsolved.

7.4 Comparison Sorting by Exchanging

Another family of comparison sorting algorithms is based on the idea that each comparison of two keys is followed systematically by interchanging the pairs of keys that are out of order, until no more such pairs are left.

Again, we first discuss a simple sorting algorithm which follows from a direct implementation of the idea, then attend to the question of the optimal utilization of the same idea. One simple algorithm based on the exchange of keys is the celebrated bubble sort.

7.4.1 The Bubble Sort

We are given a set of unordered keys, $K(1), K(2), \ldots, K(n)$ generally not necessarily all different.

We wish to sort the set into ascending order.

Algorithm Bubble Sort

Compare $K(1)$ with $K(2)$. Interchange, if out of order. Next compare $K(2)$ with $K(3)$. Interchange if out of order, etc.

The set is sorted in a finite number of steps.

The method obtained its name, the 'bubble' sort, because in the process of sorting by this method, the large keys 'bubble up' to their proper position.

The basic structure of the method is simple:

1. For $j = 1, 2, \ldots, n - 1$ do Step 2 through 4.
2. Compare the two adjacent keys:

$$K(j) \geqslant K(j + 1)?$$

If false, continue the do-loop of Step 1.
3. Exchange $K(j)$ and $K(j + 1)$.
4. Continue the do-loop of Step 1.

On the completion of scanning of all the keys once (this operation is normally referred to as a pass) it is not difficult to see that all keys above and including the last one to be exchanged must be in their final position, so they need not be examined on subsequent passes.

The passes are repeated until on the final pass (which indicates that the file is now completely ordered) no exchanges are performed.

Example 7.2

	Pass 1	Pass 2	Pass 3	Pass 4	Pass 5	Pass 6	Pass 7	Pass 8	Pass 9 (final pass)
3	(23)	23	23	23	23	23	23	23	23
4	3	(19)	19	19	19	19	19	19	19
2	4	3	(11)	11	11	11	11	11	11
1	2	4	3	(10)	10	10	10	10	10
7	1	2	4	3	(7)	7	7	7	7
6	7	1	2	4	3	(6)	6	6	6
5	6	7	1	2	4	3	(5)	5	5
10	5	6	7	1	2	4	3	(4)	4
11	10	5	6	7	1	2	4	3	3
19	11	10	5	6	6	1	2	2	2
23	19	11	10	5	5	5	1	1	(1)

7.4.2 Analysis of the Bubble Sort

We see that in the timing of the bubble sort three quantities are involved:

the number of comparisons, denoted by C,
the number of exchanges,   denoted by M,
the number of passes,   denoted by A.

To obtain complete information on the timing of the bubble sort, we need the minimum, maximum and average values of each of these three quantities.

We shall subject to a detailed analysis one quantity, M, only. The choice is partly due to the fact that the analysis of this quantity involves using an interesting mathematical modelling technique which we have not yet had an opportunity to introduce. In addition, we shall estimate the minimum and maximum values for the quantities C and A. This will enable us to form conclusions on the time complexity of the bubble sort in the worst case.

The Average Value for M

We assume that the input file is in random order and the keys are distinct. The analysis which follows uses the so called *inversion tables*.

Definition

Let $a_1 a_2 \ldots a_n$ be a permutation of the set $(12 \ldots n)$.

If $i < j$ and $a_i > a_j$, the pair (a_i, a_j) is called an 'inversion' of the permuta tion; for example, the permutation 3142 has three inversions: (31), (32), (42). Each inversion is a pair of elements that is out of order, so the only permutation with no inversions is the sorted permutation $12 \ldots n$.

The inversion table $b_1 b_2 \ldots b_n$ of the permutation $a_1 a_2 \ldots a_n$ is obtained by letting b_j be the number of elements to the left of j that are greater than j. (In other words, b_j is the number of inversions whose second component is j.)

For example the permutation

5 9 1 8 2 6 4 7 3

$b_1 b_2 b_3 b_4 b_5 b_6 b_7 b_8 b_9$

2 3 6 4 0 2 2 1 0   = 20 inversions in all

By definition we will always have

$$0 \leqslant b_1 \leqslant n - 1, \ 0 \leqslant b_2 \leqslant n - 2, \ldots 0 \leqslant b_{n-1} \leqslant 1, \ b_n = 0$$

The most important fact about inversions is the observation that an inversion table uniquely determines the corresponding permutation. This correspondence is important because we can often translate a problem stated in terms of permutations into an equivalent problem stated in terms of inversion tables, and the latter problem may be easier to solve.

For our problem of obtaining the average of M we observe that if we interchange two adjacent elements of a permutation, it is obvious that the total number of inversions will increase or decrease by unity. Hence, the number of exchanges in the bubble sort is equal to the total number of inversions in the given permutation.

Now, denote by $I_n(k)$ the number of permutations of n elements which have exactly k inversions each.

The following table lists the first few values of the function $I_n(k)$ for various n and k:

n \ k	0	1	2	3	4	5	6	7	8	9	10
1	1	0	0	0	0	0	0	0	0	0	0
2	1	1	0	0	0	0	0	0	0	0	0
3	1	2	2	1	0	0	0	0	0	0	0
4	1	3	5	6	5	3	1	0	0	0	0
5	1	4	9	15	20	22	20	15	9	4	1
6	1	5	14	29	49	71	90	101	101	90	71

In the table, the first column, $I_n(0) = 1$, corresponds to a perfectly ordered permutation, i.e. $12\ldots n$, and the second column, $I_n(1) = n - 1$, to the case when just one pair is out of order (hence $n - 1$ such permutations of n elements).

It is convenient to consider the generating function for the sequence $(I_n(k), k = 0, \ldots m)$. Denote it as follows

$$G_n(z) = \sum_{k=0}^{m} I_n(k)z^k; \tag{1}$$

say, for $n = 3$,

$$G_3(z) = 1 + 2z + 2z^2 + z^3 = (1 + z + z^2)(1 + z)$$
$$= (1 + z + z^2)G_2(z)$$
$$\text{since } G_2(z) = 1 + z,$$

for $n = 4$,

$$G_4(z) = 1 + 3z + 5z^2 + 6z^3 + 5z^4 + 3z^5 + z^6$$
$$= (1 + z + z^2 + z^3)G_3(z), \qquad \text{etc.}$$

In general,

$$G_n(z) = (1 + z + z^2 + \ldots + z^{n-1})G_{n-1}(z). \tag{2}$$

(Proof of (2) may be obtained using 'induction in reverse'.)

Now, we wish to find the average number of inversions present in a given permutation assuming that each of the $n!$ permutations of n elements is equally likely.

The probability that a given permutation has exactly k inversions is

$$p_{nk} = \frac{\text{No of permutations of } n \text{ elements with } k \text{ inversions}}{n!} = \frac{I_n(k)}{n!}. \tag{3}$$

Then the average number of inversions is

$$\sum_{k=0}^{m} kp_{nk}.$$

To determine p_{nk}, consider the generating function

$$g_n(z) = \sum_{k=0}^{m} p_{nk}z^k = \frac{1}{n!}\sum_{k=0}^{m} I_n(k)z^k = \frac{1}{n!}G_n(z)$$
$$= \frac{1 + z + z^2 + \ldots + z^{n-1}}{n} \frac{1 + z + z^2 + \ldots + z^{n-2}}{n-1} \ldots \frac{1+z}{2} 1$$
$$= h_n(z)h_{n-1}(z)\ldots h_1(z). \tag{4}$$

By virtue of the definition and expression (4), we have

$$\sum_{k=0}^{m} kp_{nk} = g_n'(1) = \frac{n-1}{2} + \frac{n-2}{2} + \ldots + \frac{1}{2} + 0 = \frac{1}{4}n(n-1). \tag{5}$$

Thus, the average value of M equals $\frac{1}{4}n(n-1)$, from which we have the important conclusion that the complexity of the bubble sort, on average, is not better than $O(n^2)$.

The analysis of the worst case of the bubble sort is straightforward and is given next.

The Maximum and Minimum Values for C, M, and A

These quantities can be obtained by the following argument. If the initial file is in order, then the first pass will also be the final pass as it will require no exchanges.

Hence, the minimum values of C, M, and A are $n-1$, 0 and 1, respectively.

Next, if the initial file is in reverse order, then the first pass will 'bubble up' the largest key into its final position, using exactly $n-1$ comparisons and $n-1$ exchanges. The second pass will 'bubble up' the next to the largest key into its final position, using exactly $n-2$ comparisons and $n-2$ exchanges, and so on. To complete the bubble sort, n passes are needed, in total, and $\frac{1}{2}n(n-1)$ each, of comparisons and exchanges. This is the worst possible case for the method, hence the maximum number of passes for the bubble sort is n and the maximum number of comparisons and exchanges is $\frac{1}{2}n(n-1)$ each.

7.4.3 Quicksort—a Comparison Sort of Complexity $n \log_2 n$, on Average

The analysis of bubble sort has shown that its time complexity is of order n^2 which is rather high for a sorting algorithm. However, a remarkably more efficient use of the basic comparison-exchange scheme is achieved by attacking the initial array at both ends simultaneously. The method, known as the Quicksort, is especially striking among the efficient comparison sorts. It has a worst case complexity proportional to n^2. On the other hand, its complexity, on average, while bounded below by $O(n \log n)$, is a fraction of the running time of other known sorting algorithms when implemented on most real computers.

Assume that an unordered array of keys, $K(1), \ldots, K(n)$, is given which we wish to sort into non-decreasing order.

Algorithm Quicksort

Keep two pointers, i and j, with $i = 1$ and $j = n$ initially. Compare $K(i)$ with $K(j)$, and if no exchange is necessary decrease j by 1 and repeat the process. After an exchange first occurs, increase i by 1, and continue comparing and increasing i until another exchange occurs. Then decrease j again, and so on, until $i = j$. At this time the original array is partitioned into two arrays and the original problem is reduced to two simpler problems, namely sorting $K(1), \ldots, K(i-1)$ and, independently, $K(i+1), \ldots, K(n)$. The same technique is, then, applied to each of the arrays.

236

Example 7.3

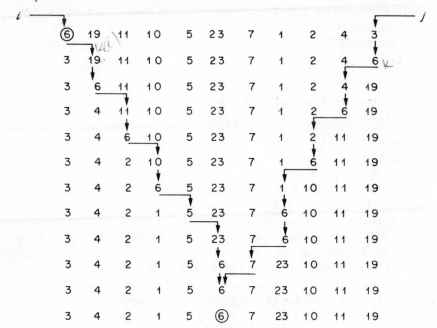

At this point, the given array of 11 keys has been partitioned into two arrays
{3 4 2 1 5} and {7 23 10 11 19}. We also observe that the original key, $K(1) = 6$,
is the partitioning boundary for the arrays as no keys on its left are greater than
6 and no keys on its right are smaller than 6. It follows that, in addition to being a
partition boundary, the key 6 has moved into its final position and does not
need to be considered in subsequent partitioning processes. This observation is a
common feature of the Quicksort, i.e. if the partitioning boundary is chosen
from among the elements of the array which is being 'quicksorted', then at the
end of the partitioning process the boundary element is in its proper final
position.

The next observation easily made is that each process of complete partitioning
of an array of length h into two arrays of length s and $h - s$ by the Quicksort,
requires h comparisons and $\leq h$ exchanges.

Hence, the number of comparisons (and, to a lesser extent, the number of
exchanges) required by the Quicksort depends on the way in which partitioning
of the arrays at intermediate stages of the sorting is achieved. Also, after each
partitioning of the current array we have temporarily to store one of the arrays
while working on the other. Thus, some auxiliary storage is needed by the
algorithm; to some extent, optimality requirements on the number of compari-

sons are governed by the optimal storage requirements of the method. We shall first study the storage requirements of Quicksort.

Quicksort Algorithm

1. Set $i \leftarrow 1$ and $j \leftarrow n$.
2. Set $K \leftarrow K(i)$.
3. Compare K with $K(j)$.
   If $K \leqslant K(j)$, decrease j by 1, then test $j = i$?
   If true go to Step 7, otherwise repeat the step.
4. Exchange $K(i)$ and $K(j)$, then increase i by 1 and test $i = j$?
   If true go to Step 7.
5. Compare K with $K(i)$.
   If $K > K(i)$, increase i by 1, then test $i = j$?
   If true go to Step 7, otherwise repeat the step.
6. Exchange $K(i)$ and $K(j)$, then decrease j by 1 and test $j = i$?
   If true go to Step 7, otherwise go to Step 3.
7. Compare the lengths of the two partitioned arrays.
   Is $i \geqslant n - i$?
   If true, put array $K(1), K(2), \ldots, K(i)$ on stack then set $i \leftarrow i + 1$ and $j \leftarrow n$; otherwise put array $K(i + 1), K(i + 2), \ldots, K(n)$ on stack then set $j \leftarrow i$ and $i \leftarrow 1$. Go to Step 2.
8. Is the stack empty?
   If false, fetch the next in line array from the stack, set i and j so as to point to the first and the last elements of the array, respectively. Go to Step 2.
9. Terminate.

7.4.4 Storage Analysis of Quicksort

In practice, the custom after each partitioning is to put the largest array on the stack and commence work on the other array, until trivially short arrays are reached. In this way, as is shown later, about $\log(n)$ arrays on average will have to be temporarily stored while awaiting their turn to be sorted. In the general case, every array put on the stack requires 2 words pointing to the ends of the array and, hence, the auxiliary storage used is about $2 \log n$ words. It may be smaller in some special cases, as, for example, when one can work on the given array in such a way that the stored arrays all come from one end. The auxiliary storage in this case is reduced to one half of the general case.

The volume of the auxiliary storage depends on the total number of arrays that are created during quicksort. In the first instance, if a perfectly ordered array is being 'quicksorted', then each partitioning operation reduces the size of the current array by only one key. Thus the maximum number of arrays created by Quicksort in this case is n, giving the auxiliary storage requirement of $2(n - 1)$ words.

Further, since each partitioning process requires a number of comparisons equal to the length of the array being currently partitioned, the maximum number of comparisons is given by

$$n + (n-1) + \ldots + 1 = \tfrac{1}{2}n(n+1) \tag{1}$$

This order of complexity immediately throws the Quicksort into the class of non-efficient sorting methods. Hence, an initially sorted file is the worst case for the Quicksort.

Now, consider a randomly unordered array of n keys which we wish to quicksort.

For simplicity, assume that at each partitioning the larger arrays all come from one end.

After the first partitioning we have

$$K(1) \ldots K(p_1) s_1 K(q_1) \ldots K(n),$$

where

$$p_1 \leqslant \frac{n}{2},$$

and the larger array is $\{K(q_1) \ldots K(n)\}$.

After the second partitioning (of the shorter array) we have

$$K(1) \ldots K(p_2) s_2 K(q_2) \ldots K(p_1)$$

where

$$p_2 \leqslant \frac{p_1}{2} \leqslant \frac{1}{2}\left(\frac{n}{2}\right),$$

and the current larger array is $\{K(q_2) \ldots K(p_1)\}$.

Repeating the process of partitioning in the above manner we eventually arrive at

$$p_k \leqslant \frac{1}{2}p_{k-1} \leqslant \ldots \leqslant \frac{n}{2^k} < 2$$

(Note: the trivial array contains one element), which yields

$$k > \log n - 1$$

where k is the maximum number of stored arrays.

Hence, in the general case of a randomly unordered initial array, the total number of subfiles created is bounded below by $\log n$. In practice, with some clever choice of the partitioning boundaries, cf. Hoare (1962), Singleton (1969), Frazer and McKellar (1970), Aho, Hopcroft, and Ullman (1974), Knuth (1969), the actual number of the arrays can be made very close to this lower bound.

7.4.5 Average Time Required by Quicksort

In order to estimate the average performance of Quicksort, it is convenient to introduce a time unit as a unit which refers equally to one operation of either comparison or exchange. Quicksort, then, lends itself to a straightforward analysis of the average time required to sort an array of n keys.

Denote this time by $T(n)$.

Assume, for simplicity, that the array is simply some permutation of the numbers $(1, 2, \ldots n)$, and that any permutation of these numbers is equally likely to occur.

Suppose that after the first partitioning we have created two arrays with the partitioning boundary s. Denote the time required to quicksort each of the arrays created, by $T(s)$ and $T(n - s)$, respectively.

Since s is equally likely to take on any value between 1 and n, and since the time expended on the first partitioning is cn, where c is a constant such that $1 < c < 2$, (the latter follows from the observation that at the first partitioning n comparisons and $\leqslant n$ exchanges are required), then we have

$$T(n) = \frac{1}{n} \sum_{s=1}^{n} \left[T(s) + T(n - s) \right] + cn, \ T(0) = T(1) = 0, \ 1 < c < 2$$

$$= \frac{1}{n} \left\{ \sum_{s=1}^{n} T(s) + \sum_{j=0}^{n-1} T(j) \right\} + cn$$

$$= \frac{1}{n} \left\{ \sum_{s=1}^{n-1} T(s) + \sum_{j=1}^{n-1} T(j) + T(n) \right\} + cn$$

$$= \frac{1}{n} \left\{ 2 \sum_{s=1}^{n-1} T(s) + T(n) \right\} + cn,$$

or

$$(n - 1)T(n) = 2 \sum_{s=1}^{n-1} T(s) + cn^2. \tag{1}$$

(1) is a recurrence relation. To solve it, we first get rid of the summation sign. Write:

$$nT(n + 1) = 2 \sum_{s=1}^{n} T(s) + c(n + 1)^2$$

$$(n - 1)T(n) = 2 \sum_{s=1}^{n-1} T(s) + cn^2$$

Subtracting the two equations we get

$$nT(n + 1) - (n - 1)T(n) = 2T(n) + c(2n + 1),$$

or

$$nT(n + 1) = (n + 1)T(n) + c(2n + 1). \tag{2}$$

It is convenient to rewrite the recurrence relation (2) in the form

$$\frac{T(n+1)}{n+1} = \frac{T(n)}{n} + c\frac{2n+1}{n(n+1)}$$

Now, recursively,

$$\frac{T(n+1)}{n+1} = \frac{T(n)}{n} + c\left[\frac{1}{n+1} + \frac{1}{n}\right]$$

$$\frac{T(n)}{n} = \frac{T(n-1)}{n-1} + c\left[\frac{1}{n} + \frac{1}{n-1}\right]$$

$$\vdots$$

$$\frac{T(2)}{2} = \frac{T(1)}{1} + c\left[\frac{1}{2} + \frac{1}{1}\right],$$

giving

$$\frac{T(n+1)}{n+1} = c\left[\left(\frac{1}{n+1} + \frac{1}{n} + \ldots + \frac{1}{2}\right) + \left(\frac{1}{n} + \frac{1}{n-1} + \ldots + 1\right)\right]$$

$$= c\left(H_{n+1} - 1 + H_{n+1} - \frac{1}{n+1}\right)$$

$$= 2cH_{n+1} - c\frac{n+2}{n+1}.$$

Hence

$$T(n) = 2cnH_n - c(n+1) \approx 2cn\log_e n - c(n+1) = O(n\ln n), \tag{3}$$

since $H_n \approx \ln n$ for large n.

7.5 Related Problem: Insertion of a Key in an Ordered Array

Suppose that an array of distinct keys, $K(1), \ldots, K(n)$ is given and that $K(1) < K(2) < \ldots < K(n)$. We wish to insert X into the ordered array by comparing it with keys of that array.

Algorithm for Sequential Insertion

Compare X with $K(1), K(2), \ldots, K(n)$ in turn, until either $K(i)$ is found such that $X < K(i)$, or else $X > K(n)$.

It is easily seen that in the worst case, the algorithm requires n comparisons.

Algorithm for Binary Insertion

Compare X with $K(\lceil n/2 \rceil)$, a key closest to the middle of the sequence $K(1), \ldots, K(n)$. If $X > K(\lceil n/2 \rceil)$, throw away $K(1), \ldots, K(\lceil n/2 \rceil)$. If $X <$

$K(\lceil n/2\rceil)$, throw away $K(\lceil n/2\rceil)$, $K(\lceil n/2\rceil + 1),\ldots,K(n)$. Repeat the process, each time throwing away half the keys, until the position for X is located.

A simple analysis shows that in the worst case, the algorithm requires $\lceil \log_2 n\rceil$ comparisons.

In fact it is readily shown that the binary insertion algorithm is optimal, that is every algorithm to insert X must require at least $\lceil \log_2 n\rceil$ comparisons in all cases.

7.6 Optimum Comparison Sorting

In this section we consider some optimal characteristics of the comparison sorting. We assume that an array of n keys to be sorted has no known structure, and that the only operation that can be used to gain information about the array is the comparison of two keys. Under these assumptions, we pose the question: what are the minimum requirements on the number of comparisons between keys in the comparative sorting algorithms?

The problem is made more precise under further assumptions that all the keys to be sorted are distinct and that the sorting methods we have in mind are based solely on an abstract linear ordering relation ' < ' between the keys. Then, all n-key comparative sorting methods which satisfy the above constraints can be represented in terms of an extended binary tree structure such as shown in Fig. 7.3.

In the tree, each internal node contains two indices '$i:j$' for comparison of $K(i)$ vs. $K(j)$. The left subtree of this node represents the subsequent comparisons to be made if $K(i) < K(j)$, and the right subtree: the subsequent comparisons if $K(i) > K(j)$. Each terminal node contains permutation $q_1,q_2,\ldots,q_n$ of

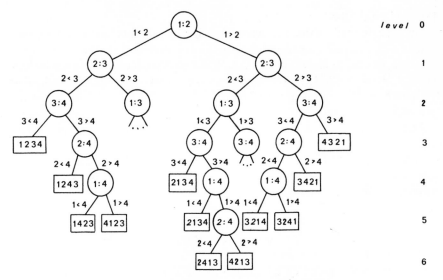

Fig. 7.3. An extended binary tree for an array of size $n = 4$

242

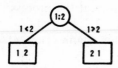

Fig. 7.4. An extended binary tree for an array of size $n = 2$

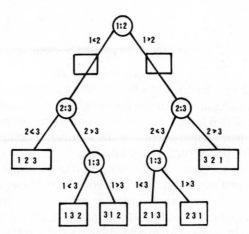

Fig. 7.5. An extended binary tree for an array
of size $n = 3$

$(1, 2, \ldots, n)$, denoting the fact that the ordering

$$K(q_1) < K(q_2) < \ldots < K(q_n)$$

has been established.

The tree is easily extended when the size of the data array is increased. Suppose that initially we consider an array of length 2, $K(1)$, $K(2)$. The tree is then constructed as shown in Fig. 7.4. When the length of the array is increased by 1, $K(1)$, $K(2)$, $K(3)$, the tree 'grows' as shown in Fig. 7.5. And, when the length of the array is again increased by 1, $K(1)$, $K(2)$, $K(3)$, $K(4)$, the tree 'grows' further as shown in Fig. 7.6.

Thus the figures represent a sorting method which first compares $K(1)$ with $K(2)$; if $K(1) > K(2)$, it goes on (via the right subtree) to compare $K(2)$ with $K(3)$, and then if $K(2) < K(3)$, it compares $K(1)$ with $K(3)$; at this stage if $K(1) < K(3)$ it 'knows' that $K(2) < K(1) < K(3)$. The method proceeds then to compare $K(3)$ with $K(4)$; if $K(3) < K(4)$ it 'knows' that $K(2) < K(1) < K(3) < K(4)$.

A comparison of $K(i)$ with $K(j)$ in our tree always means the original keys $K(i)$ and $K(j)$, not the keys which might currently occupy the ith and jth positions of the array after the keys have been shuffled around.

It is possible to make redundant comparisons. For example see Fig. 7.7. Here, there is no reason to compare $1:3$, since $K(1) > K(2)$ and $K(2) > K(3)$ implies $K(1) > K(3)$. No permutation can possibly correspond to the right

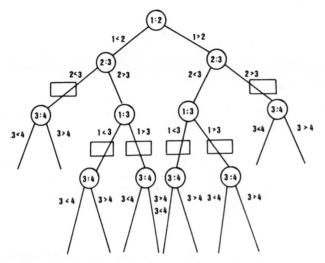

Fig. 7.6. Part of an extended binary tree for an array of size
$n = 4$

Fig. 7.7. An example of a redundant comparison,
$K(1) : K(3)$, in an extended binary tree

subtree of node $1:3$. Since we are interested in minimizing the number of comparisons, we assume that no redundant comparisons are made. All permutations of the input keys are possible and every permutation defines a unique path from the root to an external node; it follows that there are exactly $n!$ external nodes in a comparison tree which sorts n keys with no redundant comparisons.

7.6.1 Lower Bound on the Maximum Number of Comparisons

We are now able to show that any algorithm which sorts by comparisons must on some array of length n use at least $O(n \log n)$ comparisons. Indeed, if all the internal nodes of a comparison tree are at levels $< k$, then it is obvious that there can be at most 2^k external nodes in the tree. Hence

$$n! \leqslant 2^k, \tag{1}$$

where the inequality sign is explained by the fact that the number 2^k may include the redundant comparisons.

It follows from (1) that

$$k \geqslant \log_2 n! \quad \text{or the nearest upper integer.} \tag{2}$$

By Stirling's formula we have

$$k \geqslant O(n \log_2 n). \tag{3}$$

From the definition of the extended binary tree of Fig. 7.4 we observe that if no redundant comparisons are made, then k gives the minimum number of comparisons which will suffice to sort any n keys. Hence, the proof is complete.

What we have established is that under the decision tree model it is impossible to design a comparative sorting algorithm which would in any circumstances require less than $\log_2 n!$ comparisons between the keys. An extended discussion on whether this lower bound is attainable for general n, may be found in Knuth (1969).

Up-to-date there are few comparative sorts which attain the lower bound on comparisons for $n \leqslant 12$ and $n = 20, 21$. (No known sorting algorithm attains the lower bound for other than these values of n.) One such algorithm is the merge-insertion of Ford and Johnson (1959). The smallest number of keys on which the algorithm can be conveniently demonstrated is 5, namely an array of any five keys can be sorted by the merge-insertion with at most 7 comparisons.

Let $K(1)$, $K(2)$, $K(3)$, $K(4)$, $K(5)$ be an arbitrary array of five keys.

Start by comparing the two pairs $K(1):K(2)$ and $K(3):K(4)$. Leave out the key $K(5)$. The result can be visualized as follows

(2 comparisons)

where x_1, x_2 and y_1, y_2 stand for larger and smaller keys of two sorted pairs, respectively; and y_3 stands for the key which was left out.

Now compare the two larger keys of the pairs. The situation may be diagrammed as

(1 comparison)

indicating that $\quad a < b < d,$
$$c < d.$$

Next insert e among (a, b, d). This leads to one of the following situations:

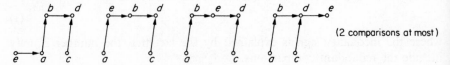

(2 comparisons at most)

Finally insert c among the keys which are less than d. This will require 2 extra comparisons at most, giving 7 comparisons in total.

We shall now apply the merge-insertion to sort in order an array of 11 keys.

1. Compare the five pairs $K(1):K(2)$, $K(3):K(4),\ldots,K(9):K(10)$. Leave out the key $K(11)$.
   We get

2. Sort the five larger keys of the pairs using merge-insertion. (Note the recursive use of the method.) A configuration obtained may be diagrammed as

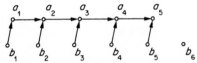

3. Insert b_3 among (b_1,a_1,a_2) then b_2 among the keys less than a_2. By now we arrive at the configuration

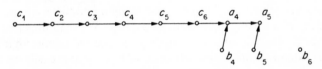

4. Insert b_5 among $(c_1,c_2,\ldots,c_6,a_4)$, e.g. first compare it to c_4, then to c_2 or c_6, etc. Then move b_4 into its proper place among the keys less than a_5. Each insertion requires three comparisons at most.
5. Insert b_6 in its place among the keys.

Summing up the comparisons required here we find that the 11 keys have been sorted in at most

$$5 + 7 + 2 + 2 + 3 + 3 + 4 = 26 \text{ comparisons.}$$

In a general case, when applied to sort an array of n keys, the merge-insertion algorithm is given as:

1. Carry out pairwise comparisons of $\lfloor n/2 \rfloor$ pairs of keys. If n is odd, leave one key out.
2. Sort the $\lfloor n/2 \rfloor$ larger keys found in Step 1, by merge-insertion.
3. Name the keys $a_1,a_2,\ldots,a_{\lfloor n/2\rfloor},b_1,b_2,\ldots,b_{\lfloor n/2\rfloor}$
   where
   $$a_1 \leqslant a_2 \leqslant a_3 \leqslant \ldots \leqslant a_{\lfloor n/2\rfloor}$$
   and
   $$b_i \leqslant a_i, \quad 1 \leqslant i \leqslant \lfloor n/2 \rfloor;$$

call b_1 and the a's the 'main chain'. Insert the remaining b's into the main chain, using binary insertion, in the following order:

$$b_1; b_3, b_2; b_5, b_4; b_{11}, b_{10}, b_9, b_8, b_7, b_6; \ldots;$$
$$b_{t_k}, b_{t_{k-1}}, \ldots, b_{t_{k-1}+1}; \ldots$$

The crucial point in the algorithm is generating of the sequence $(t_1, t_2, t_3, \ldots, t_k, \ldots) = (1, 3, 5, 11, \ldots)$ where t_k is such that each of $b_{t_k}, b_{t_{k-1}}, \ldots, b_{t_{k-1}+1}$ can be inserted with at most k comparisons. An analysis of the algorithm shows that $t_k = (2^{k+1} + (-1)^k)/3$.

The number of comparisons required to sort n keys by merge-insertion is given as

$$\sum_{k=1}^{n} \left\lceil \log_2 \left(\frac{3}{4} k \right) \right\rceil,$$

which yields the optimum value for $n \leqslant 11$ and $n = 20, 21$.

7.6.2 Lower Bound on the Average Number of Comparisons

We have considered the worst case behaviour of sorting algorithms. Another optimal measure which is of importance in many applications is, of course, the complexity of a sorting algorithm, on average. We shall now find the lower bound on the average number of comparisons required to sort an n-key array.

From the extended binary tree structure of Fig. 7.4, it is easily seen that

$$\text{the average number of comparisons} = \frac{\text{the sum of all internal nodes}}{n!}, \qquad (1)$$

averaging over all permutations.

Now, define the external path length of the tree as the sum of the distances from the root to each of the terminal nodes. Then, in general, the average number of comparisons in a sorting method is the external path length of the tree divided by $n!$

We shall prove the following theorem:

Theorem: On the assumption that all permutations of a sequence of n keys are equally likely to appear as input, the external path length of the tree is at least $n! \log n!$.

Let $D(m)$ be the external path length of an extended binary tree with m terminal nodes. We shall first show that

$$D(m) \geqslant m \log_2 m. \qquad (2)$$

The case $m = 1$ is trivial. Now, assume that statement (2) is valid for all values of m less than M. Consider an extended binary tree T with M terminal nodes. The tree T consists of a root having a left subtree T_i with i terminal nodes and a right subtree T_{M-i} with $M-i$ terminal nodes for some $1 \leqslant i \leqslant M$. And so, we have

$$D(M) = i + D(i) + (M - i) + D(M - i).$$

Therefore, the minimum sum is given by

$$D(M) \geqslant \min_{1 \leqslant i \leqslant M} [M + D(i) + D(M - i)].$$

Invoking the assumption given by (2) we may write

$$D(M) \geqslant M + \min_{1 \leqslant i \leqslant M} [i \log i + (M - i) \log (M - i)]$$

It is easily seen that the expression in square brackets attains its minimum when $i = M/2$, and, thus, we get

$$D(M) \geqslant M + 2 \frac{M}{2} \log \left(\frac{M}{2} \right) = M + M(\log M - 1) = M \log M.$$

Setting $M = n!$ we get a lower bound for the external path length of the tree, i.e.

$$D(n!) \geqslant n! \log n!$$

This completes the proof.

From the theorem it follows that the average number of comparisons in any comparative sorting algorithm

$$\geqslant \frac{n! \log_2 n!}{n!} = \log_2 n!$$

or by virtue of Stirling's formula,

$$\geqslant O(n \log_2 n).$$

7.7 Related Problem: Finding the kth largest of n.

In some applications, one is interested in selecting the kth largest element in a sequence of n elements. One special case of this problem, frequently of importance in practical work is finding the median of a sequence. The problem is closely related to sorting.

With respect to the number of comparisons involved, the 'find-kth-largest-of-n', of all sorting problems, is perhaps the most studied one. One obvious solution to the problem is to sort the sequence into non-decreasing order and then identify the kth element. Such a procedure as we have seen would require at least $n \log_2 n$ comparisons, and, hence, be an algorithm of time complexity of $O(n \log n)$.

However, for $k = 1$, we are selecting the maximum of n elements, and as we know, $n - 1$ comparisons suffice for the solution in this case. For $k = 2$, we are selecting the second-to-the-maximum, and this is equivalent to finding the first and second largest.

Schreier (1932) and Slupecki (1951) have shown that the upper bound on the number of comparisons (i.e. the worst-case number) in this case is $n - 1 + \lfloor \log_2(n - 1) \rfloor$. Kislitsyn (1964) has shown that this bound is attainable. For general $k \leqslant \lceil n/2 \rceil$, Hadian and Sobel (1969) found that $n - k + (k - 1) \times \lceil \log(n - k + 2) \rceil$ comparisons are sufficient. For $k = n/2$ this bound gives $O(n \log n)$. Until recently, it was believed that finding the median involves sorting at least half the array and, thus, the time complexity $O(n \log n)$ for this problem is asymptotically optimal. Then, in 1973, Blum, Floyd, Pratt, Rivest, and Tarjan have shown that by a careful application of the divide-and-conquer strategy, one can find the kth largest element in $O(n)$ time.

The basic idea of the approach may be formulated as follows:

> Let $K(1), \ldots, K(n)$ be the given elements. Start by selecting an arbitrary element Q, and compare it to each of the $n - 1$ others, rearranging the other elements (as, say, in Quicksort) so that all elements $> Q$ appear in positions $K(1), K(2), \ldots, K(t - 1)$, while all elements $< Q$ appear in positions $K(t + 1), \ldots, K(n)$. Thus, Q is the tth largest.
>
> If $k = t$, we are done;
>
> *if* $k < t$, we use the same method to find the *kth* largest of $K(1), \ldots, K(t - 1)$;
>
> if $k > t$, we find the $(k - t)$th largest of $K(t + 1), \ldots, K(n)$.
>
> We may assume that all elements are distinct, though this is not important. The approach works equally well for other cases.

7.7.1 Bounds on the Maximum Number of Comparisons

In order to obtain a linear time complexity of the algorithm for selecting kth largest we must be able to find a partition element in linear time such that the sizes of the arrays $K(1), \ldots, K(t - 1)$ and $K(t + 1), \ldots, K(n)$, are each no more than a fixed fraction of the size of the given array. The trick is in how the partition element Q is chosen.

We now consider a partitioning process, bearing in mind that we wish to satisfy as closely as possible the minimal requirements on the number of comparisons.

We start with choosing a 'magic' number 15. The process is then built up using the steps:

1. Partition the array into arrays of 15 elements each. Sort each of the arrays in decreasing order, noting that any 15 elements can be sorted in order with 42 comparisons (see Section 7.6.1). It follows that the number of comparisons required by this step is given as 42 $(n/15)$.

Now, denoting by $C(n)$ and $C'(n)$ the maximum number of comparisons required to find the kth largest, from scratch and from among n elements after they have been partitioned into sorted arrays of length 15, respectively, we

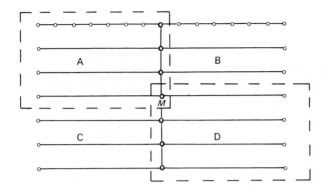

Fig. 7.8. The rows represent $n/15$ arrays of length 15, each sorted in decreasing order. The central vertical strip is the array of medians and M is its median. The elements of A are known to be greater than M. The elements of D are known to be less than M. The relations to M of the elements in blocks B and C are unknown

obtain the relation

$$C(n) = 42\left(\frac{n}{15}\right) + C'(n). \tag{1}$$

2. From each ordered array select the median to form the array of medians. The latter array contains $n/15$ elements only and its median, M, can be found 15 times faster than that of an array of n elements. Denote the number of comparisons required by this step, by $C(n/15)$.

After completion of Step 2 the situation will be as shown in Fig. 7.8. Here the rows represent $n/15$ arrays of length 15, each sorted in decreasing order. The central vertical strip is the array of medians and M is its median. The elements of A are known to be greater than M. The elements of D are known to be less than M.

The relations to M of the elements in blocks B and C are unknown.

3. To find out exactly, where among the elements of blocks B and C the median M belongs, we perform binary insertion of M into each of the rows in C and B, noting that since these rows contain seven elements each insertion requires at most 3 comparisons. This gives $3(n/15)$ comparisons in all, that are required by the step.

4. By this time, we have partitioned the array into two arrays, one containing all elements greater than M and the other all elements less than M. Let m be the number of the elements that are greater than M then if $k = m$, the solution is found;

if $k > m$, then we need to find the $(k - m)$th largest of the $(n - m)$ small elements;

and if $k < m$, then we need to find the kth largest element of the m large elements.

The observation that allows some savings in the total number of comparisons at this step, is that since we know the relative order between M and the elements of A and D, it follows that each of the arrays, of m large and of $(n - m)$ small elements, consists of no more than $\mathbf{B} + \mathbf{C} + \begin{Bmatrix} \text{either A} \\ \text{or} \quad \text{D} \end{Bmatrix}$ as the case may be $\}$ elements, that is of no more than $\frac{3}{4}n$ elements. To estimate the number of comparisons at this step we note that out of the $\frac{3}{4}n$ elements, $n/2$ represent sorted arrays of length 15 each, and the remaining $n/4$ represent sorted arrays of length 7 each. So, these shorter arrays are merged into arrays of length 15 (a dummy element being added after each pair of arrays is merged). This operation requires $13(n/60)$ comparisons since 13 comparisons at most, are needed to merge two arrays of length 7 each.

On the completion of the operation of merging, we are faced with the problem of finding some element of an arbitrary fixed rank (viz., either kth largest of the m large elements or $(k - m)$th of the $(n - m)$ small elements) from among $3n/4$ elements after they have been divided into sorted arrays of length 15. The maximum number of comparisons required by this latter operation, we denote by $C'(\frac{3}{4}n)$.

One question that remains to be explained is why the 'magic' number 15? To answer this question we first note that in each of the Steps 1, 2, and 3 of the algorithm, sorting of certain fractions of the data file is carried out. Now, since we wish to make the algorithm work in linear time these subfiles must sum to a size less than n. This is easily seen to be the case for number 15. Number other than 15 work but for certain numbers sorting the subfiles becomes expensive.

We are now ready to complete analysis of the algorithm.

From Steps 2 to 4, the following relation follows

$$C'(n) = C\left(\frac{n}{15}\right) + 3\left(\frac{n}{15}\right) + 13\left(\frac{n}{60}\right) + C'\left(\frac{3}{4}n\right) \qquad (2)$$

Substituting (1) into (2) we obtain

$$C'(n) = 42\left(\frac{n}{225}\right) + C'\left(\frac{n}{15}\right) + 3\left(\frac{n}{15}\right) + 13\left(\frac{n}{60}\right) + C'\left(\frac{3n}{4}\right)$$

giving

$$C'(n) = 0.6033n + C'\left(\frac{n}{15}\right) + C'\left(\frac{3n}{4}\right).$$

Assuming that $C'(n)$ is of the form $C'(n) = Kn$, we obtain

$$Kn = 0.6033n + \frac{Kn}{15} + \frac{3Kn}{4},$$

so that

$$K = 3.29$$

and

$$C'(n) \leqslant 3.29n.$$

Substitution of $C'(n)$ into (1) yields

$$C(n) \leqslant 2.8n + 3.29n = 6.09n.$$

As shown by Blum, Floyd, Pratt, Rivest, and Tarjan (1973) this boundary can be reduced further to $C(n) \leqslant 5.43n$ using the more scrupulous strategy of rejecting everything greater (or less) than the median M and reconstituting arrays of length 15 from the irregular arrays that remain.

A lower bound on $C(n)$ is asserted by the theorem which states that any algorithm which determines the kth largest of an unordered array of length n requires at least

$$n - 1 + \min\{k - 1, n - k\}$$

comparisons. (This gives $\frac{3}{2}n$ comparisons to find the median of the array.) An elegant proof of this theorem may be given using the concept of combinatories, see, for example, Yeh (1976).

7.7.2 Upper Bound on the Average Number of Comparisons

We assume a random order on the array and estimate the average number of comparisons required to find the kth largest of n elements.

Let $C_k(n)$ be the average number of comparisons required by the algorithm of the previous Section, to find the kth largest element. Then the following relation is asserted:

$$C_k(n) = K_1 n + \frac{1}{n} \sum_{m=k+1}^{n} C_k(m) + \frac{1}{n} \sum_{m=1}^{k-1} C_{k-m}(n - m) \tag{1}$$

where $K_1 n$ is the number of comparisons required by the algorithm to establish the rank of the 'median of medians', M, among the elements of the array i.e. Steps 1 to 3 of the algorithm.

Now, if $k > m$, we need the average number of comparisons, $C_k(m - 1)$, to find the kth largest of the array $K(1), K(2), \ldots, K(m)$ and if $k > m$ we need the average number of comparisons, $C_{k-m}(n - m)$, to find the $(k - m)$th largest of the array $K(m + 1), K(m + 2), \ldots, K(n)$. In (1), the second term on the right-hand side corresponds to the case of $k < m$ and the third term to the case of $k > m$.

Relation (1) can be rewritten as

$$C_k(n) = K_1 n + \frac{1}{n} \sum_{j=k}^{n-1} C_k(j) + \frac{1}{n} \sum_{j=n-k+1}^{n-1} C_{k+j-n}(j), \tag{2}$$

giving

$$C_k(n) \leqslant K_1 n + \max_k \left\{ \frac{1}{n} \left[\sum_{j=k}^{n-1} C_k(j) + \sum_{j=n-k+1}^{n-1} C_{k+j-n}(j) \right] \right\}. \tag{3}$$

We shall assume that $C_k(1) \leqslant K_1$, and then shall show that under this assumption

$$C_k(n) \leqslant 4K_1 n \qquad \text{for all } n \geqslant 2. \tag{4}$$

For $n = 2$ we have

$$C_k(2) \leqslant 2K_1 + \max_k \left\{ \frac{1}{2} \left[\sum_{j=k}^{1} C_k(j) + \sum_{j=3-k}^{1} C_{k+j-2}(j) \right] \right\} \leqslant 4(2K_1). \tag{5}$$

Assuming that $C_k(n) \leqslant 4K_1 n$ holds for all $n \leqslant N - 1$, we shall show that

$$C_k(N) \leqslant 4K_1 N.$$

For the induction step, write (3) as

$$C_k(N) \leqslant K_1 N + \max_k \left\{ \frac{4K_1}{N} \left[\sum_{j=k}^{N-1} j + \sum_{j=N-k+1}^{N-1} j \right] \right\}. \tag{6}$$

Performing the summation in the square brackets we get

$$C_k(N) \leqslant K_1 N + \max_k \left\{ \frac{4K_1}{N} \left[(N-1)N - \frac{(k-1)k}{2} - \frac{(N-k)(N-k+1)}{2} \right] \right\},$$

which after the maximization of the function in the square brackets gives

$$C_k(N) \leqslant K_1 N + \frac{4K_1}{N} \left[\frac{3}{4} \left(N^2 - \frac{8}{3}N + \frac{1}{3} \right) \right] \leqslant 4K_1 N.$$

The proof is complete.

Exercises

7.1   Consider an extended binary tree which sorts in order any file $K_1, \ldots, K_N$ assuming that

all keys are distinct and
the sorting is based solely on comparisons between the keys.

Using this comparison sort model and assuming that no redundant comparisons are made and that all permutations of a sequence of N keys are equally likely to appear as input

(i) show that the lower bound on the average number of comparisons required to sort N keys is of order $N \log N$,

(ii) justify the statement that the minimum number of comparisons required to sort N keys is $\geqslant \lceil \log_2 N \rceil$.

7.2 Given an arbitrary array of N distinct keys where N is a perfect square the following method is used to select the largest key:

> Divide the array into $\sqrt{N}$ groups of $\sqrt{N}$ keys each;
> Find the largest key of each group, i.e. find all the 'group leaders';
> Find the largest key of the entire array by considering the 'group leaders' file and selecting its largest key.

Extend the method to sort a complete array in order. Assuming that the largest key is obtained each time by comparison of the keys, estimate the order of the total number of comparisons required by your sorting method.

7.3 In the context of Exercise 7.2 suggest a sorting method which would use the binary tree structure. Demonstrate the method by sorting the file 13 10 7 2 1 8 4 3 12 in order.

7.4 Explain how the notion of the 'heap' is used in Heapsort.

7.5 Sort in order the array 72 39 51 81 23 92 13 5 17 33 49 11 using Heapsort.

7.6 Prove that in Heapsort a heap can be constructed in linear time.

7.7 One wishes to find both the largest and second largest elements from a file of n elements, using the notion of a 'heap'. Prove that $n + \lceil \log_2 n \rceil - 2$ comparisons are necessary and sufficient.

7.8 Find what permutations of 1 2 3 4 5 are transformed into 5 3 4 2 1 by the heap creation phase of the Heapsort.

7.9 Write a program for Heapsort.

7.10 How many inversions has the permutation 7 2 3 8 9 1 5 6 ?

7.11 Prove formula 7.4.2–(2).

7.12 Construct a binary tree of minimum height for finding the largest of 6 elements. Prove that your tree is of minimum height.

7.13 Write a program for Quicksort.

7.14 Sort an array of 13 keys using the merge-insertion algorithm of Ford and Johnson. Estimate the maximum number of comparisons required in this case.

7.15 To find the median of an arbitrary file of N keys the Blum–Floyd–Tarjan–Pratt-Rivest algorithm starts as follows:

> The input file is partitioned into subfiles of q keys each;
> Each of the subfiles of q keys is sorted into non-decreasing order;
> From each (sorted) subfile the median is selected and the subfile of medians is formed.

(i) Visualize the situation in terms of relative order of keys, after completion of steps described. Then, having in mind the minimal requirements on the number of comparisons, suggest a way to complete the process.

(ii) Assuming that $q = 9$ and that the fewest number of comparisons required to sort 9 keys is 19, estimate the lower bound on the maximum number of comparisons to find the median of N keys by the method suggested.

(iii) Analyse the algorithm under general assumption that the given file of n elements is partitioned into subfiles of length m. Determine the value of m for which the number of comparisons is minimum.

7.16 Carry out an analysis of the Blum–Floyd–Tarjan–Pratt–Rivest algorithm for the cases when the file is partitioned into subfiles of 7, 13, 15, 17 and 19 elements.

Note. The following table gives the fewest known comparisons to sort n elements

Size of file, n	1	2	3	4	5	6	7	8	9	10	11	12	13	15
No of comparisons		0	1	3	5	7	10	13	16	19	22	26	30	34	42

(Note that for $n \leqslant 12$ the number of comparisons is optimal.)

7.17 Write a program for the Blum–Floyd–Tarjan–Pratt–Rivest algorithm.

7.18 Sort the file 17 18 5 27 3 11 14 7 19 13 21 23 1 6 2 using (a) Heapsort, (b) Quicksort, (c) Merge-insertion sort of Ford and Johnson, and (d) Mergesort. In each case, how many comparisons are made?

7.19 Show that to sort an array of 15 keys by the merge-insertion algorithm requires 42 comparisons at most.

7.20 Show that in the merge-insertion algorithm of Ford and Johnson

$$t_k = (2^{k+1} + (-1)^k)/3.$$

(Hint. Derive and solve the recurrence relation $t_{k-1} + t_k = 2^k$.)

7.21 Show that $\sum_{j=1}^{n} H_j = (n+1)H_n - n$, where $H_n = 1 + \dfrac{1}{2} + \dfrac{1}{3} + \ldots + \dfrac{1}{n}$.

Chapter 8

Large Scale Data Processing: External Sorting Using Magnetic Tapes

By external sorting is meant sorting an array under the assumption that the length of the array is larger than the computer can hold in its high-speed internal memory.

External sorting is quite different from internal sorting as the data are stored on comparatively slow peripheral memory devices (viz. tapes, discs, drums) and thus the data structure has to be arranged in such a way that these devices can cope quickly with the requirements of the sorting algorithm. It is customary to call by a (sequential) file, an array which resides on disc or tape, the latter implying that at each moment one and only one of the file elements is directly accessible. Accordingly, external sorting is sometimes referred to as sorting of (sequential) files. The majority of internal sorting techniques, for example selection, exchange, insertion, are virtually useless for external sorting.

An exception is sorting by merging, where merging is a process by which two or more ordered files are combined into one ordered file. Internal sorting by merging constitutes one of the important groups of comparison sorts.

Furthermore, merging can be done without difficulty on the least expensive external memory devices as this process uses only very simple data structures, such as linear lists which are traversed in a sequential manner as stacks or as queues. (A linear list is a set of elements with linear relationships between these elements, i.e. all elements except for the last and first are preceded by exactly one and followed by exactly one element in the list. A stack and a queue are linear lists which work on the principle 'last-in-first-out' and 'first-in-first-out', respectively.)

The most common variant is an internal sorting followed by 'external merging'.

We shall first consider the basic merging process, namely merging two ordered files; this process is usually known as a 2-way merge. We shall then proceed to discuss the ways in which merging is adapted for the needs of the external sorting.

8.1 The 2-Way Merge

In the 2-way merge two ordered data files, i.e.

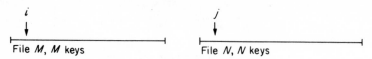

and the output file, i.e.

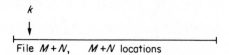

are assumed.

There are three pointers (or addresses), of which two, i and j, point to current positions in files M and N, respectively, and k points to next free position in file $M + N$. It is assumed that all keys preceding i and j have already been merged into $k - 1$ positions of file $M + N$.

Keys i and j are compared and the smaller (larger) key is transferred to position k in $M + N$. The pointers are updated and the process repeated.

In order to be aware of the situation when one of the two files becomes exhausted, the checks $i \leqslant M$? and $j \leqslant N$? are included.

We note that this process requires an additional storage of $M + N$ locations. The extra storage requirement is an essential feature of merge sorts in general. And though some reduction in the volume of additional storage may be possible due to various devices that make use of parts of files M and N which becomes vacant with the progress of merging, the merge sorts cannot be done *in situ*, since they require at least as many data moves as comparisons.

So, in merging two files, for each comparison made, at least one key goes into the output file, e.g. in the 2-way merge there are $M + N$ transmissions of keys. In general, this means that in estimating of the time complexity of merge algorithms, the operation of transmission of a key is as basic as that of comparison of keys.

There is, however, a way to reduce the volume of the additional storage required. If we add one link field to each of the $N + M$ keys, we can do everything required by the merging algorithms using simple link manipulations without moving the keys at all. This device is beneficial when the keys are long and occupy more than one location each.

8.1.1 Comparisons in the 2-Way Merge

The maximum and minimum number of comparisons required in the 2-way merge are readily obtained and are given as $M + N - 1$ and $\min\{M, N\}$, respectively.

However, estimation of the average number of comparisons calls for a more careful analysis.

Denoting by $C(N, M)$ the number of comparisons required to merge two ordered files of length M and N respectively, we can write

$$C(N, M) = M + N - S, \tag{1}$$

where S is the number of keys transmitted after exhaustion of one of the files, i.e. if the file M is exhausted after comparing $K(M)$ of the file M with $K(j)$ of the file N, then the keys $K(j), \ldots, K(N)$ of the file N are simply transmitted to the file C and thus

$$S = N - j + 1. \tag{2}$$

From (1) it follows that to find the average value of C, we have to find the average value of S.

Now, let s be some fixed integer such that $1 \leqslant s \leqslant M + N$. Then the probability that $S \geqslant s$ is given as

$$P(S \geqslant s) = p_s = \frac{\left(\begin{array}{c}\text{The number of possible} \\ \text{arrangements of } M \text{ keys} \\ \text{of file } M \text{ among first} \\ N-s \text{ keys of file } N\end{array}\right) + \left(\begin{array}{c}\text{The number of possible} \\ \text{arrangements of } N \text{ keys} \\ \text{of file } N \text{ among first} \\ M-s \text{ keys of file } M\end{array}\right)}{\left(\begin{array}{c}\text{The total number of possible} \\ \text{arrangements of } M \text{ keys of file} \\ M \text{ among } N \text{ keys of file } N\end{array}\right)}$$

or

$$p_s = \frac{\binom{M+N-s}{M} + \binom{M+N-s}{N}}{\binom{M+N}{M}}, \qquad 1 \leqslant s \leqslant M + N, \tag{3}$$

where $\binom{X}{Y}$ stands for the binomial coefficient.

The Mean and the Variance of S.

Let $q_s = P(S = s)$ for $1 \leqslant s < M + N$. \hfill (4)

Then, the mean of S, $\mu_{MN}[S]$, and the variance of S, $\sigma_{MN}^2[S]$, can be defined as

$$\mu_{MN}[S] = \sum_{s=1}^{M+N} s q_s \tag{5}$$

and

$$\sigma_{MN}^2[S] = \sum_{s=1}^{M+N} s^2 q_s - \mu_{MN}^2[S], \tag{6}$$

respectively.

Now,

$$p_s = P(S \geqslant s) = P(S = s) + P(S = s + 1) + \ldots + P(S = M + N)$$
$$= q_s + q_{s+1} + \ldots + q_{M+N}$$

and, similarly,

$$p_{s+1} = P(S \geqslant s + 1) = q_{s+1} + q_{s+2} + \ldots + q_{M+N}.$$

Subtracting the two latter expressions we get

$$p_s - p_{s+1} = q_s. \tag{7}$$

Substituting (7) into (5) and (6), we get

$$\mu_{MN}[S] = \sum_{s=1}^{M+N} s q_s = \sum_{k=1}^{M+N} p_k \tag{8}$$

and

$$\sigma_{MN}^2[S] = \sum_{s=1}^{M+N} s^2 q_s - \mu_{MN}^2[S] = \sum_{k=1}^{M+N} (2k-1)p_k - \mu_{MN}^2[S]. \tag{9}$$

Using (3) the equation (8) can be written as

$$\mu_{MN}[S] = \sum_{s=1}^{M+N} \frac{\binom{M+N-s}{M} - \binom{M+N-s}{N}}{\binom{M+N}{M}}$$
$$= \sum_{s=1}^{N} \frac{\binom{M+N-s}{M}}{\binom{M+N}{M}} + \sum_{s=1}^{M} \frac{\binom{M+N-s}{N}}{\binom{M+N}{N}}, \tag{10}$$

since

$$\binom{M+N-s}{M} = 0 \qquad \text{for } s > N,$$

$$\binom{M+N-s}{N} = 0 \qquad \text{for } s > M$$

and

$$\binom{M+N}{M} = \binom{M+N}{N}.$$

Further, using the formula

$$\sum_{k=0}^{n} \binom{r+k}{k} = \binom{r+n+1}{n},$$

relation (10) can be brought to the form

$$\mu_{MN}[S] = \frac{\binom{M+N}{N-1}}{\binom{M+N}{M}} + \frac{\binom{N+M}{M-1}}{\binom{M+N}{N}},$$

giving

$$\mu_{MN}[S] = \frac{N}{M+1} + \frac{M}{N+1}. \tag{11}$$

In a similar manner, for the variance of (9) we get

$$\sigma^2_{MN}[S] = \frac{N(2N+M)}{(M+1)(M+2)} + \frac{M(2M+N)}{(N+1)(N+2)} - \mu^2_{MN}[S]. \tag{12}$$

We conclude that the number of comparisons in the 2-way merge is given as

$$C(N,M) = \begin{cases} \min \min (N,M), \\ \text{ave } M + N - \mu_{MN}, \text{dev } \sigma_{MN}, \\ \max N + M - 1, \end{cases} \tag{13}$$

where μ_{MN} and σ^2_{MN} are determined by (11) and (12), respectively.

8.2 Merge Sorting

Two commonly known sorting methods based on the 2-way merge are the balanced (or straight) 2-way merge sort and the natural 2-way merge sort. In both methods the input file is examined from both ends, working towards the middle. The two sorts differ, however, in the way in which the itermediate subfiles are generated. In the balanced 2-way merge sort a file of n keys is sorted by starting with n subfiles of length 1 and merging these into subfiles of length 2, except possibly the last subfile, etc. The method is illustrated in Example 8.1. The first two subfiles of length 1 are the first and the last keys of the file; these are merged and placed at the first two locations of the output file. The next two subfiles of length 1 are the second and the next to the last keys of the data file; these are merged and placed at the last two locations of the output file; and so on. Referring, as usual, to the process of traversing the complete file once as a pass, we note that after k passes the subfiles generated have length 2^k, except possibly for the last subfile. The balanced 2-way merge sort is best suited for $n = 2^m$ but can be used for n other than a power of 2.

In the natural 2-way merge as distinct from the balanced merge, the lengths of the initial subfiles are decided on by the presence of 'naturally ordered' subfiles in the data file. The concept is best illustrated by an example as shown in Fig. 8.1. At the left end of the file the first j keys $K(1), K(2), \ldots, K(j)$, form an ordered sequence while the $K(j+1)$ is the first out-of-order key; hence, the initial subfile at this end is of length j. Similarly, at the right end of the file initial subfile is formed by $K(n), \ldots K(n-t)$.

260

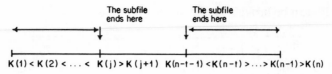

$$K(1) < K(2) < \ldots < K(j) > K(j+1) \quad K(n-t-1) < K(n-t) > \ldots > K(n-1) > K(n)$$

Fig. 8.1. Generation of 'natural' subfiles in natural 2-way merge
sort

The two 'natural' subfiles, $\{K(1), \ldots, K(j)\}$ and $\{K(n), \ldots, K(n-t)\}$ are then merged and placed appropriately in the output file. The next pair of natural subfiles is generated in a similar manner, etc.

Example 8.1

	46	2	31	54	10	25	14	72	18	39	74	91	61
Pass 1	46	61	31	74	10	18	14	72	25	54	39	91	2
Pass 2		2	46	61	91	10	18	25	72	14	74	54	39	31
Pass 3		2	31	39	46	54	61	74	91	72	25	18	14	10
Pass 4		2	10	14	18	25	31	39	46	54	61	72	74	91

8.2.1 Maximum Number of Comparisons Required by the Balanced 2-Way Merge

Let n be the length of the data file. At the start of the process there are $\lceil n/2 \rceil$ subfiles of length 1, and at the first pass these are sorted into subfiles of length 2 using

$$1 \times \left\lceil \frac{n}{2} \right\rceil \text{comparisons.}$$

At the second pass $\lceil n/2^2 \rceil$ subfiles of length 2 are 'subsorted' into subfiles of length 4 (except possibly the last subfile) using

$$3 \times \left\lceil \frac{n}{2^2} \right\rceil \text{ comparisons.}$$

In general, at the rth pass there are $\lceil n/2^r \rceil$ subfiles of length 2^{r-1} (except possibly the last subfile). These are sorted into subfiles of length 2^r using

$$(2^{r+1} - 1) \times \left\lceil \frac{n}{2^r} \right\rceil \text{ comparisons.}$$

Thus, we have

$$1 \times \left\lceil \frac{n}{2} \right\rceil + 3 \times \left\lceil \frac{n}{2^2} \right\rceil + \ldots + (2^{r+1} - 1) \times \left\lceil \frac{n}{2^r} \right\rceil + \ldots + \left(2 \left\lceil \frac{n}{2} \right\rceil - 1 \right) \times 1$$

$$= nk - \left\lceil \frac{n}{2} \right\rceil \sum_{s=0}^{k-1} \frac{1}{2^s},$$

where

$$k = \lceil \log_2 n \rceil,$$

$$= n \lceil \log_2 n \rceil - n + 1. \tag{1}$$

If n is a power of 2, then the estimate given by (1) is exact, and if n is not an exact power of 2, the estimate gives a slight overestimate of the number of comparisons required.

8.3 The Use of Merge in External Sorting

The 2-way merge can in an obvious way be extended to P-way merge when not 2 but P initially ordered subfiles are used simultaneously in a merging procedure to produce one final ordered file. The idea is illustrated in Fig. 8.2.

In this more general form, merging is extensively used in various methods of external sorting. We shall now turn to discuss some of these methods, in particular the methods associated with the use of magnetic tapes as the means for auxiliary storage.

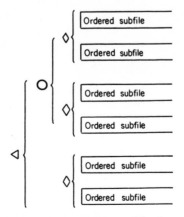

Fig. 8.2. A 6-way merge. The larger (smaller) key is placed in the output file. The merging structure is then updated

8.3.1 Main Characteristics which Effect the Performance of an External Sorting Algorithm

Suppose that the number of keys in a given file is such that the file cannot be stored in the internal memory of a computer and hence has to be placed on some peripheral device. Suppose, further, that magnetic tapes are used for this purpose.

To make the problem more specific we assume that initially data arrives from 'somewhere' and we are not concerned with the storage at this 'pre-initial' phase. At the start of the sorting process we have at our disposal a specified number of magnetic tapes. From now on, these will be referred to as 'tapes'.

There are several features which are specific to external sorting using tapes. For example, as has already been mentioned, the keys are stored on the tape sequentially and so to access them in the same order as they have been read-in the tape has first to be rewound. This operation takes time which needs to be noted accordingly.

To avoid rewinding, one can think of reading the tape backwards. This is possible in the majority of computer environments but one has to be aware that in this case the subfile generated will have the reverse ordering, i.e. if the original subfile has been read onto the tape in ascending order, then reading it from the tape backwards will produce the subfile in descending order.

The features of sorting that are associated with the use of the tapes are by-products of current computer hardware and when taken into account, make the analysis of an algorithm's efficiency machine dependent.

In other words, when searching for 'good' algorithms one is searching for the algorithms that would require shorter rewinding time and a balanced distribution of the operations of reading the tapes forward and backward. It implies that the 'best' algorithms developed will remain such only as long as the present day computer environment is valid. In order to illustrate some of the salient features of external sorting using tapes, we consider an example.

Example 8.2. (A Merge Using 4 Tapes)

Given to be sorted are 7,000 keys, $K(1), \ldots, K(7000)$.

Only 1,000 of these keys will fit in the internal computer memory. So, how shall we handle the sorting?

An obvious way is to start by sorting each of the seven subfiles $K(1), \ldots, K(1,000); \ldots; K(6,001), \ldots, K(7,000)$ internally and independently, and then merge the ordered subfiles together. For brevity, we shall denote the subfiles in this example as $K(1-1000)$, $K(1001-2000), \ldots, K(6001-7000)$.

The initial sorted subfiles are placed on the tapes alternatively so that as balanced as possible a distribution of them is ensured. In general, at the start of an external sort it is not known how many initial subfiles there will be so we cannot store 'half' of them on one tape and the other half on the other tape.

The four tapes given can be used to store the intermediate subfiles in various ways. We show three such variants. They differ by the (total) number of times that the keys are read before the sort is completed.

Variant 1

Initial $T1$ $\begin{cases} K(1-1000) \\ \end{cases}$ $\begin{cases} K(2001-3000) \\ \end{cases}$ $\begin{cases} K(4001-5000) \\ \end{cases}$ $\begin{cases} K(6001-7000) \\ \end{cases}$
(distri- $T2$ $\begin{cases} K(1001-2000) \\ \end{cases}$ $\begin{cases} K(3001-4000) \\ \end{cases}$ $\begin{cases} K(5001-6000) \\ \end{cases}$ dummy
bution) $T3$ empty
phase $T4$ empty

Rewind $T1$ and $T2$.

First $T1$ empty
$T2$ empty
merging $T3$ $\begin{cases} K(1-2000) \\ \end{cases}$ $\begin{cases} K(4001-6000) \\ \end{cases}$
phase $T4$ $\begin{cases} K(2001-4000) \\ \end{cases}$ $\begin{cases} K(6001-7000) \\ \end{cases}$

Rewind $T3$ and $T4$.

Second $T1$ $K(1-4000)$
$T2$ $K(4001-7000)$
merging $T3$ empty
phase $T4$ empty

Third $T1$ empty
$T2$ empty
merging $T3$ $K(1-7000)$
phase $T4$ empty

The sorting is now complete.

We note that given a certain number of tapes we partition them into two groups. The first group is then used for distribution of the initial subfiles (alternative order); and the second group is first used in the first merging phase; the roles of the groups are then reversed, etc.

Since reading (moving) the keys is one of the essential operations in external sorting, we shall estimate this number for Variant 1. We have

$$7,000 + 7,000 + 7,000 + 7,000 = 28,000.$$

Variant 2

Initial $T1$ $\begin{cases} K(1-1000) \\ \end{cases}$ $\begin{cases} K(3001-4000) \\ \end{cases}$ $K(6001-7000)$
(distri- $T2$ $\begin{cases} K(1001-2000) \\ \end{cases}$ $\begin{cases} K(4001-5000) \\ \end{cases}$
bution) $T3$ $K(2001-3000)$ $K(5001-6000)$
phase $T4$ empty

Rewind $T1$ and $T2$.

First $T1$ $K(6001-7000)$
$T2$ empty
merging $T3$ $\begin{cases} K(2001-3000) \\ \end{cases}$ $K(5001-6000)$
phase $T4$ $\begin{cases} K(1-2000) \\ \end{cases}$ $K(3001-5000)$

Rewind $T1$, $T3$, and $T4$.

Second
merging
phase
$T1$ empty
$T2$ $K(1-3000)$ $K(6001-7000)$
$T3$ $K(5001-6000)$
$T4$ $K(3001-5000)$

Rewind $T2$ and $T4$.

Third
merging
phase
$T1$ $K(1-5000)$
$T2$ $K(6001-7000)$
$T3$ $K(5001-6000)$
$T4$ empty

Rewind $T1$

Fourth
merging
phase
$T1$ $K(1-5000)$
$T2$ empty
$T3$ empty
$T4$ $K(5001-7000)$

Rewind $T4$

Fifth
merging
phase
$T1$ empty
$T2$ $K(1-7000)$
$T3$ empty
$T4$ empty

The sorting is now complete.
The total number of the keys read is given as

$$7{,}000 + 4{,}000 + 4{,}000 + 5{,}000 + 2{,}000 + 7{,}000 = 29{,}000.$$

We see that in this variant, aiming to have just one tape vacant at the end of each phase, we had to pay a 'fine' of an extra 1000 keys read.

In the next variant, starting with the same initial distribution as in Variant 2 and ensuring just one tape vacant at the end of each phase we shall use the 3-way merge instead of the 2-way merge.

Variant 3

Initial
(distri-
bution)
phase
$T1$ $\lceil K(1-1000)$ $\lceil K(3001-4000)$ $K(6001-7000)$
$T2$ $\lbrace K(1001-2000) \lbrace K(4001-5000)$
$T3$ $\lfloor K(2001-3000) \lfloor K(5001-6000)$
$T4$ empty

Rewind $T1$, $T2$, and $T3$.

First
merging
phase
$T1$ $K(6001-7000)$
$T2$ empty
$T3$ empty
$T4$ $K(1-3000)$ $K(3001-6000)$

Rewind $T4$.

Second
copying
phase
$T1$ $K(6001-7000)$
$T2$ $K(1-3000)$
$T3$ empty
$T4$ $K(3001-6000)$

Rewind $T2$.

Third
merging
phase

$T1$	empty
$T2$	empty
$T3$	$K(1-7000)$
$T4$	empty

In this variant, the total number of keys read is equal to

$$7{,}000 + 6{,}000 + 3{,}000 + 7{,}000 = 23{,}000,$$

which shows that Variant 3 is more 'economical' than both Variant 1 and 2.

From Example 8.2 we can see that one important problem relating to efficient external sorting is the distribution of the initial ordered subfiles on the tapes in such a way that the number of complete passes over the keys would be minimized.

Another important question is the generation of the initial subfiles in such a way that the number of complete passes required to complete the sort would be minimized.

Since the ultimate aim of the sort is to produce a single ordered file, it is clearly desirable to produce initial subfiles which are as long as possible. It will be shown later that, in fact, it is possible to produce the initial subfiles of length exceeding the internal memory of the computer. It may be noted on the other hand that increasing the length of an initial subfile requires additional processing operations during the internal sort phase. Since for all known algorithms this added cost grows faster than linearly with average subfile length, there is a limit to the 'length to which one may be willing to go.'

In the discussion which follows, we shall use the notation normally associated with external sorts using tapes:

T is the number of tapes available,
S is the number of the initial ordered subfiles produced,
P is the merge order (i.e. as in the 'P-way merge' terminology).

Now assuming first, that the initial subfiles are all of the same approximate relative length unity, we consider further examples of distribution patterns on the tapes.

Example 8.3

Let $T = 3$; $P = 2$; $S = 13$. For these parameters, three variants of merging the initial subfiles are given in Table 8.1. Here, for example, '1, 1, 1, 1, 1, 1, 1' denotes seven initial subfiles; '2, 2, 2' denotes three subfiles of relative length 2; etc.

To complete the sorting, the first two variants require 8 phases each, and the last variant requires 6 phases and it has no copying phases as opposed to the first two variants. The copying phases are 'passive' phases and an algorithm which avoids them is bound to be more efficient.

Variant 1, where at each phase the subfiles are distributed as evenly as possible on the tapes used in the current phase, is an example of the balanced P-way merge

Table 8.1

	Contents of			
	$T1$	$T2$	$T3$	Remarks
Phase 1	1,1,1,1,1,1,1	1,1,1,1,1,1	—	Initial distribution
Phase 2	—	—	2,2,2,2,2,2,1	Merge 7 subfiles to $T3$
Phase 3	2,2,2,1	2,2,2	—	Copy the subfiles alternatively to $T1$ and $T2$
Phase 4	—	—	4,4,4,1	Merge 4 subfiles to $T3$
Phase 5	4,4	4,1	—	Copy the subfiles alternatively to $T1$ and $T2$
Phase 6	—	—	8,5	Merge 2 subfiles to $T3$
Phase 7	8	5	—	Copy the subfiles alternatively to $T1$ and $T2$
Phase 8	—	—	13	Merge 1 subfile to $T3$
Phase 1	1,1,1,1,1,1,1	1,1,1,1,1,1	—	Initial distribution
Phase 2	—	—	2,2,2,2,2,2,1	Merge 7 subfiles to $T3$
Phase 3	2,2,2,1	—	2,2,2	Copy half of the subfiles onto $T1$ (Take every second subfile and copy it)
Phase 4	—	4,4,4,1	—	Merge 4 subfiles to $T2$
Phase 5	4,4	4,1	—	Copy half of the subfiles onto $T1$
Phase 6	—	—	8,5	Merge 2 subfiles to $T3$
Phase 7	—	8	5	Copy half of the subfiles onto $T2$
Phase 8	13	—	—	Merge 1 subfile onto $T1$
Phase 1	1,1,1,1,1,1,1,1 1,1,1	1,1,1,1,1	—	Initial distribution
Phase 2	—	—	2,2,2,2,2	Merge 5 subfiles onto $T3$
Phase 3	—	3,3,3	2,2	Merge 3 subfiles onto $T2$
Phase 4	5,5	3	—	Merge 2 subfiles onto $T1$
Phase 5	5	—	8	Merge 1 subfile onto $T3$
Phase 6	—	13	—	Merge 1 subfile onto $T2$

(in the example, $P = 2$, of course). Variant 2 is a specific modification of the balanced 2-way merge using 3 tapes. Variant 3 deserves a more careful analysis as it uses a very particular distribution of the initial subfiles, namely 8 on $T1$ and 5 on $T2$. This specific partitioning of 13 initial subfiles into 8 and 5 may be explained using the so called Fibonacci numbers and their properties. In fact, Variant 3 is an example of the so called polyphase sort which is based on the 'perfect Fibonacci distributions'.

To compare the variants in terms of the number of complete data file passes, we find

for Variant 1:

$$1 + 1 + 1 + 1 + 1 + 1 + 1 = 7$$

for Variant 2:

$$1 + 1 + \frac{7}{13} + 1 + \frac{8}{13} + 1 + \frac{8}{13} + 1 = 6\frac{10}{13}$$

for Variant 3:

$$1 + \frac{10}{13} + \frac{9}{13} + \frac{10}{13} + \frac{8}{13} + 1 = 4\frac{11}{13}.$$

We note that in Variant 3 only phases 1 and 6 are complete passes over all data; Phase 2 processes $\frac{10}{13}$ of the initial subfiles, Phase 3 only $\frac{9}{13}$, etc.

8.4 Number of Complete Passes Required by the Balanced P-Way Merge

Consider once again Variant 1 of Example 8.3, which is an example of the balanced P-way merge sort for $P = 2$. We wish to express the number of complete passes required by the method as a function of parameter S, the number of initial subfiles. Specifically, for Variant 1 of Example 8.3 we have:

in phase $2 (= 2^1)$ there are $\left\lceil \dfrac{S}{2} \right\rceil$ subfiles merged

in phase $4 (= 2^2)$ there are $\left\lceil \dfrac{S}{2^2} \right\rceil$ subfiles merged

$$\cdots$$

in phase 2^k there are $\left\lceil \dfrac{S}{2^k} \right\rceil = 1$ subfiles merged

wherefrom it follows that

$$k = \lceil \log_2 S \rceil \tag{1}$$

is the number of merging phases.

We also note that the sort illustrated by Variant 1 would typically have $k - 1$ copying phases.

Since each phase (the merging as well as the copying) traverses the complete file once, the total number of passes required is given by

$$k + (k - 1) + 1,\tag{2}$$

(where 1 stands for the initial distribution pass), giving

$$2\lceil \log_2 S\rceil.\tag{3}$$

When $T > 3$, we have the generalized balance $(P, T{-}P)$–way merge sort. Consider Example 8.4 for $P = 3$ and $T = 5$. At the initial distribution phase

Example 8.4 (Balanced P-way merge.)

	$P = 3$			$T - P = 2$		
	$T1$	$T2$	$T3$	$T4$	$T5$	
Phase 1	1,1,1,1,1	1,1,1,1	1,1,1,1	—	—	Initial distribution
Phase 2	—	—	—	3,3,1	3,3	3-way merge is used to merge the subfiles onto $T4$ and $T5$
Phase 3	6	6	1	—	—	2-way merge is used to merge the subfiles onto $T1, T2,$ and $T3$
Phase 4	—	—	—	13	—	3-way merge is used to merge the subfiles onto $T4$ and $T5$

the subfiles are distributed onto $T1$, $T2$, and $T3$. Then, at Phase 2 these are merged onto $T4$ and $T5$ using 3-way merge. At Phase 3 a 2-way merge is used and the subfiles generated are placed onto $T1$, $T2$, and $T3$, and so on. We note that in the general case of S initial subfiles, there are k P-way and $k - 1$ $(T - P)$-way merge phases, respectively, and k is determined from the relation

$$\left\lceil \frac{S}{P^k(T - P)^{k-1}} \right\rceil = 1,\tag{4}$$

which gives

$$k = \log_{P(T-P)} S.\tag{5}$$

Since each phase of the process requires one complete pass over the data the number of total passes is given as

$$k + (k - 1) + 1 = 2k,\tag{6}$$

or

$$2 \log_{P(T-P)} S. \tag{7}$$

Denoting

$$B = \sqrt{P(T-P)}, \tag{8}$$

we express the total number of passes as

$$\log_B S. \tag{9}$$

The formula (9) is known as the effect power of the $(P, T-P)$-way merge.

From (9) it is easily observed that the greater the base B the fewer is the total number of passes required, and in view of (8), for a fixed T, the base B achieves its maximum when $P = T - P$, giving $P = T/2$.

It follows that the balanced P-way merge (or $(P, T-P)$-way merge) sort achieves its highest efficiency in terms of the total number of processing operations in the case when $P = T/2$ and the number of complete passes is

$$\log_{T/2} S. \tag{10}$$

8.5 Polyphase Merge and Perfect Fibonacci Distributions

The polyphase merge was discovered in the late fifties (the three-tape case is due to Betz (1956) and the general pattern has been developed by Gilstad (1960), who also gave the method its present name). Analysis of the method has also been given, see, for example, MacCallum (1972) but it appears that it was Knuth (1973) who first enhanced the theory of the method by observing its relation to the so-called Fibonacci distributions.

The Fibonacci distributions are based on the Fibonacci numbers which are defined by a recurrence relation as follows

$$F_n = F_{n-1} + F_{n-2}, \quad n \geq 2, \quad F_0 = 0, \quad F_1 = 1. \tag{1}$$

For example, the first 10 Fibonacci numbers are:

$$F_0 = 0, F_1 = 1, F_2 = 1, F_3 = 2, F_4 = 3, F_5 = 5,$$
$$F_6 = 8, F_7 = 13, F_8 = 21, F_9 = 34, F_{10} = 55. \tag{2}$$

Let us now recall Variant 3 in Example 8.2. There for $T = 3$ and $P = 2$, we have

$$\begin{aligned} S = F_n &= F_{n-1} + F_{n-2} \\ &= F_7 = F_6 + F_5 \\ &= 13 = 8 + 5. \end{aligned} \tag{3}$$

It follows that the initial distribution of the subfiles onto $T1$ and $T2$, in the way defined by (3), ensures that the copying stage is completely avoided, and the method requires exactly $(n-1)$ phases to complete the sort. The distribution of initial subfiles defined by (3) and (1) is called the perfect Fibonacci distribution. If a given number of initial subfiles does not match the perfect Fibonacci distribution then the dummy subfiles can be added to make the match perfect.

The Fibonacci numbers given by (1) can be explicitly related to the quantity $P = 2$, using the following notation:

$$F_n^{(2)} = F_{n-1}^{(2)} + F_{n-2}^{(2)}, \quad n \geqslant 2, \quad F_0^{(2)} = 0, \quad F_1^{(2)} = 1. \tag{4}$$

This notation may seem somewhat artificial but it is convenient for the analysis of the general T-tape polyphase sort since form (4) can in a straightforward way be generalized to define the Fibonacci numbers for any p, i.e.

$$F_n^{(p)} = F_{n-1}^{(p)} + F_{n-2}^{(p)} + \ldots F_{n-p}^{(p)}, \quad n \geqslant p,$$
$$F_0^{(p)} = 0, \quad F_1^{(p)} = 0, \ldots, \quad F_{p-2}^{(p)} = 0, \quad F_{p-2}^{(p)} = 1. \tag{5}$$

The generalized Fibonacci sequence (5) is referred to as the pth order Fibonacci numbers.

For example, the first 11 Fibonacci numbers of fourth order are:

$$F_0^{(4)} = 0 \quad F_1^{(4)} = 0 \quad F_2^{(4)} = 0 \quad F_3^{(4)} = 1 \quad F_4^{(4)} = 1 \quad F_5^{(4)} = 2$$
$$F_6^{(4)} = 4 \quad F_7^{(4)} = 8 \quad F_8^{(4)} = 15 \quad F_9^{(4)} = 29 \quad F_{10}^{(4)} = 56 \quad F_{11}^{(4)} = 108 \tag{6}$$

Example 8.5 illustrates how the sequence (5) is used to produce the perfect Fibonacci distributions in external sorting when $P > 2$.

An important detail of the sorts based on the perfect Fibonacci distributions is the assumption that $P = T - 1$ for any given T. It means that in these sorts one tape only is made vacant at the end of each phase.

Example 8.5.

Let $T = 5$ and $P = T - 1 = 4$. The perfect Fibonacci distribution for this case is shown in Table 8.2. Here '108' denotes 108 initial subfiles (of relative length 1) placed on $T1$, '4 × 56' denotes 56 subfiles of relative length 4 placed on $T5$, etc. Phase 0 is the initial distribution phase and Phases 1–8 are merging phases.

The distribution of the subfiles shown in Table 8.2 has a direct relation to Table 8.3, based on the fourth order Fibonacci numbers.

Here $a_n = F_{n+3}^{(4)}$, the fourth order Fibonacci number, and b_n, c_n, and d_n are obtained from this number by straightforward procedures. The t_n will be useful in calculation of the total number of passes over the data.

In the general case of $P = T - 1$ we have

$$a_n = F_{n+(P-1)}^{(P)} \tag{7}$$

and

$$t_n = t_{n-1} + (P-1)a_{n-1}. \tag{8}$$

Table 8.2 A polyphase merge using five tapes

	T1	T2	T3	T4	T5	Initial subfiles processed
Phase 0	108	100	85	56		108 + 100 + 85 + 56 = 349
Phase 1	52	44	29		(56)	4 × 56 = 224
Phase 2	23	15		(29)	27	7 × 29 = 203
Phase 3	8		(15)	14	12	13 × 15 = 195
Phase 4		(8)	7	6	4	25 × 8 = 200
Phase 5	(4)	4	3	2		49 × 4 = 196
Phase 6	2	2	1		(2)	94 × 2 = 188
Phase 7	1	1		(1)	1	181 × 1 = 181
Phase 8			(1)			349 × 1 = 349

Table 8.3

Level					Total
0	1	0	0	0	1
1	1	1	1	1	4
2	2	2	2	1	7
3	4	4	3	2	13
4	8	7	6	4	25
5	15	14	12	8	49
6	29	27	23	15	94
7	56	52	44	29	181
8	108	100	85	56	349
n	a_n	b_n	c_n	d_n	t_n

where

$$a_n = a_{n-1} + a_{n-2} + a_{n-3} + a_{n-4},$$
$$b_n = a_{n-1} + a_{n-2} + a_{n-3},$$
$$c_n = a_{n-1} + a_{n-2},$$
$$d_n = a_{n-1},$$
$$t_n = t_{n-1} + 3a_{n-1}.$$

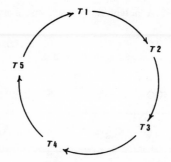

Fig. 8.3. A 'cyclic' presentation of five tapes

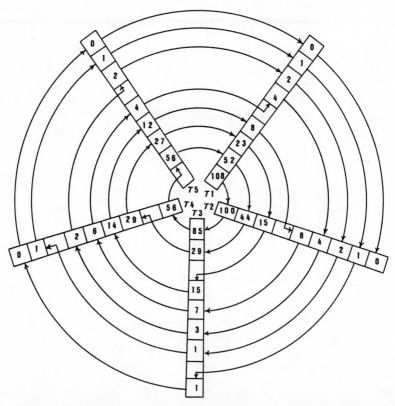

Fig. 8.4. The distribution of subfiles corresponds to Tables 8.2 and 8.3. The levels of Table 8.3 and phases of Table 8.2 have mutually reverse ordering. So, for example, the results of Phase 1 of Table 8.2 are stored on $T5$ and shown at level 7 ($= n - 1$), the results of Phase 2 are stored on $T4$ and shown at level 6 ($= n - 2$), etc. N is the total number of phases (levels). The sequence of numbers on $T1$ is the sequence of the fourth order Fibonacci numbers. This sequence governs the distribution on other tapes

The total number of initial subfiles in the sort that requires n merging phases is given by

$$S = t_n = PF_{n+P-2}^{(P)} + (P-1)F_{n+P-3}^{(P)} + \ldots + F_{n-1}^{(P)}. \tag{9}$$

In order to see the relation between Tables 8.2 and 8.3 consider Figs. 8.3 and 8.4.

The distribution illustrated in Table 8.2 can be obtained using Table 8.3. We start at the last level of Table 8.3 (i.e. the 8th level) and disperse the sequence $\{108, 100, 85, 56\}$ on the tapes shown in Fig. 8.3, starting from $T1$ and proceeding in a clockwise direction. Before moving to the next level in Table 8.3 (i.e. the 7th level) we make one (cyclic) step forward and one step 'upwards' to mark the beginning of a new phase. We then 'place' the integer on the tape and proceed as before.

The complete diagram of the process is shown in Fig. 8.4. The 'upward' shifts identify the tapes on which the results of the current phase are stored (e.g. tape $T5$ at level 8). We note that the levels of Table 8.3 and the phases of Table 8.2 have mutually reverse ordering.

8.5.1 The Number of Complete Passes Required

Given the parameters S, T, and $P = T - 1$, specifying the polyphase merge, we wish to derive the number of phases and the number of complete passes over the data required by this merge, as functions of the parameters.

We first obtain the approximate formulae for the above quantities following the approach of MacCallum (1972). To do this we consider the general case of S, T, and P, although for specific evidence where appropriate reference is made to Example 8.5.

We observe that at the kth merging phase, t_k and a_{n-k} give the relative length of the newly merged subfiles (they are all of the same relative length) and the number of these subfiles, respectively. Hence, the number of the initial subfiles processed at the kth merging phase is given as

$$t_k a_{n-k}. \tag{1}$$

The total number of the initial subfiles processed when $S = t_n$ (e.g. $S = t_8 = 349$ in Example 8.5) is obtained as

$$\sum_{k=1}^{n} t_k a_{n-k} + S, \tag{2}$$

where the second term represents the complete pass over the data required by the initial distribution phase.

In order to express relation (2) as a function of S and P, consider the recurrence relation

$$t_k = t_{k-1} + (P-1)a_{n-k+1}. \tag{3}$$

This recurrence relation can be expressed in the form which enables one readily to obtain its solution. Namely, we write for (3):

$$t_k = t_{k-1} + t_{k-2} + \ldots + t_{k-P}, \qquad k = 1, \ldots, n,$$
$$t_0 = t_{-1} = \ldots = t_{-P+1} = 1, \tag{4}$$

where n is the total number of merging phases required.

In Example 8.5, the equation corresponding to (4) is

$$t_k = t_{k-1} + t_{k-2} + t_{k-3} + t_{k-4}, \qquad k = 1, \ldots, 8,$$
$$t_0 = t_{-1} = t_{-2} = t_{-3} = 1. \tag{5}$$

The general solution of the recurrence relation (4) has the form

$$t_k = \sum_{i=1}^{P} A_i \alpha_i^k, \tag{6}$$

where the α's are roots of the characteristic polynomial equation

$$Q(x) = x^P - x^{P-1} - \ldots - x - 1 = 0. \tag{7}$$

The polynomial equation (7) has one real root α in the interval $[1, 2]$ since values of the polynomial at $x = 1$ and $x = 2$ have opposite signs and since it can be shown that all the roots except one are less than unity, in modulus. By expanding $Q(x)$ in a Taylor series about the point $x = 2$, it is easily shown that the root α is given by

$$\alpha = 2 - \frac{1}{2^{-P} - 1} + O(2^{-2P}) \tag{8}$$

which indicates that $\alpha \to 2$ when $P \to \infty$.

Thus, for reasonably large $P = T - 1$ we can assume that

$$t_k \approx A\alpha^k \qquad \text{for some constant } A, \tag{9}$$

where α is the dominant real zero of the polynomial given in (7).

Next, since $S = t_n \approx A\alpha^n$ and $1 = t_0 \approx A$, we get

$$n \approx \log_\alpha S, \tag{10}$$

that is the number of merging phases required by the polyphase sort is approximately equal to $\log_\alpha S$.

Now, the recurrence relation involving a's is given as

$$a_j = a_{j-1} + a_{j-2} + \ldots + a_{j-P}, \qquad j = 1, \ldots, n,$$
$$a_{-P} = a_{-P+1} = \ldots = a_{-1} = 0, \qquad a_0 = 1. \tag{11}$$

It can be solved in a way similar to that of relation (4). The general solution of (11) can, thus, be estimated by the approximate expression as follows:

$$a_j \approx B\alpha^j \qquad \text{for some constant } B, \tag{12}$$

where α is the dominant real root of the polynomial equation (7).

Using (12) and (9) we can now estimate the number given by (1), i.e.

$$t_k a_{n-k} \approx A\alpha^k B\alpha^{n-k} = AB\alpha^n, \tag{13}$$

which shows that at each merging phase the number of the initial subfiles processed is approximately the same. A study of Table 8.2 shows that the above result holds quite well for all the merging phases except possibly the last one, which naturally corresponds to the complete pass over the data.

To estimate the quantities given by (13) we turn to the first merging phase of the polyphase sort. At this phase we have for the (13):

$$t_1 a_{n-1} = Pa_{n-1} = P\frac{t_n - t_{n-1}}{P-1}, \tag{14}$$

where we used the recurrence relation

$$t_n = t_{n-1} + (P-1)a_{n-1}.$$

Since $t_n = S$, and by virtue of (14) and (9), the processed fraction of the initial subfiles at any given merging phase is given as

$$\frac{P}{P-1}\frac{t_n - t_{n-1}}{t_n} \approx \frac{P}{P-1}\frac{\alpha^n - \alpha^{n-1}}{\alpha^n} = \frac{P}{P-1}\frac{\alpha - 1}{\alpha}. \tag{15}$$

As it was shown earlier, the polyphase sort using S initial subfiles requires

$$n \approx \log_\alpha S = (\ln \alpha)^{-1} \ln S \tag{16}$$

merging phases. This together with (15) gives the total number of complete passes required as

$$\rho \approx \frac{P}{P-1}\frac{\alpha - 1}{\alpha}\log_\alpha S = \frac{P}{P-1}\frac{\alpha - 1}{\alpha}(\ln \rho)^{-1} \ln S, \tag{17}$$

and

$$\frac{\rho}{n} \approx \frac{P}{P-1}\frac{\alpha - 1}{\alpha} \approx 0.5, \qquad \text{for large } P.$$

The total number of passes required is given by

$$\rho + 1 \approx \frac{P}{P-1}\frac{\alpha - 1}{\alpha}\log_\alpha S + 1, \tag{18}$$

where unity represents one complete pass over the data due to the initial distribution phase.

By virtue of (8), for reasonably large P formula (18) can be expressed as

$$\rho + 1 \approx \frac{1}{2}\frac{P}{P-1}\log_2 S + 1 \tag{19}$$

or, ultimately, as

$$\rho + 1 \approx \tfrac{1}{2}\log_2 S + 1. \tag{20}$$

The estimates (19) and (20) can be used for asymptotic comparison of the polyphase merge with other external merge sorts.

A more precise analysis of the number of complete passes over the data required by the polyphase merge is given by Knuth (1973).

To estimate the quantity given by (2), the sequences

$$\{a_k = F_{k+P-1}^{(P)}, \quad k = 0, 1, \ldots n\}$$

and

$$\{t_k = t_{k-1} + (P-1)a_{k-1} = t_{k-1} + (P-1)F_{k+P-2}^{(P)}, \quad k = 1, \ldots, n, \quad t_0 = 1\}$$

are first expressed in concise form using the concept of the generating function.

It is, in particular, shown that for the two sequences the corresponding generating functions are given as

$$a(z) = \sum_{k \geq 0} a_k z^k = \frac{1}{1 - z - \ldots - z^P}, \tag{21}$$

and

$$t(z) = \sum_{k \geq 1} t_k z^k = \frac{Pz + (P-1)z^2 + \ldots + z^P}{1 - z - \ldots - z^P}. \tag{22}$$

From (21) and (22) it follows that for the general case of $S = t_n$, the number of initial subfiles processed is given as the coefficient of z^n in $a(z)t(z)$, plus t_n, for the initial distribution pass. Accordingly, the following problem is solved:

Assuming that the merging pattern has the properties which are characterised by the polynomials

$$H(z) = Pz + (P-1)z^2 + \ldots + z^P$$

and

$$Q(z) = 1 - z - \ldots - z^P$$

in the following way:

(i) The number of initial subfiles present in a 'perfect distribution' requiring n merging phases is the coefficient of z^n in $H(z)/Q(z)$,

(ii) The number of initial subfiles processed during these merging phases is the coefficient of z^n in $H(z)/Q(z)^2$,

(iii) There is a 'dominant root' α of $Q(z^{-1})$ such that $Q(\alpha^{-1}) = 0$, $Q'(\alpha^{-1}) \neq 0$, $H(\alpha^{-1}) \neq 0$ and $Q(\beta^{-1}) = 0$ implies that $\beta = \alpha$ or $|\beta| < |\alpha|$,

prove that there is a number $\varepsilon > 0$ such that, if S is the number of the initial subfiles in a perfect distribution requiring n merging phases, and if S initial subfiles are processed during those phases, we have

$$n = \ln S + b + O(S^{-\varepsilon}), \tag{23}$$

Table 8.4 Approximate number of merging phases and passes in the polyphase merge using S initial subfiles

Tapes T	Phases				Pass/phase ratio	
	$n_1 = \log_\alpha S$	$n_2 = a \ln S + b$	$\rho_1 = \dfrac{P}{P-1}\dfrac{\alpha-1}{\alpha} \times \log_\alpha S + d$	$\rho_2 = c \ln S + d$	$\dfrac{\rho_1}{n_1}$	$\dfrac{\rho_2}{n_2}$
3	$1.958 \ln S$	$2.078 \ln S + 0.672$	$1.566 \ln S$	$1.504 \ln S + 0.992$	0.80	0.72
4	$1.615 \ln S$	$1.641 \ln S + 0.364$	$1.117 \ln S$	$1.015 \ln S + 0.965$	0.69	0.62
5	$1.517 \ln S$	$1.524 \ln S + 0.078$	$0.977 \ln S$	$0.863 \ln S + 0.921$	0.64	0.57
6	$1.477 \ln S$	$1.479 \ln S - 0.185$	$0.908 \ln S$	$0.795 \ln S + 0.864$	0.61	0.54
8	$1.451 \ln S$	$1.451 \ln S - 0.642$	$0.843 \ln S$	$0.744 \ln S + 0.723$	0.58	0.51
10	$1.445 \ln S$	$1.445 \ln S - 1.017$	$0.811 \ln S$	$0.728 \ln S + 0.568$	0.56	0.50
20	$1.443 \ln S$	$1.443 \ln S - 2.170$	$0.762 \ln S$	$0.721 \ln S - 0.030$	0.53	0.50

and

$$\rho = c \ln S + d + O(S^{-\varepsilon}), \tag{24}$$

where

$$a = (\ln \alpha)^{-1},$$

$$b = a \ln\left(\frac{H(\alpha^{-1})}{-Q'(\alpha^{-1})}\right) - 1,$$

$$c = a\frac{\alpha}{-Q'(\alpha^{-1})},$$

$$d = \frac{(b+1)\alpha - H'(\alpha^{-1})/H(\alpha^{-1}) + Q''(\alpha^{-1})/Q'(\alpha^{-1})}{-Q'(\alpha^{-1})}$$

The values of n and ρ, computed for various $P = T - 1$ using the expressions (16) and (23), and (17) and (24), respectively, are compared in Table 8.4.

When the number of initial subfiles in the general T-tape polyphase merge does not match a 'perfect Fibonacci distribution', dummy subfiles are added to make the match perfect but it may be mentioned that the way in which these dummies are dispersed on the tapes affects the overall performance of the polyphase merge. Various ways of distributing the dummy subfiles have been proposed, see Knuth (1973) and Wirth (1976), although the optimal solution of this problem still awaits its discovery.

Asymptotic efficiency of the polyphase sort may be compared with that of the balanced P-way merge using an approximate estimate of the number of passes given by (20). We have

$$\frac{1}{2}\log_2 S + 1 < \log_{T/2} S = \frac{\log_2 S}{\log_2 T - 1}, \tag{25}$$

8.6 The Cascade Merge

Another important method which also makes use of a 'perfect distribution' is known as the cascade merge. It was proposed in 1959 by Betz and Carter. The method is illustrated in Table 8.5, Example 8.6, for $T = 5$. In the cascade merge the merging stages are distinguished in terms of the complete passes over the data. Pass 2, for example, is obtained by doing a four-way merge from $T1$, $T2$, $T3$ and $T4$ onto $T5$ until $T4$ is empty; then a three-way merge from $T1$, $T2$ and $T3$ onto $T4$ until $T3$ is empty; finally, a two-way merge from $T1$ and $T2$ onto $T3$ leaves $T2$ empty; for uniformity, the content of $T1$ is then copied onto $T2$. The pass is now complete, etc. The copying phases are not necessary for the process; they are included for the purpose of giving the process a fully systematic structure. Since they take a relatively small portion of the total number of moves of the data, they are not of any great inconvenience. The rules for generating a 'perfect distribution' in the cascade merge are given in

Example 8.6 (A cascade merge)

Table 8.5

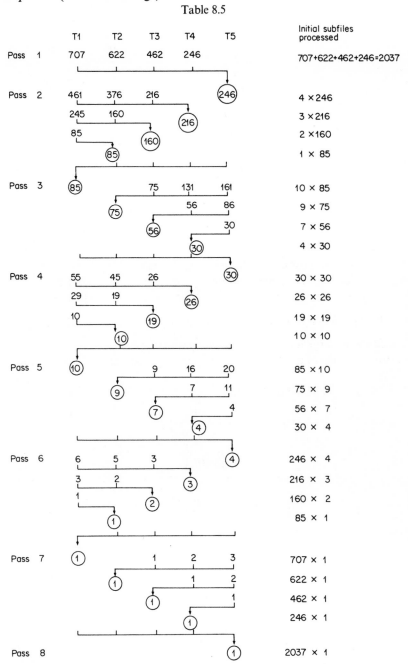

	T1	T2	T3	T4	T5	Initial subfiles processed
Pass 1	707	622	462	246		707+622+462+246=2037
Pass 2	461	376	216		(246)	4 × 246
	245	160		(216)		3 × 216
	85		(160)			2 × 160
		(85)				1 × 85
Pass 3	(85)		75	131	161	10 × 85
		(75)		56	86	9 × 75
			(56)		30	7 × 56
				(30)		4 × 30
Pass 4	55	45	26		(30)	30 × 30
	29	19		(26)		26 × 26
	10		(19)			19 × 19
		(10)				10 × 10
Pass 5	(10)		9	16	20	85 × 10
		(9)		7	11	75 × 9
			(7)		4	56 × 7
				(4)		30 × 4
Pass 6	6	5	3		(4)	246 × 4
	3	2		(3)		216 × 3
	1		(2)			160 × 2
		(1)				85 × 1
Pass 7	(1)		1	2	3	707 × 1
		(1)		1	2	622 × 1
			(1)		1	462 × 1
				(1)		246 × 1
Pass 8					(1)	2037 × 1

The initial subfiles are read 2037 × 8 = 16,296 times, in total, out of which 756 are for copying purposes.

Table 8.6

Level	a_n	b_n	c_n	d_n
0	1	0	0	0
1	1	1	1	1
2	4	3	2	1
3	10	9	7	4
4	30	26	19	10
5	85	75	56	30
6	246	216	160	85
7	707	622	462	246
n	a_n	b_n	c_n	d_n

where $a_n = a_{n-1} + b_{n-1} + c_{n-1} + d_{n-1}$
$b_n = a_{n-1} + b_{n-1} + c_{n-1}$
$c_n = a_{n-1} + b_{n-1}$
$d_n = a_{n-1}$

Table 8.6. As is seen these distributions are not exactly the Fibonacci distributions, although they are generated in the same spirit as the latter.

8.6.1 The Number of Complete Passes Required

The cascade merge has been analysed by several people.

The main results of the analysis for the general T-tape case can be summarized as follows.

The parameters in the distribution table, cf Table 8.6, are given as

$$
a_n = \frac{4}{2T-1} \sum_{-T/2 < k < \lfloor T/2 \rfloor} \cos^2 \theta_k \left(\frac{1}{2\sin\theta_k}\right)^n,
$$

$$
b_n = \frac{4}{2T-1} \sum_{-T/2 < k < \lfloor T/2 \rfloor} \cos\theta_k \cos 3\theta_k \left(\frac{1}{2\sin\theta_k}\right)^n, \tag{1}
$$

$$
c_n = \frac{4}{2T-1} \sum_{-T/2 < k < \lfloor T/2 \rfloor} \cos\theta_k \cos 5\theta_k \left(\frac{1}{2\sin\theta_k}\right)^n,
$$

where

$$
\theta_k = (4k+1)\pi/(4T-2), \qquad k = 0, 1, \ldots, n.
$$

We briefly mention that the formulae are obtained using the generating functions which happen for this problem, to be related to the Chebyshev polynomials of the second kind. Hence, the form of the solution.

In Table 8.7 are given approximate number of the passes required to merge $S = 2037$ initial subfiles by each, the polyphase and cascade merges, the latter for both, with and without copying. We note that the cascade merge without copying is asymptotically better than polyphase, on 6 or more tapes.

In a way similar to the polyphase merge, if the given number of the initial

Table 8.7 Approximate number of passes required to merge 2037 initial subfiles by each, the polyphase and cascade merges

Tapes	Passes in		
T	Polyphase merge	Cascade merge (without copying)	Cascade merge (with copying)
3	12.451	12.451	16.504
4	8.698	9.216	10.163
5	7.496	7.634	8.004
6	6.921	6.697	6.886
8	6.392	5.649	5.721
10	6.115	5.060	5.098
20	5.463	3.926	3.930

subfiles does not match the 'perfect distribution', some dummy subfiles are added to make the match perfect. Since in the cascade merge, the merging operates by complete passes, the way in which the dummies are dispersed on the tapes is unimportant.

8.7 The Oscillating Sort

This sort uses a somewhat different approach to merge sorting. It has been proposed by Sobel in 1962, and instead of assuming the initial distribution phase where all the initial subfiles are disseminated on the tapes, the algorithm

Example 8.7 (Oscillating sort)

Table 8.8

uses alternatively distribution and merging, so that much of the sorting is carried out before the input has been completely inspected.

The method uses both backward and forward tape-reading, thus producing the intermediate subfiles of ascending and descending orderings alternatively.

It is therefore important after every merging phase explicitly to emphasise the ordering of the current merged subfile. We shall denote an ascending subfile by A and a descending subfile by D. We assume that all initial subfiles are of relative length 1 (the relative lengths of intermediate subfiles can be easily deduced at every particular phase).

Example 8.7 illustrates the oscillating sort that uses 5 tapes in Table 8.8.

8.7.1 The Number of Complete Passes Required

From Example 8.5 we can see that the key which, say, was read in in the first distribution phase, is then involved in the processing only in the first and the last merging phases. In fact, it is easily observed that if the number of initial subfiles S is $(T-1)^n = 4^n$ then the oscillating sort processes each key exactly $n+1$ times (once during the distribution phase and n times during a merge). Now, assuming that the dummy subfiles are present when S is between 4^{n-1} and 4^n, bringing it up to 4^n, the total sorting time amounts to $\lceil \log_4 S \rceil + 1$ passes over all the data, and hence in the general T-tape case this time given by

$$\lceil \log_{T-1} S \rceil + 1. \tag{1}$$

Comparison with other merge sorts shows that the oscillating sort is the best method when S is a power of $T-1$.

Moreover, when S is a power of $T-1$ the oscillating sort is the 'best possible' method under an optimizing sorting model that has been developed by Karp. The model establishes the lower bound on the sorting time for any T-tape external sort as

$$\log_{T-1} S + O(1),$$

and as is seen from (1) the oscillating sort achieves this lower bound when $S = (T-1)^n$.

8.8 Generation of the Initial Subfiles

In the analysis of various external sorting techniques, it has been assumed up to now that all the initial subfiles have the same relative length which is determined by the size of the computer internal memory. However, one would generally expect that a sorting process applied to a given data file would, if starting with the 'longer' initial subfiles, require in subsequent merging phases, each key to be processed fewer times than in the case when the 'shorter' subfiles are used. In particular, it is desirable to remove the constraint on the subfile length induced by considerations of the computer internal memory.

One method which is capable of producing initial subfiles of length exceeding the computer internal memory is known as the replacement selection. The method is illustrated in Example 8.8. A 4-way merge is used on the subfiles *S*1, *S*2, *S*3, and *S*4. When the key is output, it is replaced by the next key from the data file. If the new key is smaller than the one just output, it is not considered for inclusion in the current output subfile. It is however entered into the current selection tree structure and 'suspended' until creation of the current subfile is completed (e.g. key 26 on line 4). The 'suspended' key is then considered in a normal way when generating a new subfile. This latter process starts as soon as each of the subfiles *S*1, *S*2, *S*3 and *S*4 will get a 'suspended' key.

Example 8.8

	Memory content			Output
*S*1	*S*2	*S*3	*S*4	
03	87	12	61	03
75	87	12	61	12
75	87	53	61	53
75	87	(26)	61	61
75	87	(26)	(08)	75
70	87	(26)	(08)	70
97	87	(26)	(08)	87
97	(54)	(26)	(08)	97
(09)	(54)	(26)	(08)	end subfile
09	54	26	08	08
09	54	26	12	09
etc.				

The subfiles produced in this way can contain more than four keys each even though in the selection tree structure there are never more than four keys at any time.

In practice, *P* is set fairly large so that the internal memory is essentially filled and, what is important, the unordered data file is, in fact, never partitioned explicitly. To up-date the selection tree structure, the next in line key is used to fill the place in the internal memory vacated by the output key.

The average length of the subfiles produced by the replacment selection has not yet been rigorously studied but the value 2*P* has been suggested.

It is also obvious that the minimum length of the subfile obtained equals *P* and the maximum length equals the complete data file.

8.9 Merge Trees and Optimum Merge Sorting

We have just noted that the replacement selection method produces (initial) subfiles of varying length, the average length being, with small variance, twice the capacity of the internal computer storage. Consider now a set of subfile lengths $S = (S_1, \ldots, S_k)$ which have to be merged into a sorted file, with a given merge order *P*. As we have seen in previous sections, there is a number of ways

284

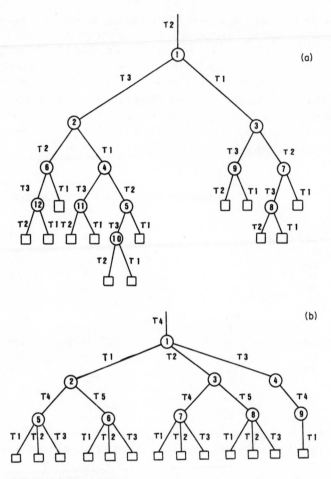

Fig. 8.5. Merge trees
(a) Tree representation of polyphase merge for $T = 3$ and $S = 13$
(b) Tree representation of balanced merge for $T = 5$, $P = 3$ and
$S = 13$

in which the merge can be performed. It is convenient to represent a P-way merge process on S subfiles by a P-ary (merge) tree with S terminal nodes. The terminal nodes of such a tree correspond to the initial subfiles, and the internal nodes to merge phases. A merge tree representing the polyphase merge given in Example 8.1, Variant 3, is shown in Fig. 8.5 (a). Here each internal node is labelled with the number of the step at which the relevant subfile was fabricated, the steps being numbered in reverse order, e.g. the last in Variant 3 step is numbered as 1st in the tree, the next-to-the-last in Variant 3 step as 2nd in the tree, etc. The branch just above each node is labelled with the name of the tape on which that subfile appears. Similarly, in Fig. 8.5(b) is shown a merge tree that represents a balanced merge for $T = 5$, $P = 3$ and $S = 13$.

The number of merge phases undergone by the keys in any subfile is indicated by the length of the path from the corresponding terminal node to the root. Allowing the subfiles of unequal length, the total number of passes over the data is determined as the sum of such path lengths each weghted by the corresponding subfile length and divided by the length of the data file. In general, every merge pattern has a tree representation, but not vice versa.

Given a set of subfiles, S, and a merge order, P, the way in which a merge can best be done depends heavily on the type of backing store available. If the backing store is tape, then its linear data topology restricts the set of subfiles to which one may economically obtain access at any given time. We wish however, to estimate bounds on optimal merge performance. One appropriate merge model for this purpose is based on the assumption that the backing store used in the process is such that it allows convenient access to any subfile at any time. Such merge processes make it possible to organise the merge so as to minimise the number of passes required. A construction which enables one to achieve such an optimal merge is that of Huffman (1952). It is defined as follows and applied recursively:

If the number of subfiles (actual and dummies, if appropriate), S, in the current list is greater than P, then replace the P subfiles of least length by a new subfile of length equal to the sum of their lengths. On the first application of this rule only m subfiles, where $S - m = O \bmod (P - 1)$, $2 \leqslant m \leqslant P$, are replaced.

The Huffman construction implies that in an optimal merge, the short subfiles are merged early and repeatedly, while the long subfiles are merged late and passed fewer times. The more short subfiles there are among a fixed number of subfiles of given expected length, the more key movement operations there are necessary in the merge. Optimal merge trees which are unsymmetrical or 'distorted' are, thus, superior to those which are not.

Now, suppose that one must choose between an internal sort algorithm which produces initial subfiles of almost constant length, and a second algorithm which produces initial subfiles of widely varying length. Assuming that both algorithms have the same expected initial subfile length defined under a certain stochastic model for generating these subfiles, one would expect the second algorithm to produce some very long subfiles as well as some very short ones, and consequently to permit a more efficient merge. This intuition was conceived and quantified by Fraser and Bennett (1972) and Bennett and Frazer (1971). Their important analysis has provided bounds on achievable merge performance which reflect directly the influence of internal sort strategy. The analysis is based on a stochastic model for generating initial subfiles which assumes that

(i) all input key permutations are equally likely,

(ii) the lengths of successive subfiles produced are independent.

(One can also assume that these lengths are dependent in a very complex and subtle way, on e.g. the input key permutation, the internal sort algorithm used, the size of the fast/backing store used and so on. It may be concluded, however, that for all practical purposes, this latter assumption gives the length distribution identical to the distribution of the independent lengths.)

Whatever algorithm is used in the initial phase, the length of a subfile produced is a random variable. This length may be taken as a dimensionless rational number if the minimum possible subfile length, S_{min}, is used as a normalizing unit. Averaging over an ensemble of applications of the same initial phase algorithm under the assumptions stated above allows one to regard the length of the ith subfile, S_i, as having been chosen according to some discrete probability distribution, $p_i(x)$, on the rational numbers. The probability distributions, $p_i(x)$, normally converge quite rapidly with i to a limiting distribution, $p(x)$, which is characteristic of the algorithm used, cf. Knuth (1963). Apart from some final subfile(s) which are almost always atypical and must be regarded as exceptions, for most of the duration of the internal sort the lengths of the subfiles produced can be regarded as a sequence of identically distributed random variables. The effect of this limiting distribution, $p(x)$, on the computation required in external merging such a set of subfiles has been studied by Frazer and Bennett.

First, assuming an optimal merge tree, T_{opt}, with a given merge order P, that operates on a set of subfiles of varying length, $S = (S_1, \ldots, S_k)$, an upper bound on the average number of (key) movement operations per key, M_{opt}, is obtained as

$$M_{opt} \leqslant 1 + \log_P k. \tag{1}$$

(When applied on a set of S equal-length subfiles formula (1) becomes

$$M_{opt} \leqslant 1 + \log_P S.)$$

This bound, however, does not take into account a length distribution function, $p(x)$, and obviously cannot be expected to be sharp.

Sharper bounds on the optimal merge are derived on the basis of the following merge model:

Given a set of k independent samples from some subfile length distribution, $p(x)$, what is the average number of merge passes per key in an optimal merge of order P?

Frazer and Bennett have shown that this average number is bounded by

$$\mu[\log_P kx] - \frac{k}{k-1} \mu[x \log_P x] (\mu[x])^{k/(k-1)} < M_{opt}$$

$$\leqslant 1 + \log_P(k\mu[x]) - \frac{k}{(k-1)\mu[x]} \mu\left[\frac{x \log_P x}{1 + \frac{x}{(k-1)\mu[x]}}\right] \tag{2}$$

where all averages, $\mu[y]$, are taken with respect to $p(x)$.

Further, it was demonstrated that the length distribution $p(x)$ which is optimal, among all distributions with mean x, in the sense of minimizing the lower bound of (2) on any finite closed interval $[1, \alpha x]$ where αx is a normalised subfile length, is indeed the distribution with maximal variance on the interval. Thus this result asserts the notion that of two distinct subfile length probability distributions with the same average value, the distribution with the greater variance permits a more efficient merge, on average. The optimal length distribution function has the form

$$
p_{\text{opt}}(x) = \begin{cases} \dfrac{\bar{x}(\alpha - 1)}{\alpha \bar{x} - 1}, & x = 1, \\[2mm] 0, & 1 < x < \alpha \bar{x}, \\[2mm] \dfrac{\bar{x} - 1}{\alpha \bar{x} - 1}, & x = \alpha \bar{x}, \quad \text{where} \quad \bar{x} = \mu[x]. \end{cases} \tag{3}
$$

Bearing in mind that the merge model used in the analysis is merely an approximation to a real process since, in the first place, the distributions, $p_i(x)$, have been replaced by the limiting distribution, $p(x)$, Bennett and Frazer (1971) have carried out a sequence of experiments on the choice of optimal strategy in a real sort, in interacting between internal sort and external merge. Experiments have proved successful, thus emphasizing the practical significance of the analytic results.

Exercises

8.1 Explain what is meant by the internal and external sorting of a given file of N keys, $K_1, \ldots, K_N$, using a computer.

8.2 There are two methods of sorting based on 2-way merge: natural 2-way merge sorting and straight 2-way merge sorting. Explain the difference between the methods.

8.3 Assuming that the average and the maximum numbers of comparisons required to merge two files with n and m keys are

$$
C_{\text{av}}(n, m) = m + n - \frac{m}{n + 1} - \frac{n}{m + 1}
$$

and

$$
C_{\text{max}}(n, m) = m + n - 1,
$$

respectively, show that in the straight 2-way merge sort of a file of $N = 2^k$ keys the average and the maximum numbers of comparisons required are

$$
C_{\text{av}} = N\left(-1 + \log_2 N - \sum_{s=0}^{k-1} \frac{1}{2^s + 1}\right)
$$

and

$$
C_{\text{max}} = N(-1 + \log_2 N) + 1.
$$

8.4   What are the additional storage requirements in the straight 2-way merge sort?

8.5   Consider the straight 2-way merge sort to sort in order a file $K_1 \ldots K_N$, $N \neq 2^k$. Derive the formula for the maximum number of comparisons assuming the following:

(i) random conditions on the file,

(ii) if $N - 1 = (b_k \ldots b_0)_2$ in binary notation, then there are $(b_k \ldots b_{j+1})_2$ merges of the ordered subfiles of equal length 2^j, and at every pass, whenever $b_j = 1$, there is one 'special' merge of the two subfiles of the lengths 2^j and $1 + (b_{j-1} \ldots b_0)_2$, respectively.

Use your formula to show that for $N = 12$ the maximum number of comparisons required is equal to 33.

8.6   Assuming that magnetic tapes are used for the auxiliary storage, describe how the replacement selection method works when used to obtain the initial subfiles in external sorting.

8.7   What happens if replacement selection is applied again to an output selection?

8.8   What is the output of the replacement selection method when the input keys are in decreasing order $K(1) \geqslant K(2) \geqslant \ldots K(n)$?

8.9   Using ordinary alphabetic order and treating each word as one key, apply the four-way replacement selection method to successive words of the following sentence:

'There was an old fellow of Trinity who solved the square root of infinity but it gave him such fidgets to count up the digits that he chucked maths and took up divinity.'

8.10   Define the pth order Fibonacci numbers and explain how these numbers are used in the polyphase sort. Make use of such terms as external sorting, distribution pattern, a complete pass over the data, and efficiency of a sort method.

8.11   Illustrate the polyphase sort process, given that the data file consists of 193 initially sorted subfiles, each of relative length unity, and that four magnetic tapes are available for the purpose.

8.12   Show that $(P - 1)n\lceil \log_2 n \rceil$ comparisons are required by the balanced P-way merge.

8.13   Assume that a sequence of 912 ordered subfiles, each of the same relative length unity, forms a file which has to be sorted 'externally' using six magnetic tapes. Carry out the sorting using (a) the balanced P-way merge; P has to be set an appropriate value, (b) the polyphase merge, (c) the cascade merge and (d) the oscillating sort. Dummy subfiles can be added where necessary. In each case, how many complete passes over the data are made?

Chapter 9

Searching

9.1 Introduction

By searching one means the operation of locating a specific item in a given sequence of n items.

As in the case of sorting, we assume that each item contains within itself a piece of information called the key. We wish then to store the sequence which is typically presented in arbitrary order, and upon demand retrieve those items whose keys match the given key K. We shall refer to the given key K as the search key.

The collection of all items stored is called a table.

Technically one can distinguish between two different tasks in information processing which are referred to as a search process, namely a storage (or entering a key into the table) and a retrieval (or accessing a key in the table) process.

The first task may be visualized as the problem where one is given an empty table and an arbitrary sequence of items. Using a particular storage algorithm, one then fills in the table with the items from the sequence. Thereafter, the same algorithm is used whenever an item has to be retrieved.

The other task is a retrieval process, when given a table already filled up, partially or fully, one searches for a match for a given key, using a particular retrieval algorithm.

In general, the search algorithms are often referred to as storage and reterieval algorithms.

We assume the case when a complete table is kept in a random-access central memory of the computer. The table may thus be either full or only partially filled in. Furthermore, the table may be static, i.e. the collection of items is given once and for all, or dynamic, i.e. occasionally one may wish to delete previously stored items or include further new items.

Basically we shall distinguish between a 'successful' search which means that the unique item containing K has been located in the table and an 'unsuccessful' search when no match for a given key is found in the table, but other interpretations of a successful/unsuccessful search are possible and will be used where appropriate.

A schematic representation of the search process may be given as follows

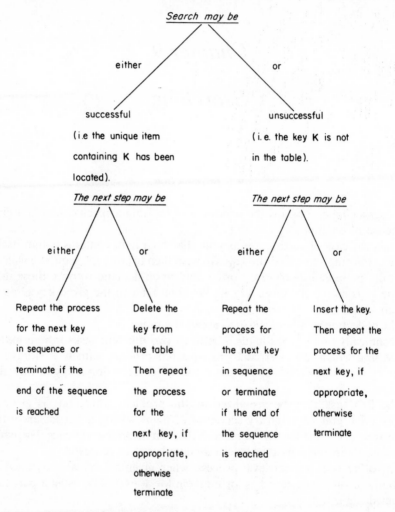

Like sorting, searching is one of the most time consuming parts of many data processing systems, hence the need for efficient search algorithms. The study of search algorithms is a recent phenomenon and a very active field of research; this to a great extent is due to a really massive 'information explosion' in recent years.

We shall now look at some basic ways in which a given sequence of items is structured in the table and the ways in which a particular item is searched for.

For a large class of searching techniques, the main operation on which a search mechanism is founded is a comparison of the given key, K, with a key

$K(j)$ from the table. For such techniques, the number of comparisons required by a particular search can be used as a measure of efficiency.

In practice the key $K(i)$ is part of a larger record of information, $R(i)$, which is being retrieved *via* its key; but for the purposes of our discussion we may concentrate solely on the keys themselves, since they are the only things which significantly enter into the search algorithms.

9.2 Classification of Search Algorithms

Linearly Ordered Tables. A large group of search methods is based on the use of linearly ordered tables. That is the table corresponding to the given sequence is obtained by sorting the sequence into a linear order, e.g.

$$K(1) \leqslant K(2) \leqslant K(3) \leqslant \ldots \leqslant K(n).$$

For the table so obtained, one rather obvious way to search is to start at the beginning of the table and to search in turn through the keys until either the key required is located or the end of the table is reached. Wood (1973) has shown that for this method the average number of comparisons required for a successful or an unsuccessful search varies between n and $\frac{1}{2}n$, depending on the degree to which the table is filled.

The degree of 'fullness' of the table is quantitatively expressed by the ratio known as the load factor, α, which is defined as

the ratio of the number of locations filled in the table, m, to the size of the table, n, i.e.

$$\alpha = \frac{m}{n}.$$

We may also note that the 'last' location of the table is normally reserved for the purpose of signalling the end of the table.

Another way to search through a table of linearly ordered keys is to start with some $K(j)$ and compare it with the search key K. Then depending on whether $K > K(j)$, $K = K(j)$ or $K < K(j)$, decide on the subsequent steps of search. Binary tree structure plays a fundamental role in analysis of this group of searching techniques. The search methods of this group are known as tree search methods.

Depending on the policy which determines how $K(j)$ is chosen and on the access probabilities or frequencies (if known) for all keys, different binary tree structures result. Static trees with known access probabilities for the keys have been studied at depth, cf. Nievergelt (1974). Here one distinguishes between the use of balanced and Fibonacci trees for uniformly distributed keys, and unbalanced trees for various non-uniform distributions of keys. Algorithms for constructing 'optimal binary trees' under assumptions of different distributions on the keys have also been developed. These algorithms are of significant theoretical interest though, perhaps, less so for practical applications.

A more realistic assumption in many applications is that of a dynamic tree with unknown access probabilities for the keys. The analysis of such trees is more difficult than in the static case and this accounts for the fact that in this area fewer results are known. But while one has obtained few theorems about dynamic trees, there are elegant and efficient insertion/deletion algorithms.

Hashing: Open Addressing and Chaining Algorithms. Another popular group of search methods uses an alternative approach which avoids sorting the given sequence altogether. Instead, for every key K of the sequence, a transformation function $h(K)$ is computed and then taken as the location of K in the table. The goal is to define the function $h(K)$ such that for any two different keys, $K(1)$ and $K(2)$, we shall get $h(K(1)) \neq h(K(2))$. It is very difficult to obtain such a function. What is usually done instead, is that the different keys are allowed to yield the same value $h(K)$ and then a special method is used to resolve any conflict situations after $h(K)$ has been computed.

A prerequisite of a good transformation function is that it distributes the keys as evenly as possible over the table. With this exception, the distribution is not bound by any pattern, and it is actually desirable if it gives the impression that it is entirely at random. This process of computing the location address, $h(K)$, for a given key has got the somewhat unscientific name hashing, i.e. 'making mess' or 'chopping the argument up', and $h(K)$ is called the hash function.

An occurrence that more than one key is hashed to the same location, i.e. that

$$h(K(i)) = h(K(j)) \quad \text{when} \quad K(i) \neq K(j),$$

is called a collision or a conflict.

Various algorithms have been developed for resolving collisions.

These can be grouped into two major sets. One is based on the approach that establishes a hash table for the storage of items and resolves collisions by somehow finding an unoccupied space for those items whose natural 'home' location is already occupied, and in such a way that the item can be later retrieved. Algorithms which use such schemes are called open addressing algorithms.

The other approach finesses the problem of collisions by using indirect addressing that allows all items which collide to maintain a claim to their 'home' location. These methods of handling collisions are commonly called chaining methods.

When discussing any search method, many complicating factors exist such as the possibility of variable length items, retrieval on secondary keys, and the possibility of duplicate keys in distinct items. We shall ignore these problems in the discussion below. Normally special circumstances can be accommodated within the framework of a particular basic algorithm.

9.3 Ordered Tables

Consider the following problem. We have an array **A** of keys that are linearly ordered by a relation '$<$', e.g.

$$K(1) < K(2) < \ldots < K(n). \tag{1}$$

We want to know whether a given key K is in **A**. From time to time we may want to insert a key or delete a key from the array.

9.3.1 Algorithm for Sequential Search

Compare K with $K(1), K(2), \ldots, K(n)$ in turn, until $K(i)$ is found such that either $K = K(i)$ in which case the search is successful or $K < K(i)$ in which case the search is unsuccessful, or else $K > K(n)$, also an unsuccessful search.

It is easily seen that in the worst case, the algorithm requires n comparisons.

9.3.2 Algorithm for Binary Search

Compare K with $K(\lceil n/2 \rceil)$, a key closest to the middle of the sequence (1). If $K = K(\lceil n/2 \rceil)$, the search terminates successfully. If $K > K(\lceil n/2 \rceil)$, discard $K(1), \ldots, K(\lceil n/2 \rceil)$. If $K < K(\lceil n/2 \rceil)$, discard $K(\lceil n/2 \rceil)$, $K(\lceil n/2 \rceil + 1), \ldots$, $K(n)$. Repeat the process, each time discarding half the keys, until the position for K is located successfully or unsuccessfully (in the sense of the search problems).

A simple analysis shows that in the worst case, the algorithm requires $\lceil \log_2 n \rceil$ comparisons.

Overhold (1973a) has shown that the binary search method is one of a class of optimal binary search methods, and the class itself may be rather large for certain values of n. In particular, he has shown that for $n = 2^m + 2^{m-1} - 1$ the number of optimal methods is of order $2^{2^m} \approx 2^{2n/3}$.

An alternative to binary search is provided by the use of the Fibonacci numbers.

9.3.3 Algorithm for Fibonaccian Search

Let n be such that $n = F_k - 1$ for some Fibonacci number F_k, $k > 1$.

Set i to F_{k-1}, p to F_{k-2} and q to F_{k-3} (p and q are consecutive Fibonacci numbers throughout the algorithm).

General Step

Compare K with $K(i)$.

If $K = K(i)$, the search terminates successfully.
If $K < K(i)$, then
        if $q = 0$, the search terminates unsuccessfully;
        otherwise set $i \leftarrow i - q$ and the pair $(p,q) \leftarrow (q, p - q)$.

If $K > K(i)$, then

if $p = 1$, the search terminates unsuccessfully;
otherwise set $i \leftarrow i + q, p \leftarrow p - 1$ then $q \leftarrow q - p$.

Repeat the step until the position of K is located successfully or unsuccessfully (in the sense of the search problem).

Efficiency of the Fibonaccian search method has been studied by Overhold (1973 b) who shows that the method is not optimal in terms of the number of comparisons required. Specifically, its search time is some 4% greater than that of the binary search and it has much greater maximum search time and standard deviation.

9.4 Search Trees

A data structure that is suited for the search processes described above is the binary search tree. Binary search trees are one of the most flexible and best understood techniques for organizing large data files.

We shall specify a binary search tree for an array **A** as a labelled binary tree in which each node i is labelled by an element $K(i) \in A$ such that

(a) for each node j in the left subtree of i, $K(j) < K(i)$,
(b) for each node j in the right subtree of i, $K(j) > K(i)$, and
(c) for each key K which is in **A**, there is exactly one node i such that $K(i) = K$.

Figs. 9.1 and 9.2 show three such trees for the size of the array $n = 7$ and $n = 12$. We number the levels in a binary search tree from zero (at the root level) upwards, k being the highest or last level. In Figs. 9.1 and 9.2, to show the connection with the corresponding search methods (e.g. standard binary search, sequential search, Fibonaccian search) we have given the relevant array segments of **A** on level zero.

We have noted earlier that the common metric for search techniques that are studied here is the number of comparisons used to find a match for a given key. In terms of the tree structure concept it implies path length measures. We define the depth of a tree as its last or maximum level (the root is placed at level 0), the length of the longest path from the root to a terminal node. Furthermore, we shall call a binary tree height-balanced if the depth of the left subtree of every node never differs by more than ± 1 from the depth of its right subtree. Examples of the balanced trees are given in Figs. 9.1(a), 9.2 and 9.3. An example of the unbalanced tree is given in Fig. 9.7. (This tree fails the height restriction on subtrees at both the 'November' and 'February' nodes.) When the access frequencies of the keys $K(1), \ldots, K(n)$ are known, it is more convenient to use the so called weighted path length which is defined taking into account the frequencies of the keys. Frequencies are then referred to as the node-weights. The weighted paths are discussed in Section 9.4.2.

9.4.1 Binary Tree Search Method and Data Structures

All the binary tree search methods have the following basic structure:

To find if a search key is in the tree we compare it with the key at the root and cases arise:

(i) there is no root (the binary tree is empty): the search key is not in the tree, and the search terminates unsuccessfully;

(ii) the search key matches the key at the root: the search terminates successfully;

(iii) the search key is less than the key at the root: the search continues by examining the left subtree of the root in the same way;

(iv) the search key is greater than the key at the root: the search continues by examining the right subtree of the root in the same way.

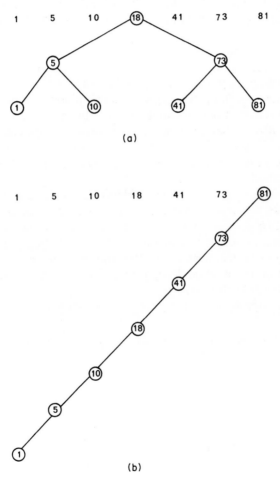

Fig. 9.1. Example of binary trees with seven nodes (a) A balanced tree corresponding to the standard binary search (b) A degenerate tree corresponding to a sequential search

296

We shall call the number of comparisons required by a particular tree search the cost of a binary search tree. An equivalent definition of the cost of a tree can be given by, first, adding $n + 1$ fictitious terminal nodes to the binary tree in a manner shown in Fig. 9.3, and then defining the following:

If a search key K is the label $K(i)$ of some node i, then the number of nodes visited when we search for K is one more than the depth of node i. If K is not in **A** and $K(i) < K < K(i + 1)$ then the number of nodes visited to search for K is equal to the depth of the fictitious terminal node i.

From the description of a binary search tree we see that the depth of such a tree and, hence, its cost is directly related to the way in which

(i) the root of the tree (or the table key for the first comparison) is selected, and
(ii) a 'divide-and-conquer' policy used in the construction of the tree, is selected.

At this stage it is convenient to point out two somewhat different roles which the binary tree structure plays in the comparison search methods. These uses can be specified by the terms a 'search tree' and a 'tree search'. Examples of the use of the binary tree in the first sense are a tree representation of a sequential search, of the standard binary search, of the Fibonaccian search, cf. Figs. 9.1, 9.2. The idea here is that given an array of keys in linear order, a particular search method is developed first. Then, for an easy illustration of this search method a binary tree structure is used. We say that a given binary tree represents the search method.

At the next stage we generalize the problem of 'searching for efficient (optimal) search methods' by considering all possible binary tree structures which can be associated with a given array of keys. We can now pose the question: what is the best (optimal) binary tree structure, that is which binary tree requires the fewest number of comparisons, on average and/or in the worst case?

The answer to this question depends crucially on the frequencies with which the search keys are likely to occur.

Assuming equiprobable search trees it can be shown that a best binary tree

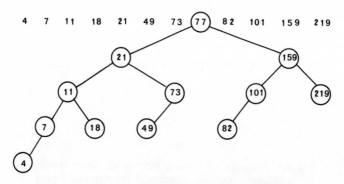

Fig. 9.2. A binary tree corresponding to the Fibonaccian search
method

from the standpoint of average search time (i.e. the number of comparisons required) is one with minimum path length. The mean search time in this case is of order $\log_2 n$ where n is the size of the data table.

9.4.2 Search Methods for Unequal Distribution Tables

The search problem under the assumption of unequal probabilities of occurrence of keys has been studied by Knuth (1971, 1973), Hu and Tucker (1971) and others. In this case the best possible binary tree will not necessarily be balanced.

Let $K(1)$, $K(2),\ldots,K(n)$ as usual be a given linearly ordered table of keys, and let

$$\alpha_0, \alpha_1, \ldots, \alpha_n \quad \text{and} \quad \beta_1, \ldots, \beta_n$$

be two sets of frequencies related to the keys as follows:

β_i is the frequency of encountering key $K(i)$, and α_i is the frequency of encountering a key which lies between $K(i)$ and $K(i+1)$; α_0 and α_n correspond to the cases when the search key is less than $K(1)$ and greater than $K(n)$, respectively.

Thus,

$$\sum_{i=1}^{n} \beta_i + \sum_{i=0}^{n} \alpha_i = 1.$$

Consider the binary tree shown in Fig. 9.4(a). Here the square nodes denote fictitious terminal nodes where no keys are stored. We denote these nodes 0, 1, 2, 3, 4, 5.

We now define the cost P of the binary search tree described as the sum of

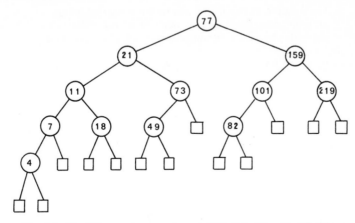

Fig. 9.3. The Fibonaccian search tree of Fig. 9.2 with added fictitious external nodes

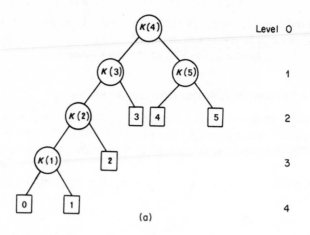

(a)

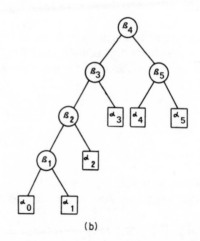

(b)

Fig. 9.4.(a) A binary search tree with $n = 6$ added fictitious external nodes (b) Binary search tree of Fig. 9.4(a) with explicitly shown probabilities associated with each node

frequencies times 'the level of the corresponding nodes plus one', i.e. we write

$$P = \sum_{i=1}^{n} \beta_i [\text{depth}(K(i)) + 1] + \sum_{i=0}^{n} \alpha_i [\text{depth}(i)]. \tag{1}$$

For example, the score for the tree shown in Fig. 9.4(a) is

$$P = 5\alpha_0 + 5\alpha_1 + 4\beta_1 + 4\alpha_2 + 3\beta_2 + 3\alpha_3 + 3\alpha_4 + 3\alpha_5 + 2\beta_3 + 2\beta_5 + \beta_4.$$

9.4.3 Optimum Cost Trees

We wish to find a minimum-cost tree, which we shall call an optimum tree. In this definition there is no need to require that the α's and β's sum to unity since we can ask for a minimum-cost tree with any given sequence of probabilities $(\alpha_0, \ldots, \alpha_n; \beta_1, \ldots, \beta_n)$.

Thus given the α's and β's, how do we find a minimum-cost tree?

We may visualize the problem as follows: if we first determine the key $K(i)$ that belongs to the root of the optimal tree then this would divide the problem into two subproblems: constructing the left subtree and constructing the right subtree.

Now, it can be observed that whichever key is an optimal choice for the root of the tree, succeeding choices in construction of both the left and right subtrees must remain optimal for the whole tree to be optimal. In other words, we observe that in the given problem all subtrees of an optimum tree are optimum. Also, any improvement to a subtree leads to an improvement to the whole tree. The difficulty in applying this principle comes from the fact that one cannot choose the root without trying many possibilities, and constructing for each one of them the optimal left- and right-subtrees. This leads to an algorithm that is based on the approach of dynamic programming.

In presenting the solution to the problem we shall use the following notation: for $0 \leqslant i \leqslant j \leqslant n$ let

T_{ij} be a minimum-cost tree for the array $\{K(i+1), \ldots, K(j)\}$;

c_{ij} be the cost of this tree;

R_{ij} be the root of this tree;

$w_{ij} = \beta_{i+1} + \ldots + \beta_j + \alpha_i + \ldots + \alpha_j$, the weight of the tree.

We can represent the tree T_{ij} as shown in Fig. 9.5.

Here, $T_{i,k-1}$ denotes the left minimum-cost subtree for the array $\{K(i+1), \ldots, K(k-1)\}$ and $T_{k,j}$ the right minimum-cost subtree for the array $\{K(k), \ldots, K(j)\}$. It is obvious that if $k = i+1$, then the left subtree is empty and if $k = j$ the right subtree is empty.

For notational convenience we assume that T_{ii} is an empty tree, its weight $w_{ii} = \alpha_i$ and its cost $c_{ii} = 0$.

An equation which relates the costs of the optimum subtrees T_{ij}, $T_{i,k-1}$ and $T_{k,j}$, when $i < j$, is derived by observing that in $T_{i,j}$ the depth of every node in the left and right subtrees has increased by one from what their depths were in $T_{i,k-1}$ and $T_{k,j}$. By virtue of formula (1) we have

$$c_{ij} = c_{i,k-1} + w_{i,k-1} + c_{k,j} + w_{k,j} + \beta_k = c_{i,k-1} + c_{k,j} + w_{ij}. \tag{2}$$

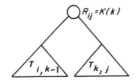

Fig. 9.5. Representation of an optimum tree T_{ij}

Equation (2) has to be solved for $k = i + 1, \ldots, j$, and the solution for k for which the sum $c_{i,k-1} + c_{k,j}$ is minimum, will give an optimum tree for $(\beta_{i+1}, \ldots, \beta_j, \alpha_i, \ldots, \alpha_j)$. Below is given an algorithm which computes a minimum-cost binary tree. Example 9.1 illustrates the method for $n = 5$.

9.4.4 Algorithm for Computing and Construction of Minimum-Cost Binary Tree

Given an array of linearly ordered keys $K(1) < K(2) < \ldots < K(n)$ and also probabilities $\beta_1, \ldots, \beta_n$ and $\alpha_0, \ldots, \alpha_n$ such that for $1 \leqslant i \leqslant n$, β_i denotes the frequency of encountering key $K(i)$, for $1 \leqslant i < n$, α_i denotes the frequency of encountering a key which lies between $K(i)$ and $K(i+1)$, and α_0 and α_n denote the frequencies corresponding to the cases when the search key is less than $K(1)$ and greater than $K(n)$, respectively, this algorithm computes and constructs a minimum cost binary search tree.

Set $w_{ii} \leftarrow \alpha_i$, $i = 0, \ldots, n$.
Set $c_{ii} \leftarrow 0$, $i = 0, \ldots, n$.
For $t = 1, \ldots, n$ do:

> For $i = 0, \ldots, n - t$ do:
> Compute $j \leftarrow i + t$.
> Compute $w_{ij} \leftarrow w_{i,j-1} + \beta_j + \alpha_j$.
> Solve $c_{i,m-1} + c_{mj} = \min_{i < k \leqslant j} \{c_{i,k-1} + c_{kj}\}$.
> Compute $c_{ij} \leftarrow w_{ij} + c_{i,m-1} + c_{mj}$.
> Set $R_{ij} \leftarrow K(m)$.

Construct the optimum tree $T(0, n)$ using:
> If $R_{ij} = K(m)$ is the root of tree $T(i,j)$ then the root of its left subtree $T(i, m-1)$ is given by $R_{i,m-1}$ and the root of its right subtree $T(m,j)$ is given by R_{mj}. If $i = j$ the tree $T(i,j)$ is null.

Example 9.1.

Given

$$K(1) < K(2) < K(3) < K(4) < K(5).$$

with $\beta_1 = 0.1, \beta_2 = 0.3, \beta_3 = 0.05, \beta_4 = 0.2, \beta_5 = 0.01$,

$\alpha_0 = 0.08, \alpha_1 = 0.04, \alpha_2 = 0.1, \alpha_3 = 0.03, \alpha_4 = 0.06$ and $\alpha_5 = 0.03$.

Results of the computation are shown in Table 9.1. For example, the values given in column $t = 4$ were obtained as follows:
For $t = 4$ consider
> $i = 0$,   giving $j = i + t = 4$.
> $w_{04} = w_{03} + \beta_4 + \alpha_4 = 0.96$.

Table 9.1

$t = j - i \longrightarrow$

$i \downarrow$

	0	1	2	3	4	5
0	$w_{00} = 0.08$ $c_{00} = 0$	$w_{01} = 0.22$ $c_{01} = 0.22$ $R_{01} = K(1)$	$w_{02} = 0.62$ $c_{02} = 0.84$ $R_{02} = K(2)$	$w_{03} = 0.70$ $c_{03} = 1.10$ $R_{03} = K(2)$	$w_{03} = 0.96$ $c_{04} = 1.80$ $R_{04} = K(2)$	$w_{05} = 1.00$ $c_{05} = 1.98$ $R_{05} = K(2)$
1	$w_{11} = 0.04$ $c_{11} = 0$	$w_{12} = 0.44$ $c_{12} = 0.44$ $R_{12} = K(2)$	$w_{13} = 0.52$ $c_{13} = 0.70$ $R_{13} = K(2)$	$w_{14} = 0.78$ $c_{14} = 1.40$ $R_{14} = K(2)$	$w_{15} = 0.82$ $c_{15} = 1.58$ $R_{15} = K(2)$	
2	$w_{22} = 0.1$ $c_{22} = 0$	$w_{23} = 0.18$ $c_{23} = 0.18$ $R_{23} = K(3)$	$w_{24} = 0.44$ $c_{24} = 0.62$ $R_{24} = K(4)$	$w_{25} = 0.48$ $c_{25} = 0.76$ $R_{25} = K(4)$		
3	$w_{33} = 0.03$ $c_{33} = 0$	$w_{34} = 0.28$ $c_{34} = 0.28$ $R_{34} = K(4)$	$w_{35} = 0.32$ $c_{35} = 0.42$ $R_{35} = K(4)$			
4	$w_{44} = 0.06$ $c_{44} = 0$	$w_{45} = 0.1$ $c_{45} = 0.1$ $R_{45} = K(5)$				
5	$w_{55} = 0.03$ $c_{55} = 0$					

Consider $0 < k \leqslant 4$, that is $k = 1, 2, 3, 4$, for which we get

$$c_{00} + c_{14} = 1.40,$$
$$c_{01} + c_{24} = 0.84,$$
$$c_{02} + c_{34} = 1.12,$$
$$c_{03} + c_{44} = 1.10,$$

and the minimum sum is clearly achieved for $k = m = 2$.
Hence, $c_{04} = w_{04} + c_{01} + c_{24} = 1.80$
and $R_{04} = K(2)$.

Next consider

$i = 1$, giving $j = i + t = 5$.
$w_{15} = w_{14} + \beta_5 + \alpha_5 = 0.82$.
Consider $0 < k \leqslant 5$, that is $k = 2, 3, 4, 5$ for which

$$c_{11} + c_{25} = 0.76,$$
$$c_{12} + c_{35} = 0.86,$$
$$c_{13} + c_{45} = 0.80,$$
$$c_{14} + c_{55} = 1.40,$$

and the minimum sum is achieved for $k = m = 2$.
Hence, $c_{15} = w_{15} + c_{11} + c_{25} = 1.58$
and $R_{15} = K(2)$.

The results of the table are specified by the root array $R\ (ij) = R_{ij} = K(m)$, which is as follows:

$$\begin{bmatrix} \text{null} & K(1) & K(2) & K(2) & K(2) & K(2) \\ & \text{null} & K(2) & K(2) & K(2) & K(2) \\ & & \text{null} & K(3) & K(4) & K(4) \\ & & & \text{null} & K(4) & K(4) \\ & & & & \text{null} & K(5) \\ & & & & & \text{null} \end{bmatrix}.$$

The optimum tree $T(0,5)$ is then obtained as shown in Fig. 9.6. It is easily seen that the algorithm requires a number of operations proportional

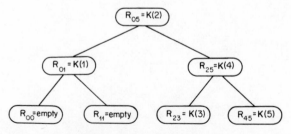

Fig. 9.6. A minimum-cost binary tree

to n^3, and memory space proportional to n^2. Knuth (1971) improved this algorithm to run in time proportional to n^2 by proving that the root of an optimal tree for $K(1),\ldots,K(n)$ need never lie outside the interval bracketed by the two roots of an optimal tree for $K(1),\ldots,K(n-1)$ and one for $K(2),\ldots,K(n)$. Hu and Tucker (1971) have presented an algorithm for constructing optimal binary search trees for the special case where all node-weights are zero. Their algorithm requires a number of operations proportional to $n \log n$ and memory space only of order n.

9.5 A Tree Search Followed by Insertion or Deletion of the Key

We shall now consider the search process where the operation of search is normally followed by insertion or deletion of the key. The corresponding binary trees in these cases are referred to as the dynamic binary trees. Consider the following example. Given is a binary tree as shown in Fig. 9.7, and we wish to insert in it the key 'July'.

To do this, we start at the root of the tree, comparing 'July' to 'May', then

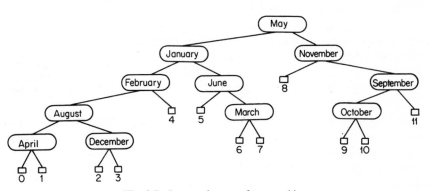

Fig. 9.7. A search tree for $n = 11$

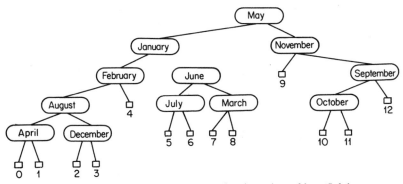

Fig. 9.8. A search tree of Fig. 9.7 after insertion of key 'July'

304

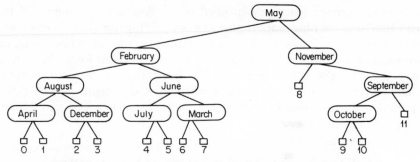

Fig. 9.9. A search tree Fig. 9.8 after deletion of node 'January'

move along the tree to the left and compare 'July' to 'January'; the search is again unsuccessful and we move further to the right and compare 'June' to 'July'; finally we move to the left and the search key is now inserted into the empty external node 5. The new tree is shown in Fig. 9.8.

At this stage we wish to delete from the tree the key 'January'. How should we go about it?

Apparently, if the right subtree of the node 'January' were empty, we could have simply deleted the node itself and joined the branch from 'February' to 'May'. But this cannot be done in our example because the right subtree in question is not empty. What we can do, though, is to find the next node for which the argument introduced holds. We then can 'delete' the latter node and reinsert it in place of the node that we really wanted to delete. In our example such a next node is 'February'. Hence, we delete 'February', join up 'August' to 'January' and then replace 'January' by 'February'. The tree obtained is shown in Fig. 9.9.

The example shown in Figs. 9.8 and 9.9 demonstrates that the operation of insertion or deletion which follows a tree search is far from a trivial task. It requires a suitable technique.

We shall now specify the two algorithms just used for insertion and deletion of a key.

It is convenient to assume that the nodes of the search tree employed in the algorithms contain the following fields:

K(i) = key stored in node i,
LL(i) = pointer to left subtree of node i,
RL(i) = pointer to right subtree of node i;

and that empty subtrees are represented by the null pointer Δ.

An example of a search tree using this notation is given in Fig. 9.10.

9.5.1 A Tree Search and Insertion Algorithm

Given an array of keys which form a binary tree, this algorithm searches for a given search key K. If K is not in the tree, a new node containing K is inserted into the tree in the appropriate place.

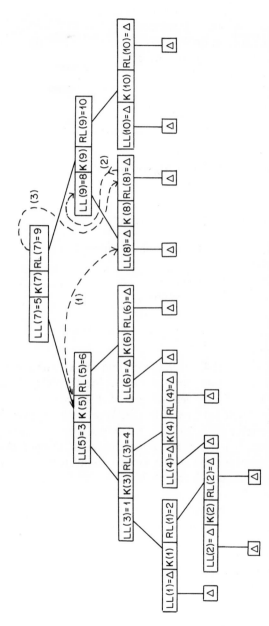

Fig. 9.10. A search tree. The dotted lines illustrate the set of operations which are carried out for deletion of key **K**(7).
Operation 1: LL(8) becomes 'equal to 5', Operation 2: LL(9) becomes 'equal to Δ', Operation 3: RL(8) becomes
'equal to 9'

Set $i = m$ where node m is the root of the given binary tree.
Compare K with $K(i)$.

If $K = K(i)$, the search terminates successfully.
If $K < K(i)$, go to the left subtree;
    if $LL(i)$ is null, insert K.
If $K > K(i)$, go to the right subtree;
    if $RL(i)$ is null, insert K.

Follow the search path until either search terminates successfully or the search key is inserted in the tree.

Next we shall give an algorithm for deletion of a key from the tree.

9.5.2 A Tree Search and Deletion Algorithm

Specifications used in this algorithm are the same as in the tree search and insertion algorithm.

(Search phase)
Set $i = m$ where node m is the root of the binary tree.
Compare K with $K(i)$.
If $K < K(i)$ go to the left subtree.
If $K > K(i)$ go to the right subtree.

(Deletion phase)
If $K = K(i)$ go to the right subtree,
$K(k)$ where $k = RL(i)$.
If $RL(i)$ is null delete $K(i)$ and join the left subtree $K(j)$ where $j = LL(i)$, to the node immediately above the deleted one and so that $K(j)$ becomes the left subtree of this node. If $K(k)$ is not empty, go to its left subtree. If this subtree is not empty, carry on down the search path testing the left subtrees in turn until the node with the empty left subtree is encountered. Denote this node by r. Our intention now is temporarily to delete the node r, i.e. to delete key $K(r)$, then to accomplish necessary changes in the tree structure and finally to replace $K(i)$, the key to be deleted, by $K(r)$. This is done performing the following sequence of statements: $LL(r)$ which is null, becomes j, i.e.

$$LL(r) \leftarrow j;$$

$LL(k)$ becomes $RL(r)$, i.e.

$$LL(k) \leftarrow RL(r),$$

then $RL(r)$ becomes $RL(i)$, i.e.

$$RL(r) \leftarrow RL(i);$$

$K(i)$ is replaced by $K(r)$.

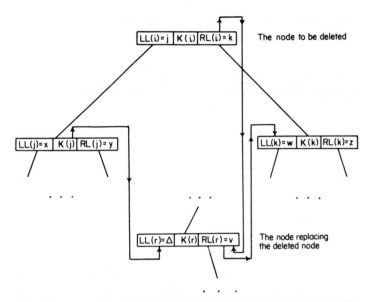

Fig. 9.11. A node deletion operation. i is the node to be deleted. The pointer LL(r) which is null becomes j. The pointer LL(k) becomes RL(r), then RL(r) becomes RL(i). Finally, K(r) replaces K(i)

This completes the tree search and deletion algorithm. A diagram illustrating the deletion sequence is shown in Fig. 9.11. Key $K(i)$ has to be deleted. Key $K(r)$ replaces the key $K(i)$.

9.6 Methods for Rebalancing the Search Trees

Suppose that an initial set of data is structured as a fairly balanced tree. This implies that its search time is close to $O(\log n)$. Now, if insertion/deletion operations are repeatedly applied to this tree, then the tree may grow very unbalanced. For example, if a sequence of deletion operations removes nodes from only the right side of the tree, we shall finish up with a tree that is unbalanced to the left. As a result, the search time of the tree will now be closer to $O(n)$. This is obviously an undesirable by-effect. To remedy it we have to make provision for some specific mechanism which would rebalance the tree after every insertion/deletion operation.

In Nievergelt (1974), a comprehensive survey is given of the methods used to create and modify binary search trees as well as some pertinent analytical results on the efficiency of these methods. A comparison of non-weighted tree balancing algorithms has been carried out by Baer and Schwab (1977).

We shall discuss one such mechanism that was developed by Adelson–Velski and Landis (1962). It is based on the use of the balanced tree structure. The tree is preserved balanced throughout the complete process of insertion/

308

deletions. The mechanism is especially elegant and was the first successful solution to the problem of efficiently handling insertion/deletion operations on the binary search trees.

9.6.1 The Balanced Trees Method of Adelson—Velski and Landis (The AVL Trees)

We shall follow Aho, Hopcroft, and Ullman (1974) in referring to the trees used in the method of Adelson–Velski and Landis as the AVL trees named after their originators.

We first recall that in a height balanced tree at each node, the depths of the left and right subtrees differ by at most one. If a subtree is missing it is deemed to be of 'depth' -1. In this section, such trees will be called the AVL trees.

The AVL trees represent a compromise between optimum binary trees for which as we know all external nodes required to be on two adjacent levels, and arbitrary binary trees. In view of such a placing of the balanced tree it is interesting to know how far from optimum a balanced tree can be. The answer to this question is given by a theorem of Adelson–Velski and Landis. It proves that the depth of a balanced tree with n internal nodes always lies between $\log_2 (n + 1)$ and $1.4404 \log_2 (n + 2) - 0.328$, and, hence, its search paths will never be more than 45 per cent longer than the optimum. We shall not dwell

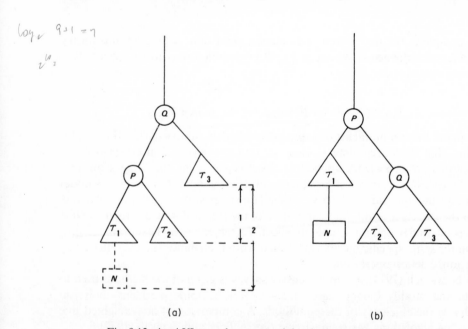

(a)   (b)

Fig. 9.12. An AVL tree, by means of the single rotation, is rearranged into a height balanced tree, i.e. into an AVL tree again, after insertion of node N (a) Node N is about to be inserted in the AVL tree (b) The tree of (a) transformed into a new AVL tree

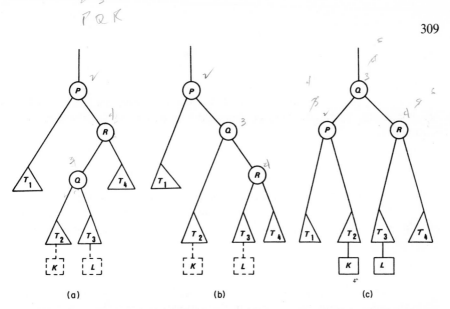

Fig. 9.13. An AVL tree, by means of double rotation, is rearranged into a height balanced tree, i.e. a new AVL tree (a) Nodes K and L are about to be inserted in an AVL tree (b) First step of the 'double rotation', a device used to transform the tree (a) into an AVL tree (c) The tree (a) transformed into a new AVL tree insertion of nodes K and L

upon proof of this theorem, and instead consider what actually happens when a new node is inserted into a balanced tree using tree insertion.

In Fig. 9.12(a) is shown a subtree of an AVL tree in which node N is about to be inserted. Before insertion, the heights of the two subtrees of P are equal, and the left subtree of Q is of greater height than the right subtree of Q. After insertion of node N the height constant is violated at node Q. Consequently, the tree is rearranged using a device which is appropriately called 'single rotation.' Namely, the subtree of Fig. 9.12(a) is 'rotated' to the right and T_2 becomes attached to Q instead of P. The new tree is again an AVL tree.

The only other case which may arise in rearranging a height balanced tree is shown in Fig. 9.13. It is handled by a device called the 'double rotation'. Here, at first the subtree of R and Q is rotated to the right and T_3 becomes attached to R. Then, the subtree of P and Q is rotated to the left and T_2 becomes attached to P.

Insertion in height balanced trees requires at most one single rotation or double rotation. Deletion, on the other hand may require as many as $h/2$ transformations where h is the height of the tree, though on average the number of transformations required is a small constant.

The AVL trees have been generalized by Foster (1973) and Bayer (1972).

9.7 Hashing

We shall now turn to examining another important family of searching methods known as hashing techniques. In hashing we first select an initial hash function, $h(K)$, and then select a method for collision resolution. A specific choice of the algorithm for computing the initial hash function may to some extent determine a method for collision resolution.

In the next section we discuss several methods for selecting the initial hash function, in the spirit of a very commendable review paper on the subject by Knott (1975).

9.8 Computing the Initial Hashing Function

We assume that the table to be searched is of dimension n from 0 to $n - 1$. The hashing function is defined on a domain of values which includes the possible keys of the items to be processed, and in our case, the range of values of a hashing function is given as

$$0 \leqslant h(K) < n - 1, \quad \text{for all keys } K.$$

Ideally one would wish to produce a hashing function which would create a perfectly random data, but because in the majority of cases the actual data are non-random, this is theoretically impossible to achieve. A more modest aim is to produce a hashing function which would lead to a reasonably small number of collisions. A good hashing function is one which minimizes the expected number of collisions.

We assume throughout that, for all keys K, the $h(K)$ can be computed in a constant amount of time.

9.8.1 Distribution-Dependent Hashing Functions

Consider a collection, G, of k integer keys which are included in a universe of possible integers, U. (Any set of keys can be assumed to be integers under an appropriate interpretation.) Now, defining a random variable X on G as

$$X(G) = G,$$

and letting F_X be its (cumulative) distribution function, we can say that the keys in G are distributed according to some discrete empirical distribution function F_X.

We wish to find a hashing function h with a domain including U and range equal to $\{0, 1, \ldots, n - 1\}$, such that the random variable $h(X)$ defined on G has a discrete uniform distribution function on $\{0, 1, \ldots, n - 1\}$. In other words, we wish to find a hashing function h which maps our k given keys to the addresses $0, 1, \ldots, n - 1$ so that the keys are uniformly distributed over the address locations. The reason for desiring a uniform distribution of the hash-values is to minimise the number of collisions which will occur.

We now recall that by definition

$$F_X(t) = P(X \leqslant t) \tag{1}$$

where $P(Y)$ denotes the probability of Y.

Furthermore, we wish to find h such that

$$P(h(X) = i) = \frac{1}{n} \quad \text{for } 0 \leqslant i \leqslant n. \tag{2}$$

In order to do this, consider the random variable $F_X(X)$ and note that

$$P\left(F_X(X) \leqslant \frac{r}{k}\right) = \frac{r}{k} \quad \text{for } 0 \leqslant r \leqslant k. \tag{3}$$

Relation (3) is strictly true only when all the keys are distinct. This assumption on the keys has been made at the beginning of the chapter and we just recall it at this point. From (3) it follows that $F_X(X)$ has a discrete uniform distribution on

$$\left\{\frac{1}{k}, \ldots, \frac{k-1}{k}, 1\right\}, \tag{4}$$

and so $nF_X(X)$ has a discrete uniform distribution on

$$\left\{\frac{n}{k}, \frac{2n}{k}, \ldots, n\right\}. \tag{5}$$

Therefore $nF_X(X) - 1$ is approximately uniform on $\{0, 1, \ldots, n-1\}$, especially when $n \leqslant k$.

Hence, we see that a good choice for h is

$$h(K) = \lceil nF_X(K) \rceil - 1. \tag{6}$$

When F_X is known and is easily computable, it can be used as shown in (6) to obtain a hashing function which is tailored to the collection of keys to be stored.

When F_X is itself a uniform distribution on $\{1, 2, \ldots, s\}$, our hashing function becomes

$$h(K) = \left\lceil \frac{nK}{s} \right\rceil - 1$$

where s is the largest key occurring in G.

One may take s as the largest key in U if the largest one occurring among the actual keys is not known.

Further variants of defining F_X can also be used, e.g. approximation of the F_X by a piecewise linear function, by a polynomial of degree higher than one; the usage of partial knowledge of F_X, etc.

9.8.2 Cluster-Separating Hashing Functions

If two or more keys in G are close to each other in some sense, they are said to form a cluster. Now, if it is known that a given collection tends to have such clusters among its keys, then a good hashing function is expected to destroy these clusters, that is the hash values of the 'clustered' keys should be not equal. In practice, however, no matter what hashing function we choose, there are notions of what constitutes a cluster for which our hashing function fails to achieve reasonable results. In this sense, there is no such thing as a general purpose hashing function.

The best we can select is a hashing function which achieves a reasonably uniform distribution of hash values given certain general assumptions about the distribution of the keys to be processed.

One common assumption is that the set of keys of collection G, is a random sample from the set of all possible keys, U.

A hashing function which is suitable under this assumption is the function with range $\{0, 1, \ldots, n-1\}$ which partitions the universe of possible keys, U, as nearly as possible into n equi-numerous sets of synonymous keys corresponding to the addresses $0, 1, \ldots, n-1$. Knott calls such a hashing function a balanced hashing function.

A more special assumption which can be made is that the keys tend to occur in arithmetic progression, i.e. if K is an occurring key, then $K + s$ for some fixed s is relatively likely to occur also. One notion for the choice of s (or the distance between two keys) is adapted from Hamming (1950). It is known as the Hamming metric and may be defined as follows:

Two keys, Q and R, considered to be t-tuples whose components are symbols of some alphabet, are separated by a distance s under the Hamming metric, $D(Q, R)$ when Q and R have different symbols in exactly s different component positions. A hashing function, h, is called d-separating with respect to the metric $D(Q, R)$ if $0 < D(Q, R) \leqslant d$ implies $h(Q) \neq h(R)$.

Based on the notion of the Hamming metric, a common clustering assumption is that if K is a key actually occurring then keys which are close to K with respect to the Hamming metric are relatively likely to actually occur also.

For example, if keys are English words of t or fewer letters then the keys are distributed in a certain manner in the set of all t-component vectors of letters, including blank. This distribution is such that our intuitive notion of clusters is indeed approximated by using the Hamming metric as a measure of closeness. If w is an English word then w changed in one or two positions is also likely to be an English word, relative to the probability that w changed in one or two positions is an English word given only that w is a random string of letters.

Methods of constructing 'cluster-destroying' hashing functions under the notion of cluster in terms of the Hamming metric, have been successfully studied using the results of algebraic coding theory. This was first suggested by Muroga (1961). One approach for the choice of the hashing function which has emerged from these studies is the use of polynomials. Let $g(x)$ be a polynomial of degree s. Also, let $y(x)$ be a polynomial of degree $t - 1$ or less, where

$$y(x) = y_0 x^{t-1} + y_1 x^{t-2} + \ldots + y_{t-1}.$$

Let K be the set of coefficients of $y(x)$, i.e.

$$K = (y_0, y_1, \ldots, y_{t-1}),$$

then the hashing function $h(K)$ is obtained as the set of coefficients of the polynomial $y(x) \bmod g(x)$.

The coefficients of $g(x)$ are defined and a suitable polynomial division algorithm is used to compute $y(x) \bmod g(x)$, where all the arithmetic among coefficients is done in the symbol field.

A care has to be exercised to ensure an appropriate choice of polynomial $g(x)$ with which one can construct d—separating functions. Such polynomials can be found but these constructions require some knowledge of the theory of finite fields, and we shall forgo their discussion here, referring the reader to some relevant literature on the subject, like Schay and Raver (1963).

9.8.3 Distribution-Independent Hashing Functions

When no specific information on the initial distribution of items is available, one turns to the ways for computing the hashing function inspired by methods of generating pseudo-random numbers.

A key may be considered as a bit-string. Based on this interpretation of the key value, some common elementary hashing functions may be listed as follows:

(i) *Division method*
   $h(K) = K \bmod n$. The range is $\{0, \ldots, n-1\}$.

(ii) *Addition method*
   $h(K) = $ a k-bit value extracted from the result of adding various segments of K together.

(iii) *Multiplication method*
   $h(K) = $ a k-bit value extracted from the value K^2 or CK for some constant C.

(iv) *Extraction method*
   $h(K) = $ a k-bit value composed by extracting k bits from K and concatenating them in some order. The range is $\{0, \ldots, 2^k - 1\}$.

(v) *Radix conversion method*
   $h(K) = $ a k-bit value extracted from the result of treating K as a binary-coded sequence of base p digits and converting this coded value into a similarly-coded base value. The range is nominally $\{0, \ldots, 2^k - 1\}$.

From the operations just given and other simple logical and arithmetic operations, reasonable hashing functions can be constructed. We shall restrict our discussion to just a few further comments on the division and multiplication methods. In respect to the division method it is usually noted that, given a suitable modulus, n, the division hashing function will have acceptable properties. For example, such functions will destroy clusters under the Hamming metric. Knuth has noted that n should not have integers 2 or 3 as factors. Whenever K is even the $K \bmod n$ is even, for instance. The choice of a prime modulus is often recommended, for these sequences of keys in arithmetic progression hash to different addresses as much as possible unless the increment of the progression is a multiple of the modulus.

However, the modulus need not necessarily be chosen to be prime, cf. Buchholz (1963), and this flexibility is one of the assets of the division method.

Another is that the quotient as well as the remainder is generally available, and this may be put to good use, see for example, the quadratic quotient method of collision resolution proposed by Bell (1970) which is discussed later in the text.

The multiplication method is sensitive to the distribution of keys and/or the choice of a constant multiplier. Below we discuss one interesting multiplicative scheme developed by Floyd (1970).

9.8.4 A Multiplicative Hashing Function

In this section we describe construction of a particular multiplicative hashing function which was proposed by Floyd. The function is given as

$$h(K) = n(cK \bmod 1), \tag{1}$$

where K is as usual a non-negative integral key and c is a constant defined in the interval $0 \leqslant c < 1$. The crucial point of the method is how to choose c.

Let us assume that keys have a tendency to occur in arithmetic progression so that we may have keys

$$K, K + s, K + 2s, \ldots \tag{2}$$

In this case it is desirable that

$$h(K), \qquad h(K + s), \qquad h(K + 2s), \ldots$$

be a sequence of distinct addresses, in so far as this is possible. Now, this goal is achieved if the set of values $\{cK \bmod 1, c(K + s) \bmod 1, \ldots\}$ is 'well-spread' in the unit interval, for then the addresses achieved by scaling will be well-spread among $\{0, 1, \ldots, n - 1\}$.

A set of points is considered to be 'well-spread' in the unit interval when the ratio of the minimum of the lengths of the subintervals defined by the given points placed in the unit interval to the maximum subinterval length is large, i.e.

$$D = \frac{\text{minimum subinterval}}{\text{maximum subinterval}} \quad \text{is} \quad \text{large} \tag{3}$$

Next in the construction proof, it is shown that for $c = p/q$ with p and q integers, the general sequence

$$\{c(K + as^i) \bmod 1, a = 0; a = 1, i = i_1; a = 2, i = i_2; a = 3, i = i_3; \ldots\}, \tag{4}$$

where the integer $s \equiv 1 \bmod q$ is the increment in arithmetic progression (2), corresponds to the sequence

$$\{(cK + ac) \bmod 1, \quad a = 0, \quad a, \ldots\} \tag{5}$$

regardless of the values of $i_1, i_2, \ldots$. This, in turn, implies that for q and s such that $s \equiv 1 \bmod q$ we have, regardless of the values of $i_1, i_2, \ldots$, that the sequence of keys

$$K, \qquad K + s^{i_1}, \qquad K + 2s^{i_2}, \ldots \tag{6}$$

corresponds to the values

$$cK \bmod 1, \qquad (cK + c) \bmod 1, \qquad (cK + 2c) \bmod 1, \ldots. \tag{7}$$

It is further assumed (without loss of generality) that K is such that

$$cK \bmod 1 = 0.$$

As a result, the problem is reduced to determining c such that the points

$$0, c \bmod 1, \quad 2c \bmod 1, \quad 3c \bmod 1, \ldots \tag{9}$$

are well spread in the unit interval.

The distribution of (9) at both, the early and asymptotic stages, has been studied by Halton (1965) and Zaremba (1966). In particular, it has been shown that in the process of successively placing the points from (2) in the unit interval, the current largest subinterval is split into two smaller subintervals by every new point placed, that is relation (3) holds throughout the process.

The analysis further shows that an appropriate choice for the integer p is such a value that the constant $c = p/q$ would be a good rational approximation to the golden ratio

$$\rho = \frac{-1 + \sqrt{5}}{2} = 0.6180339887 \tag{10}$$

For example, we may take $s = 1024 = 2^{10}$, having in mind the storage of items with 10-bit character string keys, and then we may take $q = 1023$ and $p = \lfloor q\rho \rfloor = 632$, whence the Floyd hashing function becomes

$$h(K) = \left\lfloor n\left(\frac{632}{1023} K \bmod 1\right) \right\rfloor. \tag{11}$$

Another scheme for computing (1) which involves no division is given by Knuth (1973). The parameter c is suggested as a rational fraction p/q with $q = 2^m$ where one is dealing with m-bit values in the arithmetic operations. The parameter p is a prime to q. If n is chosen to be a power of 2, say 2^k, then $\lfloor n(cK \bmod 1) \rfloor$ is just the high k bits of the low order word of the double length product pK.

We have now discussed four main families of the methods for computing the initial hashing function, $h(K)$. In addition, there are some other methods, sometimes even more elaborate than those discussed, which cater for various exotic forms of data arrays, like band matrices, triangular arrays, etc. We shall not discuss these methods, again, trusting that the interested reader can turn to appropriate sources which are available in the literature on the subject.

We finish the section by noting that in the opinion of Knuth (1973), none of the methods suggested for computing the hashing functions has proved to be superior to the simple division and multiplication methods described above.

9.9 Collision Resolution by Open Addressing

Suppose that given is a table with the load factor α, $\alpha \leqslant 1$. We wish to establish whether the key K is in the table. In the open addressing approach one looks at the various entries of the table one by one until either the key K or an empty location is found. The sequence of locations inspected during the search is called the probe sequence, and the idea is to formulate some rule by which every key K determines a probe sequence.

If using the probe sequence determined by K, an open location is encountered then one concludes that K is not in the table. The same sequence of probes will be made every time K is processed.

Let

$$h_i(K), \qquad i = 0, 1, 2, \ldots \tag{1}$$

be a sequence of relative addresses through which the search proceeds until either the key is found or an empty location is encountered. The first address, $h_0(K)$, is called the initial hash address of K. The subsequent addresses, $h_i(K)$, are evaluated by adding, modulo n, an increment d_i to the initial address, i.e.

$$h_i(K) = (h_0(K) + d_i) \bmod n, \qquad i = 1, 2, \ldots. \tag{2}$$

An illustration of the process is given in Fig. 9.14.

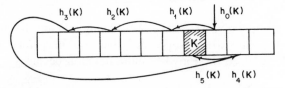

Fig. 9.14. A probe sequence in an open addressing hashing

9.9.1 The Pile-Up and Secondary Clustering Phenomena

Hashing techniques are distinguished by different rules for defining the increment d_i. Efficiency of the method is measured by the average number of probes required for a successful/unsuccessful search. To know the average behaviour of a hashing method is especially important because hashing is governed by the laws of probability. The worst case of hashing algorithms is usually bad, so one needs reassurance that the average behaviour is very good. The average number of probes required is a function of the load factor α.

We shall denote the average number of probes by $C_S(\alpha)$ and $C_U(\alpha)$ for a successful and an unsuccessful search, respectively. The functions $C_S(\alpha)$ and

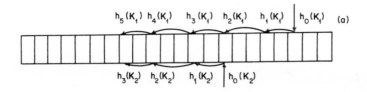

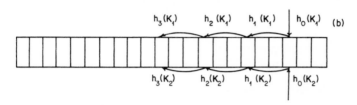

Fig. 9.15. Clustering phenomena in open addressing hashing
(a) A pile-up situation (b) A secondary clustering situation

$C_U(\alpha)$ are affected by specific phenomena occurring in searching which are known as the pile-up (or primary clustering) and secondary clustering.

By the pile-up one describes the situation which occurs in hashing when for $K_1 \neq K_2$,

if $h_i(K_1)$ and $h_j(K_2)$,   where $j \bmod n \neq i$,
designate the same location
then so will $h_{i+r}(K_1)$ and $h_{j+r}(K_2)$ for $r = 1, 2, \ldots$.

A pile-up is illustrated in Fig. 9.15(a).

The secondary clustering occurs when for $K_1 \neq K_2$,

if $h_i(K_1)$ and $h_i(K_2)$
designate the same location for some i
then so will $h_{i+r}(K_1)$ and $h_{i+r}(K_2)$ for $r = 1, 2, \ldots$.

An example of secondary clustering is shown in Fig. 9.15(b).

The two phenomena, thus, describe the situations in hashing when for two distinct keys, a partially or fully identical sequence is traced through the table. Occurrence of such situations affects the efficiency of the hashing process. It is, therefore, imperative in formula 9.9–(2) to select such a rule for computing d_i that the probing sequences will not give rise to the clustering situations. There are several algorithms for computing d_i. Each of them copes in its own way with the clustering phenomena. We intend to study these algorithms in the following section.

9.9.2 Open Addressing Algorithms

I. The Linear Search

This is historically the oldest open addressing hash method which is defined as follows:

$$h_i(K) = [h_0(K) + i] \bmod n, \qquad i = 1, 2, \ldots$$

or

$$h_i(K) = [h_{i-1}(K) + 1] \bmod n, \qquad i = 1, 2, \ldots .$$

(1)

The method was expounded by Peterson (1957) and Ershov (1958), though some primitive form of linear open addressing has been discussed as far back as 1953.

The linear search simply places colliding entries as near as possible to their nominally allocated position.

A more general form of the linear search is given by

$$h_i(K) = [h_0(K) + qi] \bmod n, \qquad i = 1, 2, \ldots,$$

(2)

where q is a constant.

It is quite obvious that linear search is affected by both pile-up and secondary clustering.

II. The Random Search

The method was developed by Morris (1968) and independently by de Balbine (1968) who named it the double hashing. The method uses two hash functions, $h_0(K)$ and $h_1(K)$, to form the probing sequence

$$h_i(K) = [h_0(K) + ih_1(K)] \bmod n.$$

(3)

As usual, $h_0(K)$ should produce a value between 0 and $n-1$, inclusive; $h_1(K)$, on the other hand, must produce a value between 0 and $n-1$ that is relatively prime to n. Say, if n is prime then $h_1(K)$ can be any value between 0 and $n-1$, and if $n = 2^r$, then $h_1(K)$ can be any odd value between 1 and $2^r - 1$. The fact that $h_1(K)$ is relatively prime to n ensures that no part of the table is examined twice, until all n locations have been probed. There are several ways to compute $h_1(K)$. Here, one has particularly to bear in mind that, for the best results, the $h_1(K)$ has to be quickly computable, yet with the property that distinct keys would tend to have different hash addresses.

In the sequel we consider five algorithms for computing hash functions $h_0(K)$ and $h_1(K)$.

Let n be prime. Compute $h_0(K) = K \bmod n$.

(4)

Now, *one* possibility for the second function is to let

$$h_1(K) = 1 + K \bmod (n - 1)$$

(5)

but since $n - 1$ is even, it would be better to let

$$h_1(K) = 1 + K \bmod (n - 2).$$

(6)

This suggests choosing n so that n and $n - 2$ are 'twin primes' like 513 and 511. The variant is due to Knuth (1973).

Another possibility is to use the quotient of the division K by n, so that

$$h_1(K) = 1 + \left\lfloor \frac{K}{n} \right\rfloor \bmod (n-2). \tag{7}$$

Computation of increment in this variant does not require any extra computer time as the quotient is a by-product of the division method used to determine $h_0(K)$. This variant is due to Bell and Kaman (1970).

Let $n = 2^r$. Compute $h_0(K)$ using the multiplicative method.
Again, *one* possibility for computing $d_i = ih_1(K)$ is to use Morris' algorithm. It is given as follows:
(Set $R \leftarrow 1$ every time the search routine is called.)
1. Set $i \leftarrow 1$.
2. Compute $5R$.
3. Shift the product r-2 bits to the left and place the result in R.
4. Set $d_i \leftarrow R/4$.
5. Increase i by 1.
6. If the hushed location is empty, terminate; otherwise, if the match is found, the algorithm terminates successfully. Otherwise go back to Step 2.

Another possibility is for the function $h_1(K)$ to take the form

$$h_1(K) = [2p(K) + 1] \bmod n, \tag{8}$$

where $p(K)$ is an integer constant that can be extracted from the key K by any hashing method. In particular, the hashing function $h_0(K)$ can be substituted for $p(K)$, and the following probing function will result

$$h_i(K) = [h_0(K) + ((2h_0(K) + 1) \bmod n)i] \bmod n. \tag{9}$$

The method is due to Luccio (1972).

The Luccio variant yields an easy analysis of the clustering phenomena. Indeed, from (9) it follows that for two distinct keys, K_1 and K_2, the same increment d_i in their respective probing sequences is used if

$$[2h_0(K_1) + 1] \bmod n = [2h_0(K_2) + 1] \bmod n. \tag{10}$$

Equation (10) is satisfied if
either

$$h_0(K_1) = h_0(K_2) \tag{11}$$

or

$$h_0(K_1) + \frac{n}{2} = h_0(K_2). \tag{12}$$

Relation (11) indicates that scheme (9) is affected by the secondary clustering, while relation (12) shows that the same increment is also used for the keys with different initial hash addresses when such addresses differ by half of the table size. In other words, the probing sequence for any K_2 traces the same locations as the probing sequence for K_1, where K_1 and K_2 are distinct and satisfy the relation

$$h_0(K_1) + \frac{n}{2} = h_0(K_2),$$

but only after one half of the table locations have been inspected. It may be concluded that for all practical purposes, the Luccio algorithm is free from pile-up.

The Luccio algorithm requires the operation of division when computing the increment. This division may be avoided if we simply set

$$h_1(K) = \begin{cases} 1 & \text{if } h_0(K) = 0 \\ n - h_0(K) & \text{otherwise.} \end{cases} \tag{13}$$

This variant, which is due to Knott (1968), is more economical than Luccio's but is still affected by secondary clustering.

III. The Quadratic Increment Search

The method was proposed by Maurer (1968) and further elaborated upon by Bell (1970). Maurer has suggested forming the d_i by using a quadratic polynomial

$$d_i = ai + bi^2 = pi + q\frac{i(i-1)}{2}, \qquad i = 1, 2, \ldots \tag{14}$$

with

$$p = a + b \quad \text{and} \quad q = 2b.$$

The probing sequence is then given by

$$h_i(K) = \left[h_0(K) + pi + q\frac{i(i-1)}{2} \right] \bmod n. \tag{15}$$

To see whether the method is affected by pile-up consider the following problem. Assuming that

$$\left[h_0(K_1) + pi + q\frac{i(i-1)}{2} \right] \bmod n = h_i(K_1)$$

$$= h_j(K_2) = \left[h_0(K_2) + pj + q\frac{j(j-1)}{2} \right] \bmod n, \tag{16}$$

find the conditions (if any) for the following to hold:

$$\left[h_0(K_1) + p(i+1) + q\frac{(i+1)i}{2} \right] \bmod n = h_{i+1}(K_1)$$

$$= h_{j+1}(K_2) = \left[h_0(K_2) + p(j+1) + q\frac{(j+1)j}{2} \right] \bmod n. \tag{17}$$

Condition (17) is equivalent to

$$(qi) \bmod n = (qj) \bmod n, \tag{18}$$

and the latter is satisfied only if

$$q \bmod n = 1. \tag{19}$$

Hence, the Maurer method is free from pile-up provided the value of q is chosen to be different from the one defined by condition (19). The method is, however, affected by the secondary clustering.

In (15) the initial hash address can be computed using the division method. In this case, the size of the table, n, is assumed to be a prime number, with the implication that the algorithm (15) covers exactly half the table after which the table is declared 'full'. The latter follows from the fact that for any prime number, Q, the quadratic remainder, R, i.e.

$$i^2 \bmod Q = R, \qquad i = 1, 2, \ldots,$$

has $(Q - 1)/2$ distinct values only, cf. Radke (1970).
These values, however, are fairly evenly distributed in the interval $[0, Q]$. The even distribution of the probed locations is a positive factor, since in hashing one always strives to obtain as even distribution as possible of items in the table.

Quadratic increment search was further studied by Ecker (1973) who has shown that under assumptions that n is not a prime number and that $q \bmod n > 1$ the whole table can be searched. The first assumption, however, precludes using the division method for computing the initial hash address.

Use of the quadratic search for the cases when the table size is 2^r and p^r with p prime, was studied by Hopgood and Davenport (1972) and Ackerman (1974). Bell has further suggested a variant of the quadratic search which is said to completely eliminate the secondary clustering. The variant is known as the quadratic quotient search and is of the form:

$$h_i(K) = [h_0(K) + pi + q(K)i^2] \bmod n \tag{20}$$

where p is a constant and $q(K) = \left\lfloor \dfrac{K}{n} \right\rfloor$.

Like the quadratic search of Maurer, the probing function (20) will search exactly half the table before declaring the table 'full'

9.10 Efficiency Analysis of the Open Addressing Algorithms

For the purpose of analysis, the hashing methods may be grouped under three distinct computational models:

Table 9.2 The average efficiencies of the three hashing processes.

Hashing method	The average number of probes required for	
	a successful search, $C_S(\alpha)$	an unsuccessful search, $C_U(\alpha)$
Linear probing	$\dfrac{1}{2}\left(1 + \dfrac{1}{1-\alpha}\right)$	$\dfrac{1}{2}\left(1 + \dfrac{1}{(1-\alpha)^2}\right)$
Double hashing with secondary clustering	$1 - \ln(1-\alpha) - \dfrac{1}{2}\alpha$	$\dfrac{1}{1-\alpha} - \ln(1-\alpha) - \alpha$
Independent double hashing	$-\dfrac{1}{\alpha}\ln(1-\alpha)$	$\dfrac{1}{1-\alpha}$

(i) the linear probing,

$$h_i(K) = [h_0(K) + i] \bmod n, \tag{1}$$

(ii) the double hashing with secondary clustering,

$$h_i(K) = [h_0(K) + f(h_0(K))] \bmod n, \tag{2}$$

where $f(h_0(K))$ is a more-or-less random function, and
(iii) the independent double hashing,

$$h_i(K) = [h_0(K) + f(h_0(K))] \bmod n, \tag{3}$$

where *each* of the possible values of the pair $(h_0(K), f(h_0(K)))$ is equally likely. In Table 9.2 are given efficiencies, on average, of the three models, in terms of the number of probes required.

We shall now derive these results.

Our aim is: given a density or load factor, α, of the table, to obtain an estimate of the average number of probes required by a particular search model. We shall use the following interpretation for a successful/unsuccessful search model.

A successful search model:

Start with an empty table. Then fill the table with keys from a data file. For each key, K, continue the probing until an empty location is encountered. Subsequently, insert the key in the location.

An unsuccessful search model:

Start with a table that is already filled to a prescribed density, α. Given keys from a random data file, probe locations of the table until match or no match (i.e. an empty location) is found for the search key. In the case of no match, the problem models an unsuccessful search.

If as before, $C_S(\alpha)$ and $C_U(\alpha)$ denote the average number of probes required for a successful and unsuccessful search, respectively, then in the open addressing hashing these quantities are related as

$$C_S\left(\frac{m}{n}\right) = \sum_{k=0}^{m-1} C_U\left(\frac{k}{n}\right).$$

9.10.1 Analysis of Linear Probing

Linear search was first analysed by Schay and Spruth (1962). They gave an approximate estimate of the average number of probes required by a successful search. Since then, Knuth (1973) has given the exact formulae for both successful and unsuccessful searches by linear probing. We shall follow the Schay and Spruth approach, while using an enhancing notation contributed to the analysis by Knuth. The approach is based on a remarkable property of linear proving which was first noticed by Peterson (1957), namely that

in a successful search by linear probing, the average number of probes does not depend on the order in which the keys were entered but only on the number of keys which hash to each location.

For example, given a table of size $n = 13$, i.e.

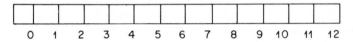

and eight keys, say, we obtain the following set of hash addresses:

$$8, 1, 5, 21 \bmod 13 = 8, 33 \bmod 13 = 7, 14 \bmod 13 = 1, 2, 8. \tag{1}$$

We shall call (1) a hashing sequence. From this sequence it follows that

$$\begin{aligned}&\text{no\quad keys hash to location 0,}\\&\text{two keys hash to location 1,}\end{aligned} \tag{2}$$

$$\text{etc.}$$

(Note that the last location is reserved for signalling the end of the table.) The resulting table looks as follows

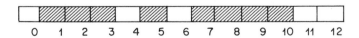

Sequence (2) given here as

$$0, 2, 1, 0, 0, 1, 0, 1, 3, 0, 0, 0, 0$$

is the only factor that determines the number of probes required.

We assume that all keys in the table have the same access frequency, though the Peterson theorem can be extended to the case when some keys are more frequently accessed than others.

Let

$$A_1 = h_0(K_1), \qquad A_2 = h_0(K_2), \ldots, \qquad A_m = h_0(K_m) \tag{3}$$

be a sequence of hash addresses computed for the given set of m keys, where $m < n$, and n is the table size. In (3), whenever a collision occurs it is resolved using the linear probing method that is the location immediately next to the hushed one is probed; we say in this case that the key is displaced by one from its hash address. The Peterson theorem states that any rearrangement of a hash sequence $A_1, A_2, \ldots, A_m$ results in a new sequence with the same average number of displacements of keys from their hash addresses. This statement will be asserted if we show that the total number of probes required to enter keys is the same for two hash sequences:

$$A_1, A_2, \ldots, A_{i-1}, A_i, A_{i+1}, A_{i+2}, \ldots, A_m$$

and

$$A_1, A_2, \ldots, A_{i-1}, A_{i+1}, A_i, A_{i+2}, \ldots, A_m, \tag{4}$$
$$1 \leqslant i \leqslant m.$$

If A_i and A_{i+1} are different then there is clearly no difference in the total number of probes in both cases; however, if the $(i+1)$st key in the second sequence hashes to the location occupied by the ith key in the first sequence, then the ith and $(i+1)$st merely exchange places, so the number of probes for the $(i+1)$st is decreased by the same amount that the number for the ith is increased. This completes the assertion.

Next, we introduce a sequence of numbers $B_0, B_1, \ldots, B_{n-1}$ associated with the hash sequence $A_1, A_2, \ldots, A_m$ and defined so that B_j is the number of A's that equal j. Clearly, $B_0 + B_1 + \ldots + B_{n-1} = m$. The value $B_j > 1$ indicates that one or more collisions have occurred at some location A_i. It means that all the keys hashed to A_i, except one, need to be 'carried over' or displaced from their hashed location.

From sequence $(B_j, j = 0, \ldots, n-1)$ we can determine the 'carry sequence' $C_0, C_1, \ldots, C_{n-1}$ where C_j is the number of keys for which both locations j and $j+1$ probed as the key is entered into the table. In other words, the value of C_j of location j is determined by the value of C_{j-1} of the previous location $j-1$ and the value of B_j of keys which have j as their hashed location.

Taking into account a 'circular' nature of the problem specified by the table size n, we deduce the relation:

$$C_{j \bmod n} = \begin{cases} 0 & \text{if } B_{j \bmod n} = C_{j-1} = 0, \\ B_{j \bmod n} + C_{j-1} - 1, & \text{otherwise.} \end{cases} \tag{5}$$

Our earlier example with $n = 13$ and $m = 8$, and

$$(B_0, B_1, B_2, \ldots, B_{12}) = (0, 2, 1, 0, 0, 1, 0, 1, 3, 0, 0, 0, 0)$$

gives

$$(C_0, C_1, C_2, \ldots, C_{12}) = (0, 1, 1, 0, 0, 0, 0, 0, 2, 1, 0, 0, 0).$$

Here, one key needs to be 'carried over' from location 1 to location 2, one from location 2 to location 3, none from locations 3, 4, 5, 6, 7, two from location 8 to location 9, one from location 9 to location 10, none from locations 10, 11, 12 and 0.

The number of probes for entering m keys can now be given in terms of the 'carry sequence':

$$\frac{C_0 + C_1 + C_2 + \ldots + C_{n-1}}{m} + 1, \tag{6}$$

where 1 stands for the first probe.

Note that both B_j and C_j do not depend on the particular location j to which i keys were hashed but only on the value i itself. This condition ensures a unique solution of equations (5) for C_j whenever $m < n$. In formula (6), to obtain the average number of probes, we need to consider the complete range of values that each C_j can take, that is the range $0 \leqslant C_j \leqslant n - 1$. Now, denoting by q_k the probability that a particular $C_j = k$, we rewrite (6) in the form

$$1 + \frac{1}{m} \underbrace{\left[\sum_{k=0}^{n-1} kq_k + \sum_{k=0}^{n-1} kq_k + \ldots + \sum_{k=0}^{n-1} kq_k \right]}_{n \text{ terms}} = 1 + \frac{n}{m} \sum_{k=0}^{n-1} kq_k. \tag{7}$$

To complete the analysis we need to compute the probability q_k. To do this, we first introduce the probability p_k that $B_j = k$ and then use equations (3) to determine the q's in terms of the p's.

The probability

$$p_k = \text{Pr}(B_j = k, \text{ i.e. exactly } k \text{ of the } A_i \text{ are equal to } j, \text{ for fixed } j)$$

is determined by the assumption that each hash sequence $A_1, A_2, \ldots, A_m$ is equiprobable, which gives

$$P_k = \binom{m}{k} \left(\frac{1}{n} \right)^k \left(1 - \frac{1}{n} \right)^{m-k}, \tag{8}$$

where the binomial coefficient $\binom{m}{k}$ stands for the number of all possible combinations of k elements of total m, the second term expresses the probability that a particular location A_i will assume a fixed value j, and the number of $A_i = j$ is k, and the third term expresses the probability that a particular location A_s from $(m - k)$ remaining will assume any fixed value between 0 and $n - 1$, except j, and the number of such A's is $m - k$.

Now, from equations (3) the relations for the probabilities $q_k = \text{Pr}(C_j = k)$ and $p_k = \text{Pr}(B_j = k)$ follow immediately:

$$\text{Pr}(C_{j \bmod n} = 0) = \text{Pr}(C_{j-1} = 0) \left[\text{Pr}(B_{j \bmod n} = 0) + \text{Pr}(B_{j \bmod n} = 1) \right]$$
$$+ \text{Pr}(B_{j \bmod n} = 0) \text{Pr}(C_{j-1} = 1),$$
$$\text{Pr}(C_{j \bmod n} = 1) = \text{Pr}(C_{j-1} = 0) \text{Pr}(B_{j \bmod n} = 2) + \text{Pr}(C_{j-1} = 1) \text{Pr}(B_{j \bmod n} = 1)$$
$$+ \text{Pr}(C_{j-1} = 2) \text{Pr}(B_{j \bmod n} = 0), \tag{9}$$
$$\vdots$$

giving

$$q_0 = q_0 p_0 + q_0 p_1 + q_1 p_0,$$
$$q_1 = q_0 q_2 + q_1 p_1 + q_2 p_0, \tag{10}$$
$$q_2 = q_0 p_3 + q_1 p_2 + q_2 p_1 + q_3 p_0,$$
$$\text{etc.}$$

Introducing the generating functions for these probability distributions,

$$B(z) = \Sigma p_k z^k \quad \text{and} \quad C(z) = \Sigma q_k z^k, \tag{11}$$

we observe that the set of equations (10) can be expressed in an equivalent form as follows:

$$B(z)C(z) = p_0 q_0 + (q_0 - p_0 q_0)z + q_1 z^2 + \ldots = p_0 q_0(1 - z) + zC(z). \tag{12}$$

(Equations (10) are obtained by equating in (12) the coefficients of like powers of z.)

Next, we note that in terms of (11) the average number of probes required, (6), is given by

$$1 + \frac{n}{m} C'(1). \tag{13}$$

Hence, the problem is reduced to determining the value $C'(1)$.

At this stage it is convenient to introduce function $D(z)$ by setting

$$B(z) = 1 + (z - 1)D(z), \tag{14}$$

where it is assumed that the initial condition $B(1) = 1$ is satisfied. Using (14), the generating functions equation (12) is rewritten as

$$[1 + (z - 1)D(z)]C(z) = p_0 q_0(1 - z) + zC(z) \tag{15}$$

giving

$$C(z) = \frac{p_0 q_0}{1 - D(z)}. \tag{16}$$

In (16) the quantity $p_0 q_0$ is determined from the initial condition $C(1) = 1$, i.e.

$$p_0 q_0 = [1 - D(z)]C(z)|_{z=1} = 1 - D(1). \tag{17}$$

Hence,

$$C(z) = \frac{1 - D(1)}{1 - D(z)}, \tag{18}$$

giving

$$C'(z) = \frac{D'(z)[1 - D(1)]}{[1 - D(z)]^2} \tag{19}$$

and

$$C'(1) = \frac{D'(1)}{1 - D(1)}. \tag{20}$$

Differentiating twice relation (14) and setting $z = 1$ we obtain

$$D(1) = B'(1) \quad \text{and} \quad D'(1) = B''(1). \tag{21}$$

Now, using (11) and (8) we get

$$B(z) = \sum p_k z^k = \sum \binom{m}{k} \left(\frac{z}{n}\right)^k \left(1 - \frac{1}{n}\right)^{m-k} = \left(\frac{z}{n} + 1 - \frac{1}{n}\right)^m. \tag{22}$$

It follows that

$$B'(1) = B'(z)\Big|_{z=1} = \frac{m}{n}\left(\frac{z-1}{n} + 1\right)^{m-1}\Big|_{z=1} = \frac{m}{n}, \tag{23}$$

and

$$B''(1) = B''(z)\Big|_{z=1} = \frac{m(m-1)}{n^2}\left(\frac{z-1}{n} + 1\right)^{m-2}\Big|_{z=1} = \frac{m(m-1)}{n}. \tag{24}$$

Finally, according to (13), (20), (23), and (24), the average number of probes required for a successful search using linear probing is given as

$$C_S(\alpha) = \frac{1}{2}\left(1 + \frac{n-1}{n-m}\right) \approx \frac{1}{2}\left(1 + \frac{1}{1-\alpha}\right), \tag{25}$$

where

$$\alpha = \frac{m}{n}.$$

To obtain the average number of probes required for an unsuccessful search we recall that in open addressing the $C_S(\alpha)$ and $C_U(\alpha)$ are related by

$$C_S\left(\frac{m}{n}\right) = \frac{1}{m}\sum_{k=0}^{m-1} C_U\left(\frac{k}{n}\right). \tag{26}$$

To solve this equation for $C_U(\alpha)$ we eliminate the summation terms in the usual way by subtracting (26) from a similar relation with m replaced by $m + 1$. This gives

$$C_U\left(\frac{m}{n}\right) = (m+1)C_S\left(\frac{m+1}{n}\right) - mC_S\left(\frac{m}{n}\right). \tag{27}$$

Substituting the expression for C_S as obtained in (25), we get

$$C_U\left(\frac{m}{n}\right) = \frac{1}{2}(m+1)\left(1 + \frac{n-1}{n-m-1}\right) - \frac{1}{2}m\left(1 + \frac{n-1}{n-m}\right) \approx \frac{1}{2}\left[1 + \frac{1}{(1-\alpha)^2}\right]. \tag{28}$$

9.10.2 Analysis of the Uniform Hashing Model and Optimality Considerations

A model of the hashing process which ignores completely the clustering phenomena was given by Peterson (1957) and is known as uniform hashing. In uniform hashing it is assumed that

(i) the keys go into random locations of the table, so that each of the $\binom{n}{m}$ possible configurations of m occupied locations and $n - m$ empty locations is equally likely;

(ii) occupancy of each location in the table is essentially independent of the others;

(iii) no pile-up or secondary clustering is present.

Consider the average number of probes required under these assumptions for an unsuccessful search. As usual, an unsuccessful search is assumed to be completed as soon as the first empty location in the table has been encountered. It is further assumed that if appropriate, the key can be inserted at this stage into the table. In the model, the probability that exactly r probes are needed to insert $(m + 1)$st key is given as

$$p_r = \frac{\text{the number of configurations where } r - 1 \text{ given locations are occupied and another is empty}}{\text{the total number of possible configurations of } m \text{ occupied locations and } n - m \text{ empty ones}} = \frac{\binom{n - r}{m - (r - 1)}}{\binom{n}{m}}, \quad (1)$$

and the average number of probes for an unsuccessful search is given by

$$C_U(\alpha) = \sum_{r=1}^{n} r p_r. \quad (2)$$

To obtain the $C_U(\alpha)$ in terms of the load factor α, the following sequence of transformations of the right-hand side of (2) can be carried out.

(a) Write

$$\sum_{r=1}^{n} r p_r = n + 1 - (n + 1) \sum_{r=1}^{n} p_r + \sum_{r=1}^{n} r p_r, \quad (3)$$

where

$$\sum_{r=1}^{n} p_r = 1 \text{ (i.e. the sum of all probabilities is equal to one)}.$$

Then

$$C_U(\alpha) = n + 1 - \sum_{r=1}^{n} (n + 1 - r) p_r = n + 1 - \sum_{r=1}^{n} (n + 1 - r) \frac{\binom{n - r}{m - (r - 1)}}{\binom{n}{m}}. \quad (4)$$

(b) Show that

$$\binom{n-r}{m-r+1} = \binom{n-r}{n-m-1}.$$

Then write

$$C_U(\alpha) = n + 1 - \sum_{r=1}^{n} (n+1-r) \frac{\binom{n-r}{n-m-1}}{\binom{n}{m}}. \tag{5}$$

(c) Show that

$$(n+1-r)\binom{n-r}{n-m-1} = (n-m)\binom{n-r+1}{n-m}.$$

Then write

$$C_U(\alpha) = n + 1 - (n-m) \sum_{r=1}^{n} \frac{\binom{n-r+1}{n-m}}{\binom{n}{m}}. \tag{6}$$

(d) By induction show that

$$\sum_{r=1}^{n} \binom{n-r+1}{n-m} = \binom{n+1}{n-m+1} \quad \text{for } 1 \leqslant m < n.$$

Then write

$$C_U(\alpha) = n + 1 - (n-m) \frac{\binom{n+1}{n-m+1}}{\binom{n}{m}} \quad \text{for } 1 \leqslant m < n, \tag{7}$$

which gives

$$C_U(\alpha) = \frac{n+1}{n-m+1} \approx \frac{1}{1-\alpha} \quad \text{for reasonably large } n. \tag{8}$$

The average number of probes for a successful search is then obtained in a usual way:

$$C_S\left(\frac{m}{n}\right) = \frac{1}{m} \sum_{k=0}^{m-1} C_U\left(\frac{k}{n}\right) = \frac{n+1}{m} \left(\frac{1}{n+1} + \frac{1}{n} + \ldots + \frac{1}{n-m+2} \right)$$

$$= \frac{n+1}{m} (H_{n+1} - H_{n-m+1})$$

$$\approx \frac{n+1}{m} (\ln(n+1) - \ln(n-m+1))$$

$$\approx \frac{1}{\alpha} \ln \frac{1}{1-\alpha}. \tag{9}$$

This completes the analysis of uniform hashing.

Effect of the secondary clustering when present, as given in 9.10–(2), increases formulae (8) and (9) to

$$C_U(\alpha) \approx \frac{1}{1-\alpha} - \ln(1-\alpha) - \alpha \tag{10}$$

and

$$C_S(\alpha) \approx 1 - \ln(1-\alpha) - \tfrac{1}{2}\alpha, \tag{11}$$

respectively.

Double hashing with secondary clustering can be described by a probing sequence

$$h_0(K),\ (h_0(K)+p_1) \bmod n,\ (h_0(K)+p_2) \bmod n, \ldots,\ (h_0(K)+p_{n-1}) \bmod n,$$

where $p_1, p_2, \ldots, p_{n-1}$ is a randomly chosen permutation of $1, 2, \ldots, n-1$ that depends on $h_0(K)$, i.e. all keys that hash to the same function $h_0(K)$ scan the same probing sequence, and the $(n-1)!^m$ possible choices of n probing sequences with this property are equally likely.

An experimental model suggested by Knuth (1973) that describes quite accurately the situation of random probing with secondary clustering applied to an initially empty table of size n, is given as follows: the key is hashed to the leftmost empty position with probability p, or else to any position except the leftmost with probability $q = 1 - p$; each of the latter $n - 1$ positions is equally likely. If the hashed position is empty the key is placed there, otherwise any empty position including the leftmost is probed and occupied, assuming that each of the empty positions is equally likely. For example, when $n = 7$ and $m = 3$, the table arrangement on completion of the above experiment will be
(occupied, empty, empty, occupied, occupied, empty, empty)
with probability

$$\tfrac{17}{3240} qqq + \tfrac{14}{180} pqq + \tfrac{1}{18} qpq + \tfrac{1}{18} qqp.$$

Here, to obtain, say, the first term of the sum we consider all possible sequences of operations with three 'q-steps' and none 'p-steps' with lead to the final arrangement of the first, fourth and fifth locations becoming occupied.
We get

(loc 4) (loc 4 → loc 1) (loc 5)	giving	$(q\tfrac{1}{6})(q\tfrac{1}{6}\tfrac{1}{6})(q\tfrac{1}{6})$,
(loc 4) (loc 5) (loc 4 → loc 1)	giving	$(q\tfrac{1}{6})(q\tfrac{1}{6})(q\tfrac{1}{6}\tfrac{1}{5})$,
(loc 4) (loc 5) (loc 5 → loc 1)	giving	$(q\tfrac{1}{6})(q\tfrac{1}{6})(q\tfrac{1}{6}\tfrac{1}{5})$,
(loc 5) (loc 4) (loc 4 → loc 1)	giving	$(q\tfrac{1}{6})(q\tfrac{1}{6})(q\tfrac{1}{6}\tfrac{1}{5})$,

(loc 5) (loc 4) (loc 5 → loc 1)　　giving　$(q\frac{1}{6})(q\frac{1}{6})(q\frac{1}{6}\frac{1}{5})$,

(loc 5) (loc 5 → loc 1) (loc 4)　　giving　$(q\frac{1}{6})(q\frac{1}{6}\frac{1}{6})(q\frac{1}{6})$.

Thus, in total we have

$$\frac{17}{3240}\,qqq.$$

Other terms of the sum are obtained in a similar way.

The model corresponds to random probing with secondary clustering when $p = 1/n$, since in this case the table locations can be renumbered so that a particular probing sequence is $1, 2, \ldots$ and all the others are random. The analysis of this model then leads to formulae (10) and (11).

Uniform hashing and its characteristics, the functions $C_S(\alpha)$ and $C_U(\alpha)$, are commonly taken as a 'rule of thumb' criteria, in relation to which the performance of various hashing methods is measured by means of empirical tests. It is intuitively accepted that uniform hashing provides a good approximation to the lower bound on hashing methods. The argument, however, depends on the way in which the uniform distribution of entries in the table is defined. For example, Ullman (1972) assigns a probability to each of the probing sequences generated instead of to each (initial) hashing address, i.e. instead of computing a hash address, $h(K)$, one maps each key K into an entire permutation of $(0, 1, \ldots, n - 1)$ that represents the probe sequence used for K and each of $n!$ permutations is assigned a probability. He then shows that it is possible to find a distribution of these probabilities of the probing sequences such that the average number of probes, i.e. the length of the probing sequence for, say, an unsuccessful search will be less than the $C_U(\alpha)$ of uniform hashing. The result, however, depends on a particular key. In fact, for one and the same distribution of probabilities of the probing sequences, the average number of probes to insert a certain key, Q, may be less than the average number of probes in uniform hashing, while the average number of probes to insert a different key, R, may be greater than the average number of probes in uniform hashing. Note that in uniform hashing the average number of probes does not depend on specific keys.

Exercises

9.1　Show that Sequential Search of Section 9.3.1 requires n comparisons in the worst case.

9.2　Show that Binary Search of Section 9.3.2 requires $[\log n]$ comparisons in the worst case.

9.3　Insert the following six keys

SUNDAY MONDAY WEDNESDAY THURSDAY
FRIDAY SATURDAY

into a binary search tree. What is the sequence of nodes visited in the tree if we wish to search for TUESDAY?

9.4   Insert the key TUESDAY into the binary tree of Exercise 9.3.

9.5   Delete the key THURSDAY from the binary tree of Exercise 9.4.

9.6   Draw the balanced binary tree with 14 nodes which has the maximum height of all 14-node balanced trees.

9.7   Write a program for the search, insertion and deletion of keys in a balanced binary tree.

9.8   Find an optimal binary search tree, given the following sequence of keys and their probabilities of occurrence:

GOOD	BAD	SWEET	BITTER	HOT	COLD
0.1	0.05	0.1	0.3	0.05	0
WET	DRY	NICE	HORRID	LIGHT	DARK
0.05	0	0	0.25	0	0

9.9   Suppose that keys are strings of letters and the following hashing function for a table of size $N = 19$ is used: Add the 'values' of the letters, where A has value 26, B has value 25, and so on. Divide the resulting sum by 17 and take the remainder. Show the contents of the hash table, given that the following keys are inserted:

ALGORITHM, DATA, STRUCTURE, PROGRAM, COMPUTER, COMPILING, LANGUAGE, DESIGN, ANALYSIS, LIST.

9.10   Explain the terms: successful and unsuccessful search, secondary clustering and pile-up, uniform probing in the context of the searching methods associated with hash coded tables.

9.11   Assuming all searches unsuccessful (successful), estimate the cost of an average search in a balanced binary search tree of height h which has minimum total path length.

9.12   Given that in the uniform probing model the average number of probes in an unsuccessful search is

$$C^{(us)}(M) = \frac{1}{1-\alpha} \quad \text{where } \alpha = \frac{M}{N},$$

is the load factor, show that the average number of probes in a successful search is

$$C^{(s)}(M) = \frac{1}{\alpha} \log_e \frac{1}{1-\alpha}.$$

9.13   Discuss the problem of the deletion of entries in the uniform probing model.

9.14   Consider the linear probing

$$h_i(K) = (h_0(K) - i) \bmod N,$$

where for a given hash sequence $h_0(K_0), h_0(K_1), \ldots, h_0(K_{M-1})$ the corresponding sequence of numbers $B_0, B_1, \ldots, B_{N-1}$ is defined with B_j being the number of $h_0(K)$'s that equal j. Derive the rule which determines the so called 'carry sequence' $C_0, C_1, \ldots, C_{N-1}$ for this process and explain why the average number of probes needed for retrieval of the M keys may be expressed as follows:

$$1 + (C_0 + C_1 + \ldots + C_{N-1})/M = 1 + \frac{N}{M} \sum_{k=1}^{N-1} kq_k,$$

where q_k is the probability that $C_j = k$.

9.15 In the context of Exercise 9.14 find the 'carry sequence' corresponding the following B_j sequence:

B_0	B_1	B_2	B_3	B_4	B_5	B_6	B_7	B_8	B_9	B_{10}	B_{11}
0	3	1	0	0	0	1	0	0	2	0	1

9.16 Discuss the factors upon which the efficiency of an open hash method depends, in particular explore such terms as the average search length, primary and secondary clustering, selection of entries and complexity of computation.

9.17 Develop an algorithm for deleting a node from a balanced tree without destroying the balance.

9.18 Write a program to perform a sequential search on an array $A[i:j]$ where $i < j$.

9.19 Assume that the linear probing method is used, how many hash sequences $\bar{a}_1, \bar{a}_2, \ldots, \bar{a}_9$ yield each of the following patterns of occupied locations in the tables:

0	1	2	3	4	5	6	7	8	9	10	11	12	13	14	15	16	17	18

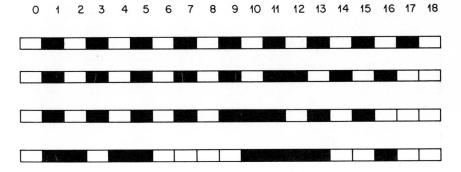

Appendix A

Some Basic Results on the Error Analysis of the Floating-Point Matrix Multiplication and the Solution of Sets of Linear Equations

Vector and Matrix Norms

In order to analyse computational processes involving matrices, it is convenient to associate with any vector or matrix a non-negative scalar that in some sense provides a measure of its magnitude. Such a scalar is normally called the norm.

Definition. The norm of a vector **x** may be defined as a real non-negative number which is denoted by $\|\mathbf{x}\|$ and which satisfies the following relations:

$\|\mathbf{x}\| > 0$, unless $\mathbf{x} = 0$,

$\|k\mathbf{x}\| = |k| \|\mathbf{x}\|$, where k is a complex number,

$\|\mathbf{x} + \mathbf{y}\| \leqslant \|\mathbf{x}\| + \|\mathbf{y}\|$.

The three most commonly known vector norms are given by

$$\|\mathbf{x}\|_p = (|x_1|^p + |x_2|^p + \ldots + |x_n|^p)^{1/p}, \tag{1}$$

where

$$\text{for } p = 1, \quad \|\mathbf{x}\|_1 = \sum_{i=1}^{n} |x_i|, \tag{2}$$

$$\text{for } p = 2, \quad \|\mathbf{x}\|_2 = \left[\sum_{i=1}^{n} |x_i|^2 \right]^{1/2} \tag{3}$$

and

$$\text{for } p = \infty, \quad \|\mathbf{x}\|_\infty = \max_i \{|x_i|\}. \tag{4}$$

Similarly, the norm of a matrix **A** may be defined as a real non-negative number which satisfies the relations:

$\|\mathbf{A}\| > 0$, unless $\mathbf{A} = \mathbf{0}$,

$\|k\mathbf{A}\| = |k| \|\mathbf{A}\|$, where k is a complex number,

$$\|\mathbf{A} + \mathbf{B}\| \leqslant \|\mathbf{A}\| + \|\mathbf{B}\|, \tag{5}$$

$$\|\mathbf{AB}\| \leqslant \|\mathbf{A}\| \|\mathbf{B}\|. \tag{6}$$

We say that the matrix norm is subordinate to the vector norm if the following relation holds

$$\|\mathbf{A}\| = \max \frac{\|\mathbf{Ax}\|}{\|\mathbf{x}\|} \tag{7}$$

This is equivalent to

$$\|\mathbf{A}\| = \max \|\mathbf{Ax}\|, \|\mathbf{x}\| = 1 \tag{8}$$

Corresponding to the three vector norms given above we have the matrix norms $\|\mathbf{A}\|_p$, where

$$\|\mathbf{A}\|_1 = \max_j \left\{ \sum_i |a_{ij}| \right\}, \tag{9}$$

$$\|\mathbf{A}\|_2 = [\text{maximum eigenvalue of } (\mathbf{A}^H \mathbf{A})]^{1/2}, \tag{10}$$

where $\mathbf{A}^H$ is the complex conjugate transpose of $\mathbf{A}$,

$$\|\mathbf{A}\|_\infty = \max_i \left\{ \sum_j |a_{ij}| \right\} \tag{11}$$

If the vector and matrix norms satisfy

$$\|\mathbf{Ax}\| \leqslant \|\mathbf{A}\| \|\mathbf{x}\| \tag{12}$$

then they are consistent.

It is properties (6) and (12) which make the vector and matrix norms so very useful in analysing the errors involved in solving linear equations. Another matrix norm, the so-called Euclidean norm,

$$\|\mathbf{A}\|_E = \left[\sum_i \sum_j |a_{ij}|^2 \right]^{1/2} \tag{13}$$

is also useful in practical error analysis. It has certain advantages over $\|\mathbf{A}\|_2$. These are

(i) the Euclidean norm of $|\mathbf{A}|$ is the same as that of $\mathbf{A}$. ($|\mathbf{A}|$ is the matrix with elements $|a_{ij}|$.)
(ii) $\|\mathbf{A}\|_E$ is easier to calculate than $\|\mathbf{A}\|_2$.

The Euclidean norm is not subordinate to any vector norm but it is consistent with $\|\mathbf{x}\|_2$.

The matrix norms provide the upper bound for the moduli of the eigenvalues of $\mathbf{A}$. Consider

$\mathbf{Ax} = \lambda\mathbf{x}$, where λ is an eigenvalue of $\mathbf{A}$

then

$$\| \mathbf{A} \| \, \| \mathbf{x} \| \geqslant \| \mathbf{A}\mathbf{x} \| = \| \lambda \mathbf{x} \| = |\lambda| \, \| \mathbf{x} \|$$

thus

$$\| \mathbf{A} \| \geqslant |\lambda| \tag{14}$$

We already know that

$$\begin{aligned}
\| \mathbf{A} \|_2^2 &= \text{maximum eigenvalue of } \mathbf{A}^H \mathbf{A} \\
&\leqslant \| \mathbf{A}^H \mathbf{A} \|_\infty \\
&\leqslant \| \mathbf{A}^H \|_\infty \| \mathbf{A} \|_\infty
\end{aligned}$$

Thus

$$\| \mathbf{A} \|_2^2 \leqslant \| \mathbf{A} \|_1 \| \mathbf{A} \|_\infty \tag{15}$$

For a more detailed review of the properties of norms see Wilkinson (1961), Wilkinson (1963), Householder (1964), Forsythe and Moler (1967).

Error Analysis of Basic Matrix Operations

(1) Multiplication by a scalar.
Let $\mathbf{B} = k\mathbf{A}$, where $\mathbf{B}$ and $\mathbf{A}$ are matrices and k is a scalar.
Then

$$b_{ij} \equiv \text{fl}(ka_{ij}) \equiv ka_{ij}(1 + e_{ij})$$

where

$$|e_{ij}| \leqslant 2^{-t}.$$

Thus

$$b_{ij} - ka_{ij} = ka_{ij}e_{ij}$$

and so

$$\| \mathbf{B} - k\mathbf{A} \|_E \leqslant 2^{-t}|k| \, \| \mathbf{A} \|_E, \tag{16}$$

giving

$$\frac{\| \mathbf{B} - k\mathbf{A} \|_E}{|k| - \| \mathbf{A} \|_E} \leqslant 2^{-t}.$$

(2) Matrix multiplication.
Consider $\mathbf{y} = \mathbf{A}\mathbf{x}$, where $\mathbf{A}$ is an $n \times n$ matrix and $\mathbf{y}$ and $\mathbf{x}$ are vectors of dimension n.
Then

$$y_i = \text{fl}\left(\sum_{j=1}^{n} a_{ij}x_j \right)$$

$$= \left[\sum_{j=1}^{n} a_{ij}x_j \right] + e_i$$

where

$$|e_i| \leqslant 2^{-t_1} [n|a_{i1}| |x_1| + n|a_{i2}| |x_2| + (n-1)$$
$$\times |a_{i3}| |x_3| + \ldots + 2|a_{in}| |x_n|] \tag{17}$$

Thus

$$\mathbf{y} = \mathbf{Ax} + \mathbf{e}, \tag{18}$$

where $|\mathbf{e}| \leqslant 2^{-t_1} |A||D||\mathbf{x}|$ and D is a diagonal matrix of the form

$$\mathbf{D} = \begin{bmatrix} n & & & & & \\ & n & & & & 0 \\ & & n-1 & & & \\ & & & \cdot & & \\ & & & & \cdot & \\ 0 & & & & & \cdot \\ & & & & & & 2 \end{bmatrix} \tag{19}$$

In terms of the norms the error $\mathbf{e}$ is bounded by

$$\|\mathbf{e}\|_2 \leqslant 2^{-t_1} \mathrm{n} \|\mathbf{A}\|_E \|\mathbf{x}\|_2 \tag{20}$$

Similarly, for the multiplication of two matrices we have

$$\mathbf{C} = \mathrm{fl}(\mathbf{AB}) \equiv \mathbf{AB} + \mathbf{E}$$

where

$$\|\mathbf{E}\|_E \leqslant 2^{-t_1} n \|\mathbf{A}\|_E \|\mathbf{B}\|_E \tag{21}$$

Since the product $\|\mathbf{A}\|_E \|\mathbf{B}\|_E$ may be very much greater than $\|\mathbf{AB}\|_E$, the computed $\mathbf{C}$ may well have a very high relative error. If we can accumulate inner products in floating-point and then write

$$\mathbf{C} = \mathrm{fl}_2(\mathbf{AB}) \equiv \mathbf{AB} + \mathbf{E}$$

we can see from the bounds given in 2.6.1–(16) that

$$|\mathbf{C} - \mathbf{AB}| \leqslant 2^{-t} |\mathbf{AB}| + \tfrac{3}{2} 2^{-t_2} |\mathbf{A}||\mathbf{D}||\mathbf{B}| \tag{22}$$

where $\mathbf{D}$ is as in (4).
Hence

$$\|\mathbf{C} - \mathbf{AB}\|_E = \| |\mathbf{C} - \mathbf{AB}| \|_E \leqslant 2^{-t} \|\mathbf{AB}\|_E + \tfrac{3}{2} 2^{-2t_2} n \|\mathbf{A}\|_E \|\mathbf{B}\|_E \tag{23}$$

The second term on the right hand side is negligible in comparison with the first and so we have

$$\frac{\|\mathbf{C} - \mathbf{AB}\|_E}{\|\mathbf{AB}\|_E} \leqslant 2^{-t} \tag{24}$$

which shows that the computed product has a very low error.

Error Analysis of the Solution of a Set of Linear Equations and of Matrix Inversion

We will consider the effect of perturbations in $\mathbf{A}$ and $\mathbf{b}$ on the solution of the set $\mathbf{Ax} = \mathbf{b}$.

(1) Let $\mathbf{k}$ be a small perturbation in $\mathbf{b}$ and $\mathbf{h}$ be the corresponding perturbation in $\mathbf{x}$.

We have

$$\mathbf{A}(\mathbf{x} + \mathbf{h}) = \mathbf{b} + \mathbf{k}, \tag{1}$$

which gives

$$\mathbf{Ah} = \mathbf{k} \text{ or } \mathbf{h} = \mathbf{A}^{-1}\mathbf{k}. \tag{2}$$

From (2) and assuming the matrix subordinate norms, we obtain:

$$\|\mathbf{h}\| = \|\mathbf{A}^{-1}\mathbf{k}\| \leqslant \|\mathbf{A}^{-1}\| \, \|\mathbf{k}\| \tag{3}$$

Also, for the matrix subordinate norms, we know that

$$\|\mathbf{b}\| = \|\mathbf{Ax}\| \leqslant \|\mathbf{A}\| \, \|\mathbf{x}\|$$

and so

$$\|\mathbf{x}\| \geqslant \|\mathbf{b}\| \, \|\mathbf{A}\|^{-1} \tag{4}$$

Therefore the relative change $\|\mathbf{h}\|/\|\mathbf{x}\|$ is given by

$$\frac{\|\mathbf{h}\|}{\|\mathbf{x}\|} \leqslant \frac{\|\mathbf{A}^{-1}\| \, \|\mathbf{k}\|}{\|\mathbf{A}\|^{-1} \, \|\mathbf{b}\|} = \frac{\|\mathbf{A}\| \, \|\mathbf{A}^{-1}\| \, \|\mathbf{k}\|}{\|\mathbf{b}\|}$$

$\|\mathbf{A}\| \, \|\mathbf{A}^{-1}\|$ is known as the condition number and usually denoted by cond (A). We, thus, have

$$\frac{\|\mathbf{h}\|}{\|\mathbf{x}\|} \leqslant \frac{\|\mathbf{k}\|}{\|\mathbf{b}\|} \text{cond}(A) \tag{5}$$

where $\text{cond}(A) = \|\mathbf{A}\| \, \|\mathbf{A}^{-1}\|$.

If cond (A) is very large then the set of equations, $\mathbf{Ax} = \mathbf{b}$, is said to be ill-conditioned and the relative error, $\|\mathbf{h}\|/\|\mathbf{x}\|$, introduced by the small perturbation $\mathbf{k}$ will be large.

(2) Let $\mathbf{E}$ be a small perturbation in $\mathbf{A}$ and $\mathbf{h}$ be again the corresponding perturbation in $\mathbf{x}$.

We have

$$(\mathbf{A} + \mathbf{E})(\mathbf{x} + \mathbf{h}) = \mathbf{b}, \tag{6}$$

which gives

$$(\mathbf{A} + \mathbf{E})\mathbf{h} = -\mathbf{Ex}. \tag{7}$$

Assuming $\mathbf{A}$ is non-singular (i.e. its inverse exists), then $(\mathbf{A} + \mathbf{E})$ may still be singular if $\mathbf{E}$ is not restricted. We shall, thus, assert the conditions on $\mathbf{E}$, under which matrix $(\mathbf{A} + \mathbf{E})$ will be non-singular if $\mathbf{A}$ is non-singular. Note that a matrix is called non-singular if its determinant is not equal to zero.

$(\mathbf{A} + \mathbf{E})$ can be expanded as

$$\mathbf{A} + \mathbf{E} = \mathbf{A}(\mathbf{I} + \mathbf{A}^{-1}\mathbf{E}), \tag{8}$$

giving

$$(\mathbf{A} + \mathbf{E})^{-1} = (\mathbf{I} + \mathbf{A}^{-1}\mathbf{E})^{-1}\mathbf{A}^{-1}. \tag{9}$$

From (9) it follows that, provided $\mathbf{A}$ is non-singular, for the $(\mathbf{A} + \mathbf{E})$ to be non-singular, the inverse $(\mathbf{I} + \mathbf{A}^{-1}\mathbf{E})^{-1}$ must exist, i.e. the determinant $|\mathbf{I} + \mathbf{A}^{-1}\mathbf{E}|$ must be non-zero.

Consider the matrix $(\mathbf{I} + \mathbf{A}^{-1}\mathbf{E})$. Its eigenvalues have the form $1 + \lambda_i$, where $\lambda_i, i = 1, \ldots, n$, are the eigenvalues of the matrix $\mathbf{A}^{-1}\mathbf{E}$. Now, if we assume the condition

$$\|\mathbf{A}^{-1}\mathbf{E}\| < 1, \tag{10}$$

then it follows that all $|\lambda_i|$ are less than unity, since by virtue of the matrix norms properties,

$$\|\mathbf{A}^{-1}\mathbf{E}\| \geqslant |\lambda_i| \tag{11}$$

holds for any matrix norm.

But if all $|\lambda_i|$ are less than unity, then $1 + \lambda_i$ are all non-zero, and thus the determinant $|\mathbf{I} + \mathbf{A}^{-1}\mathbf{E}| \neq 0$, since the determinant of a matrix is equal to the product of its eigenvalues. We conclude that provided A is non-singular, $(\mathbf{A} + \mathbf{E})$ will be non-singular if condition (10) holds.

Now, assuming that condition (10) is satisfied, consider (7) again. We have

$$\mathbf{h} = -(\mathbf{A} + \mathbf{E})^{-1}\mathbf{E}\mathbf{x} \tag{12}$$

giving

$$\mathbf{h} = -(\mathbf{I} + \mathbf{A}^{-1}\mathbf{E})^{-1}\mathbf{A}^{-1}\mathbf{E}\mathbf{x}, \tag{13}$$

where we used the relation (9).

From (13) we, further, have

$$\|\mathbf{h}\| \leqslant \frac{\|\mathbf{A}^{-1}\mathbf{E}\| \|\mathbf{x}\|}{\|\mathbf{I} + \mathbf{A}^{-1}\mathbf{E}\|} \leqslant \frac{\|\mathbf{A}^{-1}\| \|\mathbf{E}\| \|\mathbf{x}\|}{1 - \|\mathbf{A}^{-1}\| \|\mathbf{E}\|}, \tag{14}$$

since

$$\|\mathbf{A}^{-1}\mathbf{E}\| \leqslant \|\mathbf{A}^{-1}\| \|\mathbf{E}\|,$$

$$\|\mathbf{I} + \mathbf{A}^{-1}\mathbf{E}\| \geqslant \|\mathbf{I}\| - \|\mathbf{A}^{-1}\mathbf{E}\| \geqslant \|\mathbf{I}\| - \|\mathbf{A}^{-1}\| \|\mathbf{E}\|$$

and

$$\|\mathbf{I}\| = 1.$$

The relative error is given by

$$\frac{\|\mathbf{h}\|}{\|\mathbf{x}\|} \leqslant \frac{\|\mathbf{A}\|\,\|\mathbf{A}^{-1}\|\,\|\mathbf{E}\|/\|\mathbf{A}\|}{1 - \|\mathbf{A}\|\,\|\mathbf{A}^{-1}\|\,\|\mathbf{E}\|/\|\mathbf{A}\|}$$

$$\leqslant \frac{\|\mathbf{E}\|}{\|\mathbf{A}\|}\,\frac{\mathrm{cond}\,(A)}{1 - \dfrac{\|\mathbf{E}\|}{\|\mathbf{A}\|}\,\mathrm{cont}\,(A)} \tag{15}$$

Note that for the Euclidean norms we have $\|\mathbf{I}\|_E = \sqrt{n}$, which means that unity on the right-hand side of (14) and (15) has to be replaced by n.

From (15) we can see that the relative error, $\|\mathbf{h}\|/\|\mathbf{x}\|$ will be large if cond (A) is large.

Appendix B

Some Basic Preliminaries on Laws of Probability and Statistical Analysis

Probability

The most basic notion encountered in the theory of probability is that of an *event*, by which one means the occurrence of a specified outcome of an experiment, and the notion of probability has to do with the frequency of occurrence of an event X in repeated independent trials of an experiment.

If an experiment is repeated n times and if the event X occurs $n(X)$ times, then one expects the ratio $n(X)/n$ to cluster about a unique number, the probability of X. More precisely, the probability is defined as a real-valued function, $P(X)$, on the events of an experiment, satisfying $0 \leqslant P(X) \leqslant 1$.

Random Variable and the Distribution Function

The numerical equivalent of an event is called the random variable. A set of probabilities for the different possible values of the random variable is called a probability distribution.

If ξ is a random variable then associated with ξ is the (cumulative) distribution function which is defined as the real-valued function of a real variable and given as

$$F(x) = P(\xi \leqslant x),$$

i.e. it is equal to the probability that the variable assumes a value not greater than x.

The function $f(x) = F'(x)$, if it exists for all x, is called the distribution density of the random variable ξ.

For ξ continuous on an interval $[a, b]$, which may be finite or infinite, its (continuous) distribution function is given by

$$F(x) = \int_a^x f(x) \mathrm{d}x, \qquad a \leqslant x \leqslant b.$$

For ξ discrete, i.e. defined as a set of discrete values, $\{x_i, \ i = 1, 2, \ldots\}$,

341

the probability is given as

$$P(\xi = x_i) = p_i, \qquad \sum_i p_i = 1,$$

and the distribution function has the form

$$F(x) = \sum_{x_i \leqslant x} p_i.$$

Mean Value and Standard Deviation

Associated with each random variable is its mean value defined as

$$\mu[\xi] = \begin{cases} \int\limits_a^b xf(x)\mathrm{d}x = \int\limits_a^b x\,\mathrm{d}F(x), & \text{if } \xi \text{ is continuous in } [a,b], \\ \sum_i x_i p_i, & \text{if } \xi \text{ is defined as a set } \{x_i\}. \end{cases}$$

Another important characteristic is the variance

$$\mathrm{var}[\xi] = \sigma^2[\xi] = \begin{cases} \int\limits_a^b (x - \mu)^2 f(x)\mathrm{d}x = \int\limits_a^b (x - \mu)^2\,\mathrm{d}F(x) & \text{if } \xi \text{ is continuous in } [a,b] \\ \sum_i (x_i - \mu)^2 p_i & \text{if } \xi \text{ is given as a set } \{x_i\}. \end{cases}$$

The standard deviation is then defined as

$$\sigma = (\mathrm{var}[\xi])^{1/2}.$$

Uniform Distribution

The standard uniform distribution is defined for $0 \leqslant x \leqslant 1$ as follows: the distribution density is given by

$$f(x) = \begin{cases} 0, & x < 0, \\ 1, & 0 \leqslant x \leqslant 1, \\ 0, & x > 1, \end{cases}$$

and the (cumulative) distribution function is obtained as

$$F(x) = \int\limits_{-\infty}^x f(x)\mathrm{d}x = \begin{cases} 0, & x < 0 \\ x, & 0 \leqslant x \leqslant 1. \\ 1 & x > 1 \end{cases}$$

In Figure B.1 are shown graphical illustrations of both functions.
The mean value and the variance of the ξ are, then, given as

$$\mu[\xi] = \int\limits_{-\infty}^{\infty} xf(x)\mathrm{d}x = \frac{1}{2},$$

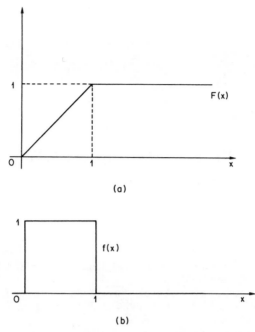

Fig. B.1. (a) The standard (cumulative) distribution function (b) The standard density function

and

$$\text{var}[\xi] = \sigma^2[\xi] = \int_{-\infty}^{\infty} \left(x - \frac{1}{2}\right)^2 f(x)\mathrm{d}x = \frac{1}{12},$$

respectively.

In the general case of the uniform random variable y given in $a \leqslant y \leqslant b$, y is generated from the standard distribution using the relation

$$y = a - \xi(a - b).$$

Permutation

A permutation of n objects is an arrangement or ordering of the objects; the number of permutations is the maximum number of ways the n objects can be arranged or ordered. This number is given as

$$p_n = n(n - 1)(n - 2)\ldots 1 = n!$$

For large n, the number of permutations can be estimated approximately using Stirling's formula, i.e.

$$n! \approx \sqrt{2\pi n}\left(\frac{n}{e}\right)^n.$$

This formula helps to avoid laborious calculations of factorials for large n.

Bibliography and References

Ackerman (1974).
　A. F. Ackerman, Quadratic search for hash tables of size p, *CACM*, **17**, No. 3, 164.
Adelson-Velski and Landis (1962).
　G. H. Adelson-Velski and E. H. Landis, An algorithm for the organization of information, *Doklady Akademii Nauk SSSR*, **146**, 263–266 (in Russian). English translation in *Soviet Math. Dokl.*, **3**, 1259–1262 (1962).
Aho, Hopcroft, and Ullman (1974).
　Aho, A. V., Hopcroft, J. E., and Ullman, J. D., *The Design and Analysis of Computer Algorithms*, Addison-Wesley Publ. Co.
Baer and Schwab (1977).
　Baer, J.-L., and Schwab, B., A comparison of tree-balancing algorithms, *CACM*, **20**, No. 5, 322–330.
Bakhvalov (1971).
　Bakhvalov, N. S., On the stable evaluation of polynomials (in Russian), *J. of Comp. Math. and Math. Phys.*, **11**, No. 6, 1568–1574.
de Balbine (1968).
　de Balbine, G., *Ph. D. Thesis*, Calif. Inst. of Technology, 1968, see in Knuth (1973), p. 521.
Bayer (1972).
　Bayer, R., Symmetric Binary B-trees: Data structure and maintenance algorithms, *Acta-Informatica*, **1**, No. 4, 290–306.
Belaga (1958).
　Belaga, E. C., Some problems in the computation of polynomials, *Dokl. Akad. Nauk SSSR*, **123**, 775–777 (Russian).
Belaga (1961).
　Belaga, E. C., On computing polynomials in one variable with initial conditioning of the coefficients, *Problemy Kibernetiki*, **5**, 7–15 (Russian). English translation in *Problems of Cybernetics*, **5**, 1–13.
Bell (1970).
　Bell, J. R., The quadratic quotient method: A hash code eliminating secondary clustering, *CACM*, **13**, No. 2.
Bell and Kaman (1970)
　Bell, J. R., and Kaman, C. H., The linear quotient hash code, *CACM*, **13**, No. 11.
Bellman and Dreyfus (1961).
　Bellman, R. E., and Dreyfus, S. E., *Applied Dynamic Programming*, Princeton Univ. Press, Princeton, New Jersey.

Benneth and Frazer (1971).
  Bennett, B. T., and Frazer, W. D., *Approximating Optimal Direct-Access Merge Performance*, IFIP.
Bergland (1968).
  Bergland, G. D., A fast Fourier transform algorithm using base eight iterations, *Math. Computation*, **22**, 275–279.
Betz (1956).
  Betz, B. K., *Unpublished Memorandum*, Minneapolis–Honeywell Regulator Co. (Reference is taken from Knuth, Vol. 3, p. 268.)
Bingham, Godfrey, and Tukey (1967).
  Bingham, C., Godfrey, M. D., and Tukey, J. W., Modern techniques of power spectrum Estimation, *IEEE Trans.*, AU-15 (June 1967), 56–66.
Blum, Floyd, Pratt, Rivest, and Tarjan (1973).
  Blum, M., Floyd, R. W., Pratt, V. R., Rivest, R. L., and Tarjan, R. E., Time bounds for selection, *Journal of Computer and System Sciences*, **1**, No. 4, 448–461.
Borodin (1971).
  Borodin, A., Horner's Rule is uniquely optimal. In *Theory of Machines and Computations*, Kohavi and Paz (Eds.), Academic Press, New York, 45–58.
Borodin (1973).
  Borodin, A., Computational complexity: theory and practice, in *Currents in the Theory of Computing*, by Aho, A. V. (Ed.), Prentice-Hall.
Borodin and Munro (1975).
  Borodin, A., and Munro, I., *The Computational Complexity of Algebraic and Numerical Problems*, Elsevier Publ. Co.
Brent (1970a).
  Brent, R. P., Algorithms for matrix multiplication, *Tech. Report CS157*, March 1970, Computer Sci. Dept., Stanford University.
Brent (1970b).
  Brent, R. P., Error analysis of algorithms for matrix multiplication and triangular decomposition using Winograd's Identity, *Num. Math.*, **16**, 145–156.
Brent (1971a).
  Brent, R. P., *Alogorithms for Finding Zeros and Extrema of Functions Without Calculating Derivatives*, STAN-CS-71–198.
Brent (1971b).
  Brent, R.P., An algorithm with guaranteed convergence for finding a zero of a function. *Comp. J.*, **14**, 422–425.
Brent, Winograd, and Wolfe (1973).
  Brent, R., Winograd, S., and Wolfe, P., Optimal iterative processes for root-finding, *Numer. Math.*, **20**, 327–341.
Brigham (1974).
  Brigham, E. O., *The Fast Fourier Transform*, Prentice-Hall.
Brockett and Dobkin (1973).
  Brockett, R., and Dobkin, D., On the optimal evaluation of a set of bilinear forms, *Proc. 5th Annual SIGACT Symp. on Theory of Computing*, Austin 1973.
Buchholz (1963).
  Buchholz, W., File organization and addressing, *IBM Systems Journal*, **2**, 86–111.
Bus and Dekker (1974).
  Bus, J. C. P., and Dekker, T. J., Two efficient algorithms with quaranteed convergence for finding a zero of a function, *Afdeling Numericke Wiskunde*, NW 13/74, Stichting Mathematisch Centrum, Amsterdam.
Businger and Golub (1971).
  Businger, P., and Golub, G. H., Linear least squares solutions by Householder transformations, Contribution I/8 in *Handbook for Automatic Computation*, Volume II, edited by Wilkinson, J. H., and Reinsch, C., Springer–Verlag.

346

Cheney (1962).

Cheney, E. W., *Algorithms for the Evaluation of Polynomials Using a Minimum Number of Multiplications*, Technical Note 2, Computation and Data Processing Center, Aerospace Corporation, El Segundo, California.

Chernous 'ko (1968).

Chernous'ko, F. L., An optimal algorithm of the root search for the function which is evaluated approximately, *Journal of Comp. Math. and Math. Physics*, **8**, No. 4. pp. 705–724 (Russian).

Clenshaw (1955).

Clenshaw, C. W., A note on the summation of Chebyshev Series, *MTAC*, **9**, 118–120.

Conte and de Boor (1972).

Conte, S. D., and de Boor, C., *Elementary Numerical Analysis—An Algorithmic Approach*, McGraw-Hill, Kogakusha, Second Edition.

Cook (1966).

Cook, S. A., On the minimum computation time of functions, *Doctoral Thesis*, Harvard University Cambridge, Massachusetts, 1966.

Cooley, Lewis, and Welch (1977).

Cooley, J. W., Lewis, P. A. W., and Welch, P. D., The fast Fourier transform and its application to time series analysis, in *Statistical Methods for Digital Computers*, edited by Enslein, K., Ralston, A., and Wilf, H. S., J. Wiley, London.

Danielson and Lanczos (1942).

Danielson, G. C., and Lanczos, C., Some improvements in practical Fourier analysis and their application to X-ray scattering from liquids, *J. Franklin Inst.*, **233**, 365–380; 435–452.

Dekker (1969).

Dekker, T. J., Finding a zero by means of successive linear interopolation, In: Dejon, B., and Henrici, P. (eds.), *Constructive Aspects of the Fundamental Theorem of Algebra*, Wiley Interscience, London.

Ecker (1973).

Ecker, A., The period of search for the quadratic and related hash methods, *The Computer Journal*, **17**, No. 4, 340–343.

Ershov (1958a).

Ershov, A. P., On programming arithmetic operations, *CACM*, **1**, No. 8, 3–6.

Ershov (1958b).

Ershov, A. P., *Doklady Akad. Nauk SSSR*, **118**, 427–430.

Eve (1964).

Eve, J., The evaluation of polynomials, *Num. Mathematik*, **6**, 17–21.

Feldstein and Firestone (1969).

Feldstein, A., and Firestone, R. M., A study of Ostrowski efficiency for composite iteration algorithms, *Proceedings of ACM 1969*.

Fiduccia (1971).

Fiduccia, C.M., Fast matrix multiplication, in *Proceedings 3rd Annual ACM Symposium on Theory of Computing*, pp. 45–49.

Fiduccia (1972).

Fiduccia, C. M., On obtaining upper bounds on the complexity of matrix multiplication, In: Miller, R. E., and Thatcher, J. W. (eds.), *Complexity of Computer Computations*, Plenum Press.

Fike (1967).

Fike, C. T., Methods of evaluating polynomial approximations in function evaluation routines, *CACM*, **10**, 175–178.

Flanagan (1967).

Flanagan, J. L., Spectrum analysis in speech coding, *IEEE Trans. AU-15* (June 1967), 66–69.

Floyd (1964).
   Floyd, R. W., Algorithm 245: Treesort 3, *CACM*, **7,** No. 12, 701.
Floyd (1970).
   Floyd, R. W., The reference to personal communication in Knott (1975).
Ford and Johnson (1959).
   Ford, L. R., and Johnson, S. M., A tournament problem, *American Math. Monthly*, **66,** 387–389.
Forsythe and Moler (1967).
   Forsythe, G. E., and Moler, C. B., *Computer Solution of Linear Systems*, Prentice-Hall.
Foster (1973).
   Foster, C. C., A generalization of AVL trees, *CACM*, **16,** 8, 513–517.
Fox (1954).
   Fox, L., Practical solution of linear equations and inversion of matrices, *Appl. Math. Ser. Nat. Bur. Stand.*, **39,** 1–54.
Fox and Mayers (1968).
   Fox, L., and Mayers, D. F., *Computing Methods for Scientists and Engineers*, OUP.
Frazer and Bennett (1972).
   Frazer, W. D., and Bennett, B. T., Bounds on optimal merge performance, and a strategy for optimality, *JACM*, **19,** 4, 641–648.
Frazer and McKellar (1970).
   Frazer, W. D., and McKellar, A. C., Samplesort: A sampling approach to minimal storage tree sorting, *JACM*, **17,** No. 3, 466–507.
Gentleman (1969).
   Gentleman, W. M., Off-the-shelf black boxes for programming, *IEEE Trans. Education*, E-12, 43–50.
Gentleman (1973).
   Gentleman, W. M., On the relevance of various cost models of complexity. In: Traub, J. F. (ed.), *Complexity of Sequential and Parallel Numerical Algorithms*.
Gilstad (1960).
   Gilstad, R. L., in *Proceed. AFIPS Eastern Jt. Computer Conf.*, 18, 143–148.
Givens (1954).
   Givens, W., Numerical computation of the characteristic values of a real symmetric matrix, *Rep. ORNL-1954*, Oak Ridge Associated Universities, Oak Ridge, Tennessee.
Good (1958, 1960).
   Good, I. J., The interaction algorithm and practical Fourier series, *J. Roy. Static. Soc.*, *Series B*, **20,** 361–372 (1958). *Addendum*, **22,** 372–375 (1960).
Gross and Johnson (1959).
   Gross, O., and Johnson, S. M., Sequential minimax search for a zero of a convex function, *MTAC* (now *Math. Comp.*), **13,** 44–51.
Hadian and Sobel (1969).
   Hadian, A., and Sobel M., Selecting the tth largest using binary errorless comparisons, *Tech. Rept. 121*, Dept. of Statistics, University of Minnesota, Minneapolis.
Halton (1965).
   Halton, J. H., The distribution of the sequence $\{n, n = 0, 1, 2, \ldots\}$, *Cambridge Philosophical Society Proceedings*, **61,** 665–670.
Hamming (1950).
   Hamming, R. W., Error detecting and error correcting codes, *BSTJ*, **26,** No. 2, 147–160.
Hart (1968).
   Hart, J. F., *et al.*, *Computer Approximations*, John Wiley. Harter (1972).
Harter (1972).
   Harter, R., The optimality of Winograd's formula, *CACM*, **15,** No. 5, 352.
Hartmanis and Hopcroft (1971).
   Hartmanis, J., and Hopcroft, J. E., An overview of the theory of computational complexity, *JACM*, **18,** No. 3, 444–475.

348

Helms (1967).
Helms, H. D., Fast Fourier transform method of computing difference equations and simulating filters, *IEEE Trans.*, AU-15, June 1967, 85–90.
Hoare (1962).
Haore, C. A. R., Quicksort, *Computer Journal*, **5**, No. 1, 10–15.
Hopcroft and Kerr (1971).
Hopcroft, J. E., and Kerr, L. R., On minimizing the number of multiplications necessary for matrix multiplication, SIAM, *J. Applied Math.*, **20**:**1**, 30–36.
Hopcroft and Musinski (1973).
Hopcroft, I., and Musinski, J., Duality in determining the complexity of noncommutative matrix multiplication, *Proc. 5th Symposium on Theory of Computing*, Austin 1973.
Hopgood and Davenport (1972).
Hopgood, F. R. A., and Davenport, J., The quadratic hash method when the table size is a power of 2, *The Computer Journal*, **15**, No. 4, 314–315.
Householder (1958a).
Householder, A. S., A class of methods for inverting matrices, *Journal of SIAM*, **6**, 189–195.
Householder (1958b).
Householder, A. S., Unitary triangularization of a non-symmetric matrix, *Journal of A.C.M.*, **5**, 339–342.
Householder (1964).
Householder, A. S., *The Theory of Matrices in Numerical Analysis*, Blaisdell, 1964, Chapter 2.
Hu and Tucker (1971).
Hu, T. C., and Tucker, A. C., Optimal computer search trees and variable-length alphabetic codes, *SIAM J. Appl. Math.*, **21**, 4, 514–532.
Huffman (1952).
Huffman, D. A., A method for the construction of minimum redundancy codes, *Proc. IRE*, Sept. 1952, 1098–1101.
Jarratt (1966a).
Jarratt, P., Multipoint iterative methods for solving certain equations, *Comp. J.*, **8**, 398–400.
Jarratt (1966b).
Jarratt, P., A rational iteration function for solving equations, *Comp. J.*, **9**, 304–307.
Jarratt (1966c).
Jarratt, P., Some fourth order multi-point iterative methods for solving equations, *Math. Comp.*, **20**, 434.
Jarratt (1969).
Jarratt, P., Some efficient fourth order multipoint methods for solving equations, BIT, **9**, 2, 119–124.
Jarratt (1973).
Jarratt, P., A review of methods for solving non-linear algebraic equations in one variable, in Rabinowitz, P. (ed.), *Numerical Methods for Nonlinear Algebraic Equations*, 1973, Gordon and Breach Science Publ., New York.
Jarratt and Nudds (1965).
Jarratt, P., and Nudds, D., The use of rational functions in the iterative solution of equations on a digital computer, *Comp. J.*, **8**, 62–65.
Kaneko and Liu (1970).
Kaneko, T., and Liu, B., Accumulation of round-off error in fast Fourier transforms, *JACM*, **17**, No. 4, 637–654.
Karatsuba and Ofman (1962).
Karatsuba, A., and Ofman, Yu., Multiplication of multiple numbers by means of Automata, *Doklady Akad. Nauk USSR*, **145**, No. 2, 293–294 (Russian).

349

Kirkpatrick (1972).
Kirkpatrick, D., On the additions necessary to compute certain functions, *Proc. 4th Annual ACM Symposium on Theory of Computing*, 94–101.
Kislitsyn (1964).
Kislitsyn, S. S., On the selection of the kth element of an ordered set by pairwise comparison, *Sibirsk. Mat. Zh.*, **5**, 557–564.
Klinkhamer (1968).
Klinkhamer, J. F., *On Key-to-Address Transformation for Mass Storage*, University of Utah, Computer Science Department Report.
Klyuev and Kokovkin-Shcherbak (1965).
Klyuev, V. V., and Kokovkin-Shcherbak, H. I., Minimization of the number of arithmetic operations in the solution of linear algebraic systems of equations, *Zh. Vychisl. Mat. i Mat. Fiz.*, **5**, 1, pp. 21–33. English translation in *U.S.S.R. Computational Mathematics and Mathematical Physics*, **5**, 25–43.
Knott (1975).
Knott, G. D., Hashing functions, *The Computer Journal*, **18**, No. 3.
Knuth (1963).
Knuth, D. E., Length of strings for a merge sort, *ACM*, **6**, No. 11, 685–688.
Knuth (1965).
Knuth, D. E., Problem No. 47, *BIE*, **5**, 142.
Knuth (1968).
Knuth, D. E., *The Art of Computer Programming*, Vol. 1.
Knuth (1969).
Knuth, D. E., *The Art of Computer Programming*, Vol. 2.
Knuth (1971).
Knuth, D. E., Optimum binary search trees, *Acta Informatica*, **1**, 14–25.
Knuth (1973).
Knuth, D. E., *The Art of Computer Programming*, Vol. 3.
Kogan (1967).
Kogan, T. I., Construction of the high order iterative processes, *Journal of Comp. Math. and Math. Physics*, **7**, No. 2, 423–424 (in Russian).
Krautstengel (1968).
Krautstengel, R., On one iterative method of calculating a single root of equation $f(x) = 0$, *Journal of Comp. Math. and Math. Physics*, **8**, No. 6, 1327–1329 (Russian).
Kung and Traub (1973).
Kung, H. T., and Traub, J. F., Computational complexity of one-point and multi-point iteration. In: Karp, R., (ed.), *Complexity of Real Computation*, American Mathematical Society, Providence, R. I.
Kung and Traub (1974).
Kung, H. T. and Traub, J. F., Optimal order of one-point and multi-point iteration, *Journal of ACM*, **21**, No. 4.
Le Bail (1972).
Le Bail, R. C., Use of fast Fourier transforms for solving partial differential equations in physics, *J. of Comp. Physics*, **9**, 440–465.
Lohmann and Paris (1967).
Lohmann, A. W., and Paris, D. P., Binary Fraunhofer holograms generated by computer, *Appl. Optics*, **6**, 1739–1748.
Luccio (1972).
Luccio. F., Weighted increment linear search for scatter tables, *CACM*, **15**, No. 12, 1045–1047.
MacCallum (1972).
MacCallum, I. R., A simple analysis of the nth order polyphase sort, *Comp., J.*, **16**, No. 1, 16–18.

350

Makarov (1975a).

Makarov, O. M., On relation of two multiplication algorithms (Russian), *Journal of Comp. Math. and Math. Physics*, **15**, No. 1, 227–231.

Makarov (1975 b).

Makarov, O. M., On relation of the algorithms of the fast Fourier transform and Hadamard with the algorithms of Karatsuba, Strassen, and Winograd (Russian), *Journal of Comp. Math. and Math. Physics*, **15**, No. 5, 1095–1105.

Martin, Peters, and Wilkinson (1965).

Martin, R. S., Peters, G., and Wilkinson, J. H., Symmetric decomposition of a positive definite matrix, *Num. Math.*, **7**, 362–383.

Maurer (1968).

Maurer, W. D., An improved hash code for scatter storage, *CACM*, **11**, No. 1.

Mesztenyi and Witzgall (1967).

Mesztenyi, C. K., and Witzgall, C., Stable evaluation of polynomials, *Journal of Research of the National Bureau of Standards*, **71B**, 11–17.

Miller (1973).

Miller, W., Computational complexity and numerical stability, *Thomas J. Watson Research Report RC-4480*, Yorktown Heights.

Moler (1967).

Moler, C. B., Iterative refinement in floating point, *Journal of ACM*, **14**, 316–321.

Moler (1969).

Moler, C. B., *Numerical Solution of Matrix Problems*, Presented at the Joint Conference on Mathematical and Computer Aids to Design, Anaheim, 1969.

Morris (1968).

Morris, R., Scatter storage techniques, *CACM*, **11**, No. 1.

Motzkin (1955).

Motzkin, T. S., Evaluation of polynomials, *Bull. Amer. Math. Soc.*, **61**, 163.

Muller (1956).

Muller, D. E., A method of solving algebraic equations using an automatic computer, *MTAC*, **10**, 208–215.

Murega (1961).

Murega, S., Unpublished report, IBM.

Newbery (1974).

Newbery, A. C. R., Error analysis for polynomial evaluation, *Math. of Comp.*, **28**, No. 127, 789–793.

von Newman and Goldstine (1947).

von Newman, J., and Goldstine, H. H., Numerical inverting of matrices of high order, *Bull. Amer. Math. Soc.*, **53**, 1021–1099.

Nievergelt (1974).

Nievergelt, J., Binary search trees and file organization, *Computing Surveys*, **6**, No. 3, 195–207.

Ofman (1962).

Ofman, Yu., On the algorithmic complexity of discrete functions, *Doklady Akad. Nauk USSR*, **145**, No. 1, 48–51 (in Russian).

Ostrowski (1954).

Ostrowski, A. M., On two problems in abstract algebra connected with Horner's Rule, *Studies in Mathematics and Mechanics*, Presented to R. von Mises, Academic Press, New York, pp. 40–48.

Ostrowski (1960).

Ostrowski, A. M., *Solutions of Equations and Systems of Equations*, Academic Press, New York.

Ostrowski (1966).

Ostrowski, A. M., *Solutions of Equations and Systems of Equations*, Second Edition. Academic Press, New York, Chapter 12, p. 338.

Overholt (1973a).
Overholt, K. J., *Optimal Binary Search Methods*, BIT, **13**, 84–91.
Overholt (1973b).
Overholt, K. J., *Efficiency of the Fibonacci Search Method*, BIT, **13**, 92–96.
Pan (1966).
Pan, V. Ya., Methods of computing values of polynomials, *Uspekhi Math. Nauk*, **21**, 103–134 (Russian). English translation in *Russian Math. Surveys*, **21**, 105–136.
Pan (1959).
Pan, V. Ya., Schemes for computing polynomials with real coefficients, *Dokl. Akad. Nauk SSSR*, **127**, 266–269 (in Russian). English translation in Math. Reviews, **23**, 1962, Review Number B560.
Paterson (1972).
Paterson, M. S., Efficient iterations for algebraic numbers in *Complexity of Computer Computations*, ed. by Miller, R. E., and Thatcher, J. W., Plenum Press.
Peters and Wilkinson (1969).
Peters, G., and Wilkinson, J. H., Eigenvalues of Ax = Bx with band symmetric A and B, *Comp. J.*, **12**, 398–404.
Peterson (1957).
Peterson, W. W., Addressing for random access storage, *IBM J. Research and Development*, **1**, No. 2, 130–146.
Radke (1970).
Radke, C. E., The use of quadratic residue research, *CACM*, **13**, No. 2.
Ramos (1971).
Ramos, G. U., Round-off error analysis of the fast Fourier transform, *Math. Comp.*, **25**, No. 16, 757–768.
Reingold and Stocks (1972).
Reingold, E. M., and Stocks, A. I., Simple proofs of lower bounds for polynomial evaluation, *Complexity of Computer Computations*, Plenum Press, 1972, pp. 21–30.
Rice (1965).
Rice, J. R., On the conditioning of polynomial and rational forms, *Num. Math.*, **7**, 426–435.
Rice (1972).
Rice J., *A Seminar given at Purdue University*, Spring 1972.
Rabin (1974).
Rabin, M. O., Theoretical impediments to artificial intelligence, in *Information Processing 74* (Invited Paper). North Holland.
Rabin and Winograd (1971).
Rabin, M., and Winograd, S., *Fast Evaluation of Polynomials by Rational Preparation*, IBM Technical Report RC 3645, Dec. 1971.
Runge and König (1924).
Runge, C., and König, R., Die Grundlehren der Mathematischen Wissenschaften, in *Vorlesung über Numerisches Rechnen*, 11, Springer, Berlin, 1924.
Schay and Raver (1963).
Schay, G., Jr., and Raver, N., A method for key-to-address transformation, *IBM Journal of Research and Development*, **7**, No. 2, 121–126.
Schay and Spruth (1962).
Schay, G., and Spruth, W. G., Analysis of a file addressing method, *CACM*, **5**, 459–462.
Schönhage (1966).
Schönhage, A., Multiplikation Grosser Zahlen, *Computing (Arch. Elektron. Rechnen)*, **1**, 182–196.
Schönhage and Strassen (1971).
Schönhage, A., and Strassen, V., Schnelle Multiplikation Grosser Zahlen, *Computing*, **7**, 281–292.

Schreier (1932).

Schreier, J., On Tournament Eliminations Systems, *Mathesis Polska*, **7**, 154–160 (Polish).

Shaw and Traub (1974).

Shaw, M., and Traub, J. F., On the number of multiplications for the evaluation of a polynomial and some of its derivatives, *J. ACM*, **21**, No. 1, 161–167.

Singleton (1969a).

Singleton, R. C., An algorithm for computing the mixed radix fast Fourier transform, *IEEE Trans. Audio and Electroacoustics*, AU-17, **2**, 93–103.

Singleton (1969 b).

Singleton, R. C., Algorithm 347: An algorithm for sorting with minimal storage, *CACM*, **19**, No. 3, 185–187.

Slupecki (1951).

Slupecki, J., On the system S of tournaments, *Colloq. Math.*, **2**, 286–290.

Strassen (1969).

Strassen, V., Gaussian elimination is not optimal, *Num. Math.*, **13**, 354–356.

Strassen (1972).

Strassen, V., Evaluation of rational functions, in *Complexity of Computer Computations*, ed. by Miller, R. E., and Thatcher, J. W., Plenum Press.

Sukharev (1976).

Sukharev, A. G., Optimal root search for the function which satisfies the Lipschitz condition, *Journal of Comp. Math. and Math. Physics*, **16**, No. 1, 20–29 (in Russian).

Todd (1955).

Todd, J., Motivations for working in numerical analysis, *Comm. Pure Appl. Math.*, **8**.

Toom (1963).

Toom, A. L., The complexity of a scheme of functional elements realizing the multiplication of integers, *Dokl. Akad. Nauk SSSR*, **150**, 496–498.

Traub (1964).

Traub, J. F., *Iterative Methods for the Solution of Equations*, Chapter 5. Prentice-Hall.

Ullman (1972).

Ullman, J. D., A note on the efficiency of hashing functions, *JACM*, **19**, No. 3, 569–575.

Weinstein (1969).

Weinstein, C. J., Round-off noise in floating point fast Fourier transform computation, *IEEE Trans. Audio and Electroacoustics*, AU-17, 209–215.

Welch (1969).

Welch, P. D., A fixed point fast Fourier transform error analysis. *IEEE Trans.* AU-17 (June 1969), 151–157.

van Wijngaarden, Zonneveld, and Dijkstra (1963).

van Wijngaarden, A., Zonneveld, J. A., and Dijkstra, E. W., Programs AP200 and AP230 de serie AP200, in Dekker, T. J. (ed.), 1963, *The Series AP200 of Procedures in ALGOL 60*, The Mathematical Centre, Amsterdam.

Wilkinson (1961).

Wilkinson, J. H., Error analysis of direct methods of matrix inversion, *Journal ACM*, **8**, 281–330.

Wilkinson (1963).

Wilkinson, J. H., *Rounding Errors in Algebraic Processes*. Prentice-Hall.

Wilkinson (1965).

Wilkinson, J. H., *The Algebraic Eigenvalue Problem*. OUP.

Wilkinson (1967).

Wilkinson, J. H., Two algorithms based on successive linear interpolation, *Report CS 60*, Computer Sci. Dept., Stanford Univ.

Williams (1964).

Williams, J. W. J., Algorithm 232; Heapsort, *CACM*, **7**, No. 6, 347–348.

Winograd (1967).
Winograd, S., On the number of multiplications required to compute certain functions, *Proc. Natl. Acad. Sci. U.S.A.*, **58**, 1840–1842.

Wingrad (1968).
Winograd, S., A new algorithm for the inner product, *IEEE Trans. on Comp.*, C17, 693–694.

Winograd (1970a).
Winograd, S., On the multiplication of 2×2 matrices, *IBM Research Report*, RC 267, Jan. 1970.

Winograd (1970b).
Winograd, S., On the number of multiplications necessary to compute certain functions, *Comm. Pure and Appl. Math.*, **23**, 165–179.

Winograd (1973).
Winograd, S., Some remarks on fast multiplication of polynomials, in *Complexity of Sequential and Parallel Numerical Algorithms* ed. by Traub, J. F., Academic Press.

Wirth (1976).
Wirth, N., *Algorithms + Data Structures = Programs*, Prentice-Hall series in Automatic Computation.

Wood (1973).
Wood, D., A note on table look-up, *BIT*, **13**, 245–246.

Wozniakowski (1974a).
Wozniakowski, H., Rounding error analysis for the evaluation of a polynomial and some of its derivatives, *SIAM J. Numer. Anal.*, **11**, No. 4.

Wozniakowski (1974b).
Wozniakowski, H., Maximal stationary iterative methods for the solution of operator equations, *SIAM J. Numer. Anal.*, **11**, No. 5.

Wozniakowski (1975a).
Wozniakowski, H., *Maximal Order of Multipoint Iterations Using n Evaluations*, Department of Computer Science, Carnegie-Hellon University.

Wozniakowski (1975b).
Wozniakowski, H., *Numerical Stability for Solving Nonlinear Equations*, Department of Computer Science, Carnegie-Mellon University.

Wozniakowski (1975c).
Wozniakowski, H., *Numerical Stability of the Chebyshev Method for the Solution of Large Linear Systems*, Department of Computer Science, Carnegie-Mellon University.

Yeh (1976).
Yeh, R. T. (ed.), *Applied Computation Theory: Analysis, Design, Modeling*, Prentice-Hall.

Zaremba (1966).
Zaremba, S. K., Good lattic points, discrepancy, and numerical integration, *Annali de Matematica Pura ed Applicata*, Series 4, **73**, 293–317.

Index